Aristotle's Modal Syllogistic

Aristotle's Modal Syllogistic

Marko Malink

HARVARD UNIVERSITY PRESS
Cambridge, Massachusetts & London, England
2013

Library of Congress Cataloging-in-Publication Data

Malink, Marko, 1980–
Aristotle's modal syllogistic / Marko Malink.
pages cm
Includes bibliographical references and index.
ISBN 978-0-674-72454-9 (alk. paper)
1. Aristotle. 2. Logic, Ancient. 3. Modality (Logic) 4. Syllogism.
5. Predicate (Logic) 6. Aristotle. Prior analytics. I. Title.
B491.L8M35 2013
160.92—dc23 2012049482

To the memory of my brother,
Jurij

Contents

Acknowledgments

In the course of writing this book, I have incurred many debts, both intellectual and personal. The project originated in my master's studies at the University of Leipzig and evolved during my doctoral and postdoctoral work at Humboldt University of Berlin. I am especially indebted to my supervisor, Christof Rapp, for his invaluable guidance and support. Without him, it is unlikely that I would have written this book. He and my fellow graduate students created an ideal environment for studying Aristotle at Humboldt University. It was an exhilarating experience. During most of that time, my work was supported by funding from the *Studienstiftung des deutschen Volkes.*

I was also very fortunate to work with Paolo Crivelli and Benjamin Morison at Oxford in 2005–6. They read several drafts of parts of my dissertation, and their comments did much to help shape my views. Moreover, I benefited from a stimulating seminar they gave on Aristotle's assertoric syllogistic in the same year. Ben and Paolo taught me, each in his own way, how to approach Aristotle's logic.

I am grateful to Alan Code, who kindly hosted me at Rutgers University during a sabbatical semester in 2008–9 (funded by the Berlin *Topoi* cluster). This helped me turn the dissertation into a book manuscript. Parts of the manuscript were presented in a graduate seminar I co-taught with Jonathan Beere in Berlin. I thank Jonathan and the participants in that seminar for much interesting discussion. I have also learned from conversations with the ancient philosophers from the *Topoi* community. More recently, I am grateful to my colleagues and students at the University of Chicago for the supportive and invigorating community they help to create.

Papers based on drafts of parts of this book have been presented to audiences at Berkeley, Berlin, Bloomington, Chicago, Harvard, Milwaukee,

Munich, Oxford, Rostock, Rutgers, Syracuse, Toronto, UCLA, and Wellington. I am grateful for the many helpful comments I received on these occasions. David Charles, Christian Pfeiffer, and Barbara Vetter offered helpful suggestions on parts of earlier drafts. I thank Nic Koziolek, Robin Smith, and an anonymous referee, who read the whole manuscript and generously provided pages of valuable comments. This book has been much improved by their efforts.

Parts of the interpretive framework presented here, especially in Appendix B, were derived from a paper entitled "A Reconstruction of Aristotle's Modal Syllogistic" (*History and Philosophy of Logic* 2006). In general, this book follows the same approach outlined in the paper, although it deviates from it in a number of ways. Also, Chapters 1–6 of this book are informed by discussions presented in two earlier papers entitled "A Non-Extensional Notion of Conversion in the *Organon*" (*Oxford Studies in Ancient Philosophy* 2009) and "*TΩI* vs *TΩN* in *Prior Analytics* 1.1–22" (*Classical Quarterly* 2008). Chapter 9 relies on "Categories in *Topics* I.9" (*Rhizai* 2007).

I thank my colleague and friend Jacob Rosen for countless hours of enjoyable conversations. They have been a constant source of inspiration and motivation. More specifically, I have greatly benefited from our collaboration on some portions of Aristotle's modal logic which are not treated in this book ("A Method of Modal Proof in Aristotle," *Oxford Studies in Ancient Philosophy* 2012; and "Proof by Assumption of the Possible in *Prior Analytics* 1.15," forthcoming in *Mind*).

Finally, I thank my parents for their support. While I was working on the manuscript, my brother, Jurij, struggled with his health and died in 2010 at the age of 21. This book is dedicated to his memory.

Abbreviations of Aristotle's Works

APost.	*Posterior Analytics*
APr.	*Prior Analytics*
Cael.	*de Caelo*
Cat.	*Categories*
de An.	*de Anima*
EE	*Eudemian Ethics*
EN	*Nicomachean Ethics*
GA	*de Generatione Animalium*
HA	*Historia Animalium*
Int.	*de Interpretatione*
MA	*de Motu Animalium*
Met.	*Metaphysics*
Meteorol.	*Meteorology*
PA	*de Partibus Animalium*
Phys.	*Physics*
Top.	*Topics*

Aristotle's Modal Syllogistic

Introduction

Aristotle was the first to undertake a systematic study of deductive inference. He is therefore considered the founder of logic. At the heart of his mature logical theory lies the assertoric syllogistic, presented in chapters 1.1–2 and 1.4–7 of the *Prior Analytics.* The assertoric syllogistic deals with inferences that consist of nonmodal propositions such as 'A belongs to all B' or 'A does not belong to some B'. Now, Aristotle is the founder not only of logic but also of modal logic. He developed a system of modal syllogistic, presented in *Prior Analytics* 1.3 and 1.8–22. There he is concerned with modalized propositions, that is, with propositions that contain modal qualifiers such as 'necessarily' and 'possibly'. Typical examples would be 'A necessarily belongs to all B' and 'A possibly belongs to no B'. Aristotle examines a large number of inferences involving such modalized propositions and determines which of these inferences are valid and which not.

One of the first things to note about the modal syllogistic is that it is considerably more complex than the assertoric one. It gives rise to a number of interpretive problems and has been the subject of controversy since antiquity. Commentators have identified various mistakes in the modal syllogistic, especially when reading it in the light of modern logic. Thus Łukasiewicz, in his influential 1957 book *Aristotle's Syllogistic from the Standpoint of Modern Formal Logic,* writes, "Aristotle's modal syllogistic is almost incomprehensible because of its many faults and inconsistencies" (133).

In subsequent years, commentators have attempted to challenge Łukasiewicz's verdict and to vindicate the modal syllogistic. However, these attempts have generally not been successful, which has led to some skepticism about the modal syllogistic. Smith aptly writes,

> In recent years, interpreters have expended enormous energy in efforts to find some interpretation of the modal syllogistic that is consistent and nevertheless preserves all (or nearly all) of Aristotle's results; generally, the outcomes of such attempts have been disappointing. I believe this simply confirms that Aristotle's system is incoherent and that no amount of tinkering can rescue it. (Smith 1995: 45)[1]

The view that the modal syllogistic is incoherent is widely shared, for example, by the Kneales, Hintikka, and Striker.[2] Similarly, it is often thought that Aristotle's claims about the validity and invalidity of inferences in the modal syllogistic include substantive mistakes.[3] However, there is no consensus as to exactly where the supposed mistakes or incoherences lie. Depending on the interpretive framework employed, commentators offer different diagnoses of what went wrong and why.

In view of this situation, the aim of this book is to explore the prospects for understanding the modal syllogistic as a coherent and consistent logical system. I argue that the obstacles to such an interpretation can eventually be overcome, albeit at the cost of some interpretive complexity. I develop a model that matches all of Aristotle's claims about the validity and invalidity of inferences in the modal syllogistic. Thus it will be shown that the set of these claims is consistent and that, with respect to the proposed model, these claims do not contain mistakes. Of course, this does not mean that the modal syllogistic is free of problems. Far from it. But the model will help us to see more clearly precisely where the problems lie, and where not.

In addition to proposing a model that matches Aristotle's claims of validity and invalidity, I seek to explain why Aristotle made the claims

1. Smith makes the same point elsewhere (1989: xxviii and 2011: 240–1).

2. Authors who hold this view include Łukasiewicz (1957: 133, 181, and 198), Kneale & Kneale (1962: 86–91), Hintikka (1973: 140–1), van Rijen (1989: 195–9), Striker (1994: 39; 2009: xv, 115, 146, and 166–7), Thom (1996: 123–49), Mignucci (1998: 52), and Mueller (1999b: 8).

3. For example, Henle (1949: 99), Brogan (1967: 57–61), Patterson (1995: 171–85 and 194–8), Nortmann (1996: 133, 266–82, and 376), and Ebert & Nortmann (2007: 667–8).

he did make. In some cases, the proposed model itself provides such an explanation, by adequately representing Aristotle's reasons for judging a given inference valid or invalid. I pursue this kind of explanatory project for some central parts of the modal syllogistic, especially for *Prior Analytics* 1.8–12. However, I am not able to do so for the whole modal syllogistic. With regard to chapters 1.13–22, my focus is more narrowly on gathering his claims of validity and invalidity, examining their various consequences, and organizing them into a coherent model. Correspondingly, I do not attempt to analyze and reconstruct all of Aristotle's proofs of these claims. Some of his more complicated proofs remain unaccounted for in this book, due to both limitations of space and specific problems with them.

Before launching into the modal syllogistic, however, it will be helpful to have a look at the assertoric syllogistic. We will do so in Part I of the book. Part II and Part III are then devoted to the modal syllogistic. In the remainder of this introduction, I give an overview of each of the three parts.

THE ASSERTORIC SYLLOGISTIC. The assertoric syllogistic deals with non-modal propositions such as 'Every man is an animal' or 'Not every man is walking'. Aristotle usually represents these propositions by means of a somewhat artificial construction using the verb 'belong to'. For example, he would use the phrase 'A belongs to all B' instead of 'Every B is A', and 'A does not belong to some B' instead of 'Not every B is A'.[4] Aristotle focuses on four kinds of assertoric propositions, which are usually indicated by the letters 'a', 'e', 'i', and 'o'. In the secondary literature on the modal syllogistic, the lack of modal qualifiers in these propositions is often indicated by the letter 'X'. Thus the four common kinds of assertoric propositions can be written as follows:

Aa_XB	A belongs to all B
Ae_XB	A belongs to no B
Ai_XB	A belongs to some B
Ao_XB	A does not belong to some B

4. Thus, 'A does not belong to some B' does not mean 'it is not the case that A belongs to some B', but is equivalent to 'A does not belong to all B'.

If an a_X-proposition 'A belongs to all B' is true, let us say that A is a_X-predicated of B. In the same way, we will speak of e_X-predication, and so on.

In Part I of this book, we briefly consider the syntax of Aristotle's assertoric propositions (Chapter 1), and then examine their semantics (Chapters 2–5). The focus is on the semantics of a_X-propositions. Specifying their semantics means giving an account of the relation of a_X-predication. It is often thought that this relation is determined by the sets of individuals that fall under its argument-terms, as follows: A is a_X-predicated of B if and only if every individual that falls under B falls under A. Thus a_X-predication is taken to be definable by means of the relation of an individual's falling under a term.

I argue that Aristotle did not accept such a definition but that he treated a_X-predication as a primitive and undefined relation. To substantiate this claim, I examine what is known as Aristotle's *dictum de omni et de nullo,* found in the opening chapter of the *Prior Analytics.* As we will see, the *dictum de omni* can be taken to state that the primitive relation of a_X-predication is reflexive and transitive. This relation is then used in the *dictum de nullo* to specify the semantics of assertoric universal negative propositions, as follows: A is e_X-predicated of B just in case A is not a_X-predicated of anything of which B is a_X-predicated. Similar definitions will be given for i_X- and o_X-predication. The result will be an adequate semantics for Aristotle's assertoric syllogistic, based on the primitive relation of a_X-predication. In this semantics, the truth of assertoric propositions is not determined by the sets of individuals that fall under the terms involved.

MODALIZED PROPOSITIONS. As mentioned above, the modal syllogistic deals with propositions that contain modal qualifiers such as 'necessarily' and 'possibly'. In the secondary literature, the former qualifier is usually indicated by the letter 'N'. Propositions that contain this qualifier are called necessity propositions or N-propositions. There are four common kinds of these propositions:

Aa_NB	A necessarily belongs to all B
Ae_NB	A necessarily belongs to no B

$\mathrm{Ai_NB}$	A necessarily belongs to some B
$\mathrm{Ao_NB}$	A necessarily does not belong to some B

Although the label "necessity proposition" is somewhat artificial, it is preferable to the more natural "necessary proposition," for this latter label suggests that the proposition in question is necessarily true, whereas Aristotle's N-propositions need not be necessarily true. For example, 'Horse necessarily belongs to all man' is an N-proposition, since it contains the qualifier 'necessarily'; but it is not necessarily true (in fact, it is necessarily false). To avoid such potential misunderstandings, I use the label "necessity proposition." Likewise, propositions that contain the modal qualifier 'possibly' are called possibility propositions.

The modal syllogistic is concerned with inferences that consist of assertoric and modalized propositions. Aristotle first considers inferences that consist exclusively of assertoric propositions and necessity propositions (*Prior Analytics* 1.8–12). This portion of the modal syllogistic is traditionally called the apodeictic syllogistic. We will examine it in Part II of this book. Aristotle goes on to treat of inferences that involve possibility propositions (*Prior Analytics* 1.13–22). This portion is called the problematic syllogistic, and we will examine it in Part III.

THE APODEICTIC SYLLOGISTIC. Aristotle begins the apodeictic syllogistic by briefly discussing inferences that consist exclusively of necessity propositions (*Prior Analytics* 1.8). His treatment of these inferences is strictly parallel to that of assertoric inferences in the assertoric syllogistic. Aristotle next considers inferences from mixed premise pairs consisting of an assertoric proposition and a necessity proposition (*Prior Analytics* 1.9–11). In his view, some of these premise pairs yield a necessity proposition as conclusion. For example, he takes the following schema, known as Barbara NXN, to be valid:

Major premise:	$\mathrm{Aa_NB}$	A necessarily belongs to all B
Minor premise:	$\mathrm{Ba_XC}$	B belongs to all C
Conclusion:	$\mathrm{Aa_NC}$	A necessarily belongs to all C

Barbara NXN allows us to infer a necessity proposition from a necessity major premise and an assertoric minor premise. On the other hand,

Aristotle denies that a necessity proposition can be inferred when the major premise is assertoric; he takes Barbara XNN to be invalid:

Major premise:	Aa_XB	A belongs to all B
Minor premise:	Ba_NC	B necessarily belongs to all C
Conclusion:	Aa_NC	A necessarily belongs to all C

Aristotle's treatment of these two schemata has been the subject of controversy. Especially controversial is his endorsement of the former schema. Theophrastus and Eudemus, both pupils of Aristotle's, and several later Platonists denied the validity of Barbara NXN.[5] Aristotle himself offers only little explanation as to why this schema should be valid. On the other hand, its validity is fundamental to the whole modal syllogistic. Barbara NXN is arguably the most important among Aristotle's modal schemata, comparable to assertoric Barbara—that is, the transitivity of a_X-predication—in the assertoric syllogistic. Understanding why he took it to be valid is therefore key to understanding the modal syllogistic.

I argue that Aristotle's endorsement of Barbara NXN is motivated by the account of predication that he develops in the *Topics*. More specifically, I suggest that his endorsement of this schema relies on the *Topics*' theory of predicables and categories. Let me illustrate this claim by outlining some elements of that theory. The presentation is bound to be somewhat preliminary and sketchy here. A full exposition will be given in Part II.

The *Topics*' theory of predicables provides a classification of predicative relations between terms. According to it, A is predicated of B just in case A is a definition, genus, differentia, proprium, or accident of B. For example, 'animal' is predicated as a genus of 'man', 'biped' is predicated as a differentia of 'man', 'walking' as an accident, and so on. On the other hand, 'horse' is not predicated of 'man', since it is not a definition, genus, differentia, proprium, or accident of 'man'. If someone falsely asserts that every man is a horse, or that 'horse' is a genus of 'man', this does not count as a predication in the relevant sense. When A is a definition, genus, or differentia of B, then the predication is

5. See Alexander *in APr.* 124.8–30, Ammonius *in APr.* 38.38–39.2, and Philoponus *in APr.* 123.15–17, 124.9–125.18.

essential, which means that A specifies (part of) the essence of B. Otherwise the predication is accidental.

In addition, the *Topics*' theory of categories introduces, among other things, a distinction between two kinds of terms. The first group contains substance terms like 'animal' and 'man', and nonsubstance terms like 'color', 'redness', and 'motion'. Call these *essence terms.* The second group contains nonsubstance terms like 'colored', 'red', and 'moving'. Call these *nonessence terms.*

Essence terms:	'animal', 'man', 'color', 'redness', 'motion', ...
Nonessence terms:	'colored', 'red', 'walking', ...

Aristotle imposes restrictions on the predications in which essence terms can occur. Several passages from the *Topics* imply that essence terms can be predicated only of terms of which they are predicated essentially. For example, 'animal' is predicated only of terms like 'man', 'horse', and 'Socrates', and it is predicated essentially of all of them. It is not predicated of a term like 'walking', on the grounds that this term does not stand for a proper species of animal. Likewise, 'color' is predicated only of terms like 'redness', 'scarlet redness', and 'crimson redness', and it is predicated essentially of all of them. Thus essence terms are predicated essentially of everything of which they are predicated. In other words, they are a definition, genus, or differentia of everything of which they are predicated; they are not a proprium or accident of anything. On the other hand, nonessence terms are predicated nonessentially of subjects; for example, both 'red' and 'walking' may be predicated as accidents of 'man'. Of course, all of this requires further justification and explanation, which will be provided in Chapters 8–10.

The interpretation of the modal syllogistic pursued here starts from the idea that the relations of a_X- and a_N-predication are closely connected with the *Topics*' relations of predication and essential predication. They are connected in such a way, I suggest, that essence terms are a_X-predicated only of terms of which they are predicated essentially. For example, 'animal' is a_X-predicated only of terms like 'man' and 'horse', but not of terms like 'walking'. Although every individual that falls under 'walking' also falls under 'animal', the latter term is not

a_X-predicated of the former. Thus, a_X-predication is not determined by the sets of individuals that fall under the terms involved.

On this account, essence terms are predicated essentially of everything of which they are a_X-predicated. Given that essential predication implies a_N-predication, the following holds for any B:

Thesis 1: If B is an essence term, then B is a_N-predicated of everything of which it is a_X-predicated

In addition, I argue that Aristotle imposes a restriction to the effect that only essence terms can serve as the subjects of a_N-predications.[6] The reason why nonessence terms cannot serve as subjects of a_N-predications is, roughly speaking, that a_N-predications need to be grounded in the definable essence of their subjects and that nonessence terms lack such an essence. Thus, every subject of an a_N-predication is an essence term. For example, 'color' is a_N-predicated of 'redness', but 'colored' is not a_N-predicated of 'red'. The following holds for any A and B:

Thesis 2: If A is a_N-predicated of B, then B is an essence term

Given the above two theses, Aa_NB and Ba_XC imply Ba_NC. So the premise pair of Barbara NXN implies the premise pair of Barbara NNN:

Major premise:	Aa_NB	A necessarily belongs to all B
Minor premise:	Ba_NC	B necessarily belongs to all C
Conclusion:	Aa_NC	A necessarily belongs to all C

This does not mean that the premise pair of Barbara NXN and that of Barbara NNN are identical. They are distinct because their minor premises are two distinct linguistic items, differing in the presence or absence of the qualifier 'necessarily'. Nevertheless, the former premise pair implies the latter.

Now, the validity of Barbara NNN is undisputed and generally accepted. Hence the above two theses suffice to justify the validity of

6. This means that nonessence terms cannot serve as the subjects of true a_N-propositions, although they can serve as the subjects of false a_N-propositions.

Barbara NXN: the premise pair of Barbara NXN implies that of Barbara NNN, and therefore yields an a_N-conclusion. Of course, Aristotle does not explicitly state the two theses in the modal syllogistic. Nonetheless, I argue that he endorsed them, based on the *Topics*' theory of predication, and that he took them to justify the validity of Barbara NXN (Chapters 8–10). On the other hand, the two theses do not entail the validity of Barbara XNN. Thus they help explain Aristotle's treatment of the two Barbaras in the apodeictic syllogistic.

THE SEMANTICS OF E_N-PROPOSITIONS. Let us consider another example of how the two theses can help explain the apodeictic syllogistic. The example concerns Aristotle's treatment of universal negative necessity propositions (e_N-propositions). On the one hand, Aristotle claims that Celarent NXN is valid (*Prior Analytics* 1.9):

Major premise:	Ae_NB	A necessarily belongs to no B
Minor premise:	Ba_XC	B belongs to all C
Conclusion:	Ae_NC	A necessarily belongs to no C

On the other hand, he takes the following rule of e_N-conversion to be valid (*Prior Analytics* 1.3):

Premise:	Ae_NB	A necessarily belongs to no B
Conclusion:	Be_NA	B necessarily belongs to no A

It is not immediately clear how e_N-propositions can be interpreted in such a way that both of these inferences are validated. Becker (1933) and others think that there is no such interpretation. They hold that Aristotle's endorsement of the two inferences is a sign of incoherence, resulting from a confusion between a *de re* and a *de dicto* reading of e_N-propositions.[7] As we will see in Chapter 11, the former reading validates Celarent NXN but not e_N-conversion, whereas the latter reading validates e_N-conversion but not Celarent NXN. Hence it is often thought

7. On the *de re* reading, A is e_N-predicated of B just in case for every individual that falls under B, it is necessary that this individual does not fall under A. On the *de dicto* reading, the e_N-predication obtains just in case it is necessary that no individual falls under both A and B.

that Aristotle's use of e_N-propositions is ambiguous: when asserting e_N-conversion, he has in mind the *de dicto* reading, and when asserting Celarent NXN, he has in mind the *de re* reading.[8] Since in some passages Aristotle simultaneously relies on the validity of Celarent NXN and e_N-conversion, his modal syllogistic seems to be incoherent.

This view assumes that the *de re* and *de dicto* readings are the only options available for an interpretation of e_N-propositions. However, there are alternative readings of e_N-propositions that validate both of Aristotle's inferences. Within the framework introduced above, such a reading can be specified as follows:

Ae_NB if and only if (i) A and B are essence terms, and
(ii) A is e_X-predicated of B

On this reading, e_N-predications are exactly those e_X-predications in which both terms are essence terms. For example, 'man' is e_N-predicated of 'horse', and 'redness' of 'whiteness', but 'red' is not e_N-predicated of 'white'. As we will see, this account is confirmed by Aristotle's treatment of e_N-predication in *Prior Analytics* 1.34.

Given the rule of e_X-conversion (whose validity is uncontroversial), the above definition of e_N-predication is clearly symmetric and hence validates the rule of e_N-conversion. Moreover, given the above two theses, it also validates Celarent NXN. To see this, suppose that A is e_N-predicated of B, and B is a_X-predicated of C. Then conditions (i) and (ii) are satisfied. In order to establish that A is e_N-predicated of C, it remains to show that

(iii) C is an essence term, and
(iv) A is e_X-predicated of C

The first of these two conditions can be established as follows. Given (i), B is an essence term. Since B is a_X-predicated of C, it follows that B is a_N-predicated of C (Thesis 1), and hence C is an essence term (Thesis 2).

8. Becker (1933: 42), Kneale & Kneale (1962: 89–91), Hintikka (1973: 139–40), Sorabji (1980: 202), Striker (1994: 40–1; 2009: xvi–xvii and 115), and Mignucci (1998: 50–2 and 61–2). A similar ambiguity of e_N-propositions is diagnosed by Patterson (1989: 15–16; 1990: 155–66; 1995: 41–87); see p. 16n11 below.

Condition (iv) follows straightforwardly from (ii) and the premise that B is a_X-predicated of C, by means of purely assertoric Celarent. Thus Celarent NXN is valid on the proposed reading of e_N-propositions.

In sum, both inferences are validated by a single reading of e_N-propositions. Contrary to what is often thought, there is no need to attribute to Aristotle an ambiguity or incoherence in his use of e_N-propositions. Thus, the appeal to the *Topics*' theory of predication can help us understand the modal syllogistic as a coherent system within the wider context of Aristotle's logic.

THE PROBLEMATIC SYLLOGISTIC. The problematic syllogistic deals with inferences that involve possibility propositions. Aristotle distinguishes two kinds of these propositions, traditionally referred to as one-sided and two-sided possibility propositions. Being two-sided possible means being neither impossible nor necessary, and being one-sided possible simply means being not impossible. Thus, two-sided possibility precludes necessity, whereas one-sided possibility does not. For example, 'Possibly no man is a horse' is true if understood as a one-sided possibility proposition, but not if understood as a two-sided possibility proposition. Two-sided possibility propositions prevail in Aristotle's modal syllogistic. In the secondary literature, they are usually indicated by the letter 'Q'. Aristotle distinguishes four kinds, which can be represented as follows:

Aa_QB	A two-sided-possibly belongs to all B
Ae_QB	A two-sided-possibly belongs to no B
Ai_QB	A two-sided-possibly belongs to some B
Ao_QB	A two-sided-possibly does not belong to some B

Likewise, one-sided possibility propositions are indicated by the letter 'M', and represented by formulae such as Aa_MB and so on.

Aristotle's treatment of possibility propositions raises several questions and difficulties. One of them concerns their logical relation to necessity propositions. It is clear from some of his proofs in the problematic syllogistic that Aristotle endorses the following principles of incompatibility (see pp. 199–201 below):

Aa_QB	is incompatible with	Ae_NB
Ai_QB	is incompatible with	Ae_NB

At the same time, however, Aristotle commits himself to denying some other such principles. Most strikingly, he is committed to denying that e_Q-propositions are incompatible with the corresponding a_N-propositions. This follows from his claim that the premise pair Be_QA, Ba_NC does not yield any conclusion (1.19 38a26–b4); for if e_Q-propositions were incompatible with a_N-propositions, it would be easy to prove that this premise pair yields the conclusion Ao_XC.[9] Aristotle is therefore committed to the view that

Ae_QB is compatible with Aa_NB

If we are to verify Aristotle's claims of invalidity in the modal syllogistic, we have to accept that some e_Q-predications coincide with a_N-predications. This consequence is counterintuitive and constitutes a major difficulty in interpreting the problematic syllogistic. It is often taken to show that the modal syllogistic is inconsistent.[10] By contrast, I argue that the modal syllogistic can be viewed as consistent if we refrain from attributing to Aristotle certain principles of modal opposition, that is, principles about the incompatibility of possibility propositions and necessity propositions. Aristotle's treatment of these principles is unusual and asymmetric, but nevertheless coherent; for, as we will see, he consistently commits himself to denying principles of modal opposition for affirmative necessity propositions while accepting them for negative necessity propositions (Chapters 13 and 14).

This approach to modal opposition will enable us to construct an adequate deductive system for the entire modal syllogistic (Chapter 15). Every schema held to be valid by Aristotle in the modal syllogistic will

---

9. Suppose Be_QA and Ba_NC. In order to establish the conclusion Ao_XC, assume for *reductio* its contradictory, Aa_XC. From Be_QA and Aa_XC we infer Be_QC by means of the syllogism Celarent QXQ (which Aristotle accepts as valid). So if Be_QC were incompatible with Ba_NC, the proof by *reductio* would be successful.

10. McCall (1963: 93), Patterson (1995: 194–8), Thom (1996: 128–9), Nortmann (1996: 279), Ebert & Nortmann (2007: 655 and 667–8), and Striker (2009: 161); similarly, Johnson (2004: 303).

be deducible in this system, but no schema held to be invalid by him will be deducible in it.

THE PREDICABLE SEMANTICS OF THE MODAL SYLLOGISTIC. Another characteristic feature of the problematic syllogistic is that Aristotle is committed to a rather strong principle about the realization of certain Q-predications. As we will see in Chapter 16, Aristotle's endorsement of the schema Ferio XQM implies that the following holds for any A and B:

If B is a substance term, and $\mathrm{Ai_QB}$, then $\mathrm{Ai_XB}$

In other words, every term that is $\mathrm{i_Q}$-predicated of a substance term is also $\mathrm{i_X}$-predicated of it. For example, if 'sitting' is $\mathrm{i_Q}$-predicated of the substance term 'man', then it is also $\mathrm{i_X}$-predicated of it. Thus, every $\mathrm{i_Q}$-predication whose subject is a substance term is realized or actualized by the corresponding $\mathrm{i_X}$-predication. On the face of it, this principle seems implausible. Why cannot some man be two-sided-possibly sitting while no man is actually sitting? Surely Aristotle should accept that some man is two-sided-possibly sitting. At the same time, there are several passages in the modal syllogistic in which Aristotle assumes that every man is moving, and hence presumably that no man is sitting. This and other problems with the above principle about the realization of $\mathrm{i_Q}$-predications are discussed in Chapter 16. Nevertheless, the principle follows from Aristotle's endorsement of Ferio XQM, and must therefore be taken on board for an adequate interpretation of his modal syllogistic.

Recognition of this fact, along with the asymmetric approach to modal opposition outlined above, will put us in a position to specify a semantic interpretation of Aristotle's possibility propositions (Chapters 17 and 18). This interpretation, which I call the predicable semantics of the modal syllogistic, will be based on three primitive relations:

(i) $\mathrm{a_X}$-predication
(ii) $\mathrm{a_N}$-predication
(iii) strong $\mathrm{a_N}$-predication

The third relation, that of strong a_N-predication, picks out those a_N-predications whose subject is a substance term and which do not coincide with an o_M-predication (and hence also not with an e_Q-predication). The three primitive relations suffice to introduce into the predicable semantics some notions of the *Topics*' theory of predication. For example, they can be used to define the notions of substance term and essence term as follows:

B is an essence term	if and only if	there is an A that is a_N-predicated of B
B is a substance term	if and only if	there is an A that is strongly a_N-predicated of B

With the help of these definitions, all of Aristotle's assertoric and modalized propositions will be given a semantic interpretation within the predicable semantics. Thus, Aristotle's N-, Q-, M-, and X-propositions are interpreted solely by means of the above three primitive relations.

As we will see, some of these interpretations are complex and technical, especially those of Q- and M-propositions. When such interpretations are involved, the predicable semantics does not likely reflect the reasons Aristotle had for his claims of validity and invalidity, and hence does not explain why he made the claims he made. In some cases, I argue, the predicable semantics does provide such an explanation, particularly for the apodeictic syllogistic (see p. 270). But the predicable semantics is not intended to provide this kind of explanation for the problematic syllogistic. Moreover, the predicable semantics cannot account for all of Aristotle's proofs in the modal syllogistic (see pp. 268–269). Nevertheless, it is useful as a model of the modal syllogistic in that it verifies all of Aristotle's claims of validity and invalidity: every inference held to be valid by Aristotle in the modal syllogistic will be valid in the predicable semantics, and every inference held to be invalid by him will be invalid in it. In this sense, the predicable semantics establishes the consistency of the modal syllogistic.

PREVIOUS RECONSTRUCTIONS OF THE MODAL SYLLOGISTIC. As mentioned above, a lot of work has been done over the past decades to provide a formal reconstruction of the modal syllogistic. I cannot enter here into a detailed discussion of this work. But I want to give a brief overview of its main directions and to indicate how my own approach fits into it.

The project of interpreting the modal syllogistic from the perspective of modern formal logic was initiated by the pioneering work of Becker (1933). He pointed out a number of problems faced by attempts to translate Aristotle's modal syllogistic into modal first-order logic. For instance, Becker diagnosed the supposed ambiguity between *de re* and *de dicto* readings of necessity propositions that we saw above. This diagnosis has been widely influential ever since.

Thirty years later, McCall (1963) put forward a syntactic calculus for the modal syllogistic, without giving a semantic interpretation for it. His calculus is adequate for the apodeictic syllogistic, being in accordance with all of Aristotle's claims of validity and invalidity there. However, it is not adequate in this way for the problematic syllogistic, partly because McCall attributes to Aristotle some of the troublesome principles of modal opposition mentioned above. The deductive system for the modal syllogistic given in Chapter 15 builds on McCall's calculus but avoids its difficulties with modal opposition.

In the last three decades, scholars adopted a more semantic approach than McCall's, interpreting Aristotle's modalized propositions within various logical frameworks. Although these interpretations are quite diverse, we may identify three kinds of frameworks that have been used to interpret the modal syllogistic. First, some authors employ a set-theoretic framework in which terms are assigned several sets of individuals; this approach is taken by Johnson (1989, 2004), Thom (1991, 1996), and Thomason (1993, 1997). Second, others employ the framework of modern modal first-order logic; for example, Brenner (1993, 2000), Nortmann (1996), Schmidt (2000), and Rini (2011). Third, Patterson's (1995) semiformal interpretation employs the framework of Aristotle's predicables.

My own approach belongs to the third group, relying on the *Topics*' theory of predicables. The present book has benefited a lot from Patterson's inspiring work. It adopts some of his ideas, although it develops them in a different way. One difference is that Patterson attributes to Aristotle an ambiguity of necessity propositions, much the same as Becker's ambiguity between *de re* and *de dicto* readings.[11] Consequently, Patterson's account is nonuniform in the sense that necessity propositions are given different interpretations in different contexts. By contrast, the predicable semantics developed in the present book is uniform in that every kind of modal and assertoric proposition is given the same interpretation throughout the whole modal syllogistic.[12]

My approach also has some similarities with Rini's work (2011). She argues that the modal syllogistic is based on a distinction between two kinds of terms, which she calls red and green terms: the former are predicated necessarily of everything of which they are predicated, the latter not. In the present book, this corresponds to the distinction between essence terms and nonessence terms introduced above: the former are a_N-predicated of everything of which they are a_X-predicated, the latter not. Like Rini, I will emphasize the importance of this distinction for the modal syllogistic, but I will put it to use in a different way than she does.[13]

Some of the studies mentioned above offer formal models that match all of Aristotle's claims of validity and invalidity in the apodeictic

---

11. Patterson (1989: 15–16; 1990: 155–66; 1995: 41–87) diagnoses an ambiguity between what he calls weak *cop* and strong *cop* readings. Brennan (1997: 229–30) and Rini (2011: 55) argue that this distinction is basically equivalent to that between *de re* and *de dicto* readings.

12. Further examples of interpretations that fail to be uniform in this sense are Nortmann (1996: 51–4 and 125–6) and Schmidt (2000: 31–4).

13. Rini does not make the distinction explicit in the semantic interpretation of modalized propositions. Instead, she introduces it implicitly in external restrictions on the kinds of terms that are allowed to occur in certain contexts. Thus, she assumes that some inferences held to be valid by Aristotle are valid not for every choice of terms, but only when certain terms in them are taken to be red terms (Rini 2011: 43–4, 79–80, and 153–5). This raises questions about the status of the modal syllogistic as a system of formal logic. In the present book, by contrast, the distinction will be made explicit in the semantic interpretation of modalized propositions, so that inferences will be valid for every choice of terms without restriction.

syllogistic.[14] Several studies offer formal models covering both the apodeictic and the problematic syllogistic.[15] Unlike the predicable semantics, however, none of these models matches all of Aristotle's claims of validity and invalidity.[16]

14. McCall (1963), Johnson (1989, 1993, 1995, 2004), Thomason (1993, 1997), and Brenner (1993, 2000).

15. McCall (1963), Patterson (1995), Nortmann (1996), Thom (1996), Schmidt (2000), Johnson (2004), and Rini (2011).

16. For the places where the models diverge from Aristotle's claims, see McCall (1963: 93), Patterson (1995: 194–8), Nortmann (1996: 133, 266–82, and 376), Thom (1996: 123–49), Schmidt (2000: 58–9, 174, and 178), Johnson (2004: 303), and Rini (2011: 100–2).

I

The Assertoric Syllogistic

Aristotle's syllogistic is concerned with categorical propositions. These have a tripartite syntax, consisting of a subject term, a predicate term, and a copula. For example, the assertoric proposition 'A belongs to all B' consists of the subject term B, the predicate term A, and the copula 'belongs to all'. Assertoric propositions are not modalized; that is, they contain no modal qualifiers. In the first seven chapters of the *Prior Analytics,* Aristotle develops a deductive system of these propositions, based on conversion rules and perfect first-figure schemata (Chapter 1).

Aristotle does not describe the semantics of his assertoric propositions in any detail. Nevertheless, it is important for an interpretation of his syllogistic to give an account of their semantics. In what follows, we will focus on the semantics of a_X-propositions, that is, of assertoric universal affirmative propositions (Chapters 2–5). When an a_X-proposition 'A belongs to all B' is true, let us say that A is a_X-predicated of B. Aristotle characterizes the relation of a_X-predication in his *dictum de omni,* found in the opening chapter of the *Prior Analytics.* In formulating this *dictum,* he seems to assume that every argument-term of a categorical proposition is assigned a certain plurality of items (without specifying what these items are). Based on this, his *dictum de omni* states that A is a_X-predicated of B just in case every member of the plurality associated with B is a member of the plurality associated with A. Aristotle also mentions a *dictum de nullo,* characterizing the semantics of e_X-propositions, that is, of assertoric universal negative propositions. It states that A is e_X-predicated of B just in case no member of the plurality associated with B is a member of the plurality associated with A. This reading of the *dictum de omni et de nullo* leads to what I call the abstract *dictum* semantics of the assertoric syllogistic (Chapter 2).

The abstract *dictum* semantics is abstract in that it does not specify what the plurality associated with a term is. It is often thought that this plurality should be identified with the set of individuals that fall under a given term. For example, the plurality associated with the term 'animal' would be a set consisting of individuals such as Socrates, Odysseus, and his dog Argos. Accordingly, the *dictum de omni* is taken to read: A is a_X-predicated of B just in case every individual that falls under B also falls under A. Similarly, the *dictum de nullo* is taken to read: A is e_X-predicated of B just in case no individual that falls under B falls under A. This reading leads to what may be called the orthodox *dictum* semantics of the assertoric syllogistic. I argue that Aristotle did not endorse this reading of the *dictum de omni et de nullo* and that the orthodox *dictum* semantics should be rejected (Chapter 3).

I go on to argue that the plurality associated with a term A should instead be identified with the set of those items of which A is a_X-predicated. For example, the plurality associated with the term 'animal' consists of terms such as 'man', 'horse', and 'dog' (or of the species signified by these terms). Thus, the *dictum de omni* reads: A is a_X-predicated of B just in case A is a_X-predicated of everything of which B is a_X-predicated. The *dictum de nullo* reads: A is e_X-predicated of B just in case A is not a_X-predicated of anything of which B is a_X-predicated. This leads to what may be called the heterodox *dictum* semantics of the assertoric syllogistic. In this semantics, unlike in the orthodox one, a_X-predication is not defined by means of more primitive notions. Rather, it is regarded as a primitive relation that is reflexive and transitive. The relations of e_X-, i_X-, and o_X-predication are defined in terms of this primitive relation. I defend the heterodox *dictum* semantics against potential objections. One of its advantages over the orthodox *dictum* semantics is that it provides a solution to the well-known problem of existential import (Chapter 4).

The heterodox *dictum* semantics naturally determines a class of first-order models. This class of models may be called the preorder semantics of the assertoric syllogistic. I discuss the preorder semantics in detail, comparing it to the set-theoretic semantics—the natural class of models determined by the orthodox *dictum* semantics (Chapter 5). Finally, we will consider Aristotle's proofs by ecthesis in the assertoric syllogistic. I argue that these proofs are based on Aristotle's *dictum de omni et de nullo* and that they can be adequately reconstructed within the abstract and heterodox *dictum* semantics (Chapter 6).

Chapters 5 and 6 are not presupposed by the remainder of the book. Readers primarily interested in the modal syllogistic may omit them.

1

Categorical Propositions

THE TRIPARTITE SYNTAX OF CATEGORICAL PROPOSITIONS. In the first chapter of the *Prior Analytics*, Aristotle begins his investigation of deductive inference by clarifying what a proposition (πρότασις) is:

> A proposition is a sentence affirming or denying something of something.[1] (*APr.* 1.1 24a16–17)

Propositions are sentences, that is, linguistic expressions of a certain language.[2] Every proposition contains two constituents, one of which is affirmed or denied of the other. Aristotle calls these constituents 'terms' (ὅροι):

---

1. πρότασις μὲν οὖν ἐστὶ λόγος καταφατικὸς ἢ ἀποφατικός τινος κατά τινος. Here and in what follows, translations of passages from the *Prior Analytics* are taken, with some modifications, from Smith (1989) or Striker (2009).

2. Aristotle characterizes propositions as λόγοι. In doing so, he seems to rely on his discussion of λόγοι in *de Interpretatione* 4 and 5 (see Alexander *in APr.* 10.13–12.3, Smith 1989: xvii). There, Aristotle defines a λόγος as a significant utterance, that is, a linguistic expression of a certain kind (*Int.* 4 16b26). Hence the propositions introduced in *Prior Analytics* 1.1, too, seem to be linguistic expressions rather than nonlinguistic items (Crivelli & Charles 2011: 194, Crivelli 2012: 114). This justifies translating λόγος at *APr.* 1.1 24a16 as 'sentence' (Smith 1989: 1, Striker 2009: 1, Crivelli 2012: 113).

> I call a term that into which a proposition may be broken up, that is, both what is predicated and what it is predicated of, with the addition of 'to be' or 'not to be'.[3] (*APr.* 1.1 24b16–18)

Terms are parts of propositions. Hence given that propositions are linguistic items, terms seem to be linguistic items as well.[4] Typical examples of terms used in the *Prior Analytics* are 'man', 'stone', and 'walking'. The term that is affirmed or denied of something in a given proposition is traditionally called the predicate of that proposition. The term of which it is affirmed or denied is called the subject of the proposition.

In order to compose a proposition out of two terms, a third constituent needs to be added. In the passage just quoted, this constituent is referred to as 'to be' or 'not to be' (τὸ εἶναι ἢ μὴ εἶναι); commentators usually call it the copula. The alphabet of Aristotle's language may then be taken to consist of two kinds of expressions: terms and copulae. Propositions are obtained by adding a copula to an ordered pair of terms, which serve as the predicate and as the subject. Since antiquity, these propositions have been called categorical propositions.[5] So categorical propositions are simple declarative sentences with a tripartite syntax, consisting of a subject term, a predicate term, and a copula.

THE QUALITY OF CATEGORICAL PROPOSITIONS. Aristotle classifies categorical propositions according to what is known as their quantity, quality, and modality. This classification is as follows:

> [Modality:] Every proposition expresses either belonging, or belonging of necessity, or being possible to belong; [quality:] and some of these are affirmative and others negative, for each prefix respectively; [quantity:] and of the affirmative and negative premises, in turn, some are universal, others are particular, and others indeterminate. (*APr.* 1.2 25a1–5)

---

3. ὅρον δὲ καλῶ εἰς ὃν διαλύεται ἡ πρότασις, οἷον τό τε κατηγορούμενον καὶ τὸ καθ' οὗ κατηγορεῖται, προστιθεμένου τοῦ εἶναι ἢ μὴ εἶναι. This text incorporates Ross's (1949: 290) widely accepted excision of ἢ διαιρουμένου after προστιθεμένου.

4. Crivelli (2012: 115).

5. See Bobzien (2002: 364n18).

Let us first consider the distinction in quality, that is, the distinction between affirmative and negative propositions. In affirmative propositions, the predicate is affirmed of the subject; examples are 'All pleasure is good' and 'Some pleasure is good'. In negative propositions, the predicate is denied of the subject; examples are 'No pleasure is good' and 'Some pleasure is not good'.

In the tripartite syntax of categorical propositions, negative propositions are not obtained by applying a negative constituent to an affirmative proposition. Instead, they are obtained by applying a negative copula (τὸ μὴ εἶναι) instead of an affirmative one (τὸ εἶναι) to two terms.[6] In the above examples, the words "not" and "no" do not correspond to an independent constituent in the tripartite syntax of categorical propositions. Rather, they are part of the copula, indicating that the copula applied to the two argument-terms is a negative instead of an affirmative one.

Things are different when it comes to describing the semantics of categorical propositions (a task we will embark upon in Chapters 2–5). Trained in twentieth-century logic, we may wish to describe their semantics in a language that recognizes negations as independent syntactic constituents. For example, the semantic interpretation of the syllogistic to be developed in this study will use sentential negation operators of modern propositional logic. Aristotle's language of categorical propositions, on the other hand, does not recognize such operators as independent negative constituents.

THE QUANTITY OF CATEGORICAL PROPOSITIONS. According to quantity, Aristotle distinguishes between universal, particular, and indeterminate propositions. Indeterminate propositions differ from universal and particular ones in the absence of quantifying expressions such as 'all', 'no', or 'some'; for instance, 'Pleasure is good' and 'Pleasure is not good'. Examples of particular propositions are 'Some pleasure is good' and 'Some pleasure is not good'. Examples of universal propositions are 'All pleasure is good' and 'No pleasure is good'.

6. See Smith (1989: xvii), Patterson (1995: 17–18), Whitaker (1996: 80–1), and Charles (2000: 380). Aristotle indicates the difference between affirmative and negative copulae at *APr.* 1.1 24b18, 1.8 30a1, and *Int.* 12 21b27.

The presence or absence of quantifying expressions does not affect the tripartite syntax of categorical propositions. Like universal and particular propositions, indeterminate propositions are categorical propositions consisting of two terms and a copula. Quantifying expressions do not correspond to an independent constituent in the tripartite syntax of categorical propositions. Rather, they seem to be part of the copula, indicating that the copula is universal or particular.[7] Thus, universal and particular propositions are not obtained by applying a quantificational constituent to an indeterminate proposition. They are obtained by applying a universal or particular copula, instead of an indeterminate one, to two terms.

Again, in order to describe the semantics of categorical propositions, one may wish to use a language that recognizes independent quantificational constituents. For instance, the semantic interpretation to be developed in this study makes use of quantifiers of modern first-order logic. Aristotle's language of categorical propositions, however, does not recognize such quantifiers.

In the passage quoted above, indeterminate propositions are put on an equal footing with particular and universal ones. Nevertheless, they play only a minor role in Aristotle's syllogistic. Throughout chapters 1.2–22, Aristotle consistently focuses on particular and universal propositions. Where he does mention indeterminate propositions, he treats them as similar in logical force, or as equivalent, to particular propositions.[8] Following Aristotle's lead, we will concentrate on particular and universal propositions in this study.

THE MODALITY OF CATEGORICAL PROPOSITIONS. According to modality, Aristotle distinguishes between assertoric propositions, necessity propositions, and possibility propositions. As mentioned in the Introduction (p. 11), he makes a further distinction within possibility propositions, differentiating between one- and two-sided possibility propositions.

Assertoric propositions are nonmodal propositions. They differ from necessity and possibility propositions by the absence of modally qualifying expressions like 'necessarily' and 'possibly'. Examples of assertoric

7. See Patterson (1995: 19–21).

8. See pp. 284–285 below.

propositions are 'No man is a horse' and 'Some horse is beautiful'. Although the former proposition is true by necessity, it is not a necessity proposition, because it does not contain a modally qualifying expression. For the same reason, the latter proposition is not a possibility proposition, although it is possibly true. On the other hand, 'Some horse necessarily is a man' is a necessity proposition although it is not necessarily true (in fact, it is necessarily false).

The presence of modally qualifying expressions does not affect the tripartite syntax of categorical propositions. Necessity and possibility propositions have the same tripartite syntax as assertoric ones. Modally qualifying expressions do not correspond to an independent constituent in the tripartite syntax of categorical propositions. Rather, they seem to be part of the copula, indicating that the copula is not assertoric but modally qualified.[9] Consequently, necessity and possibility propositions are not obtained by applying a modally qualifying constituent to an assertoric proposition. In particular, they are not obtained by adding a modal sentential operator to an assertoric proposition. Rather, necessity and possibility propositions are obtained by applying a necessity or possibility copula, instead of an assertoric one, to two terms. Thus, modalized propositions result from applying a modally qualified copula to two terms, whereas assertoric propositions result from applying a modally unqualified copula to two terms.

This is confirmed by a passage from chapter 1.8 of the *Prior Analytics,* in which Aristotle states that necessity propositions differ from assertoric ones "in the addition of 'necessarily belonging' or '[necessarily] not belonging' to the terms."[10] In chapter 1.1, recall, Aristotle stated that categorical propositions result from two terms by "the addition of 'to be' or 'not to be.'"[11] In this passage, "to be" refers to an affirmative copula, whereas "not to be" refers to a negative copula. Likewise, "necessarily belonging" in the passage from chapter 1.8 seems to refer

9. See Patterson (1995: 15–17), Charles (2000: 381–7), and Raymond (2010: 195–9). That Aristotle takes modal expressions to be part of the copula seems to be confirmed by *Int.* 12 21b26–30 and 22a8–10; see Charles (2000: 381–3); similarly, Whitaker (1996: 159).

10. διοίσει τῷ προσκεῖσθαι τοῖς ὅροις τὸ ἐξ ἀνάγκης ὑπάρχειν ἢ μὴ ὑπάρχειν, *APr.* 1.8 29b39–30a2.

11. προστιθεμένου τοῦ εἶναι ἢ μὴ εἶναι, *APr.* 1.1 24b17–18; see p. 24 above.

to an affirmative necessity copula, whereas "[necessarily] not belonging" seems to refer to a negative necessity copula. Thus, the passage from chapter 1.8 suggests that necessity propositions result from the addition of a necessity copula to two terms.[12] Correspondingly, necessity propositions differ from assertoric ones in that a necessity copula is added to two terms instead of an assertoric copula.

Again, this is not to deny that one might wish to use sentential modal operators to describe the semantics of categorical propositions. As a matter of fact, however, I will not make use of such operators to specify a semantics for the modal syllogistic.

THE TRADITIONAL NOTATION OF CATEGORICAL PROPOSITIONS. The quality and quantity of categorical propositions is traditionally indicated by the following four letters:

a	universal affirmative
e	universal negative
i	particular affirmative
o	particular negative

(There are no special letters for indeterminate propositions, since these are of minor importance in the syllogistic.) In the more recent secondary literature, the modality of categorical propositions is often indicated by the following four letters:

X	assertoric
N	necessity
Q	two-sided possibility
M	one-sided possibility

Thus, for example, the four major kinds of assertoric propositions are written as follows:

$\mathrm{Aa_XB}$	A belongs to all B
$\mathrm{Ae_XB}$	A belongs to no B

12. See Patterson (1995: 17). In both passages the addition of the copula to the two terms is indicated by the same verb, namely, προστίθεσθαι.

$\mathrm{Ai_X B}$	A belongs to some B
$\mathrm{Ao_X B}$	A does not belong to some B

In these formulae, 'A' represents the predicate term, 'B' the subject term, and expressions like '$\mathrm{a_X}$' and '$\mathrm{e_X}$' represent the copula.[13] Likewise, the four kinds of necessity propositions are represented as follows:

$\mathrm{Aa_N B}$	A necessarily belongs to all B
$\mathrm{Ae_N B}$	A necessarily belongs to no B
$\mathrm{Ai_N B}$	A necessarily belongs to some B
$\mathrm{Ao_N B}$	A necessarily does not belong to some B

The representations of one- and two-sided possibility propositions are obtained from these formulae by replacing the subscripted 'N' by 'M' and 'Q' respectively.

When an $\mathrm{a_X}$-proposition is true, we say that the predicate term is $\mathrm{a_X}$-predicated of the subject term. In the same way, we will speak of $\mathrm{e_N}$-predication, $\mathrm{i_Q}$-predication, $\mathrm{o_M}$-predication, and so on.[14] For example, A is $\mathrm{e_N}$-predicated of B just in case the proposition $\mathrm{Ae_N B}$ is true.

DEDUCTIVE ARGUMENTS. After having explained what propositions and terms are, Aristotle proceeds to define what a συλλογισμός is (1.1 24b18–20). The central idea in his definition is that something follows necessarily from something. While some details of the definition have been controversial, it is now generally agreed that a συλλογισμός is a valid deductive argument in which a conclusion follows necessarily from some premises. Thus, a συλλογισμός is a sequence of categorical propositions constituting a valid deductive argument from given premises to a certain conclusion. At the same time, the term 'συλλογισμός' is also used by Aristotle to refer to schemata of such arguments, that is, to patterns in which concrete terms like 'man' and 'walking' are replaced by schematic placeholders like 'A' and 'B'.[15]

13. The subscripted 'X' is often omitted in the secondary literature, especially when only assertoric propositions are under consideration.

14. For this terminology, see Barnes (2007: 140–2).

15. See Striker (2009: 67).

In *Prior Analytics* 1.1–22, Aristotle focuses on argument-schemata that consist of two premises and a conclusion. More specifically, he focuses on schemata which fall into one of his three figures. Using 'x', 'y', and 'z' as placeholders for a copula, the three figures are:

	first figure	second figure	third figure
Major premise:	AxB	BxA	AxB
Minor premise:	ByC	ByC	CyB
Conclusion:	AzC	AzC	AzC

By replacing 'x', 'y', and 'z' with concrete copulae, we obtain schemata of deductive arguments. In the secondary literature, a schema that falls into one of the three figures is often called a 'mood'. Some of these moods are valid according to Aristotle, while others are invalid. In the assertoric syllogistic, the valid moods in the first figure are the following:

	Barbara	Celarent	Darii	Ferio
Major premise:	$\mathrm{Aa_XB}$	$\mathrm{Ae_XB}$	$\mathrm{Aa_XB}$	$\mathrm{Ae_XB}$
Minor premise:	$\mathrm{Ba_XC}$	$\mathrm{Ba_XC}$	$\mathrm{Bi_XC}$	$\mathrm{Bi_XC}$
Conclusion:	$\mathrm{Aa_XC}$	$\mathrm{Ae_XC}$	$\mathrm{Ai_XC}$	$\mathrm{Ao_XC}$

Aristotle regards these four moods not only as valid but also as perfect (*Prior Analytics* 1.4). That is, he takes their validity to be evident and not in need of proof. By contrast, the moods that he identifies as valid in the second and third figures are not perfect, but stand in need of proof. In order to prove their validity, Aristotle makes use of the perfect first-figure moods and the following rules of conversion:

	$\mathrm{e_X}$-conversion	$\mathrm{i_X}$-conversion	$\mathrm{a_X}$-conversion
Premise:	$\mathrm{Ae_XB}$	$\mathrm{Ai_XB}$	$\mathrm{Aa_XB}$
Conclusion:	$\mathrm{Be_XA}$	$\mathrm{Bi_XA}$	$\mathrm{Bi_XA}$

Aristotle states these conversion rules in *Prior Analytics* 1.2, and uses them in chapters 1.5–6 to establish the validity of the following moods:

Second figure: Cesare, Camestres, Festino, Baroco
Third figure: Darapti, Disamis, Datisi, Felapton, Ferison, Bocardo

Aristotle's proofs of these moods take the form of direct and indirect deductions. The rest of this chapter explains briefly how these deductions proceed.

DIRECT DEDUCTIONS. In order to represent Aristotle's proofs of second- and third-figure moods, it is useful to make use of a deductive system such as the ones proposed by Corcoran, Smiley, and others.[16] The deductive system can be taken to be based on seven deduction rules: the three conversion rules mentioned above, and the four perfect first-figure moods considered as deduction rules.

Within such a system, direct deductions start with a number of propositions assumed as premises. Each of the subsequent propositions is derived from the propositions preceding it by means of one of the seven deduction rules. The last proposition is the conclusion of the deduction. For example, Aristotle's direct deduction establishing the validity of the second-figure mood Camestres proceeds as follows (1.5 27a9–14):

1. Ba_XA (major premise)
2. Be_XC (minor premise)
3. Ce_XB (from 2; by e_X-conversion)
4. Ce_XA (from 1, 3; by Celarent)
5. Ae_XC (from 4; by e_X-conversion)

Each of the steps leading from the premises in lines 1 and 2 to the conclusion in line 5 is performed by means of a deduction rule of the system. Thus the validity of Camestres is established.

INDIRECT DEDUCTIONS. Among Aristotle's valid moods in the assertoric syllogistic, there are two whose validity cannot be established by direct deductions, namely, Baroco and Bocardo. Aristotle instead establishes their validity by indirect deductions, that is, by the method of *reductio ad absurdum.* Aristotle does not explicitly formulate a rule for indirect

16. Corcoran (1972: 697–8; 1974: 109–10), Smiley (1973: 141–2), and Smith (1983: 226; 1989: xix–xxi). In what follows, I rely on the system given by Smith.

deductions.[17] It is, however, clear that indirect deductions involve a step of assuming for *reductio* the contradictory of the intended conclusion. Aristotle determines the contradictories of assertoric propositions as follows:[18]

$\mathrm{Aa_XB}$	is the contradictory of	$\mathrm{Ao_XB}$,	and vice versa
$\mathrm{Ae_XB}$	is the contradictory of	$\mathrm{Ai_XB}$,	and vice versa

Moreover, in some of his indirect deductions, Aristotle avails himself of the following principle concerning the incompatibility of $\mathrm{a_X}$- and $\mathrm{e_X}$-propositions:[19]

$\mathrm{Aa_XB}$ is incompatible with $\mathrm{Ae_XB}$, and vice versa

Given these principles of contradictoriness and incompatibility, Aristotle's method of indirect deduction can be described as follows. First some premises are assumed. Then the contradictory of the intended conclusion is assumed for *reductio* as an additional premise. Based on the resulting extended set of premises, we begin to construct a direct deduction. We try to go on until the direct deduction contains two propositions that are contradictory to or incompatible with each other. If successful, we have given an indirect deduction of the intended conclusion from the original premises. Thus, for example, Aristotle's indirect deduction establishing the validity of Baroco can be represented as follows (1.5 27a36–b1):

1. $\mathrm{Ba_XA}$	(major premise)
2. $\mathrm{Bo_XC}$	(minor premise)
3. $\mathrm{Aa_XC}$	(assumption for *reductio;* contradictory of $\mathrm{Ao_XC}$)
4. $\mathrm{Ba_XC}$	(from 1, 3; by Barbara)

17. Ross (1949: 372), Smith (1989: 140), Barnes (2007: 366), and Striker (2009: 70).

18. See *APr.* 2.8 59b8–11, 2.15 63b23–30, and *Int.* 7 17b16–20.

19. He states this principle at *APr.* 2.15 63b28–30 and *Int.* 7 17b20–3; see p. 41 below.

Line 3 contains the assumption for *reductio,* which is the contradictory of the intended conclusion of Baroco. The step in line 4 is justified by Barbara, which is among the deduction rules of the deductive system. Since the proposition in line 4 is contradictory to that in line 2, the indirect deduction is complete. Thus, it is proved that the conclusion $\mathrm{Ao_XC}$ follows from the two premises, and the validity of Baroco is established. The indirect deduction establishing the validity of Bocardo is exactly parallel to that for Baroco.

The system of direct and indirect deductions just described captures the entirety of moods held to be valid by Aristotle in the assertoric syllogistic: all these moods can be proved to be valid in this system. Chapter 15 discusses how the system can be extended to capture the whole of Aristotle's modal syllogistic.

A SEMANTICS FOR THE ASSERTORIC SYLLOGISTIC? As we have seen, the assertoric syllogistic can be reconstructed as a deductive system of categorical propositions. Given such a purely syntactical system, two questions naturally arise. First, does the deductive system match Aristotle's claims about invalidity? In other words, is it weak enough so that it does not establish the validity of any mood held to be invalid by Aristotle? Second, what, if anything, justifies the deduction rules of the system? In other words, on what grounds did Aristotle assert the validity of the three conversion rules and the four perfect assertoric moods?

Both questions can be answered by appealing to an account of the meaning of Aristotle's assertoric propositions and their copulae. The purpose of Chapters 2, 3, and 4 is to provide such an account by developing a semantics for the assertoric syllogistic. In fact, a major goal of this study is to develop a semantics for the whole syllogistic, including the modal syllogistic. This project is hampered by the fact that Aristotle says very little about the semantics of categorical propositions in the *Prior Analytics.* A notable exception is a brief but important passage in which he characterizes the semantics of $\mathrm{a_X}$-propositions: the *dictum de omni* in the opening chapter of the *Prior Analytics.* In Chapter 2, I examine this *dictum de omni* and determine how it contributes to justifying the seven deduction rules of the deductive system.

2

The *dictum de omni*

INTRODUCING THE *DICTUM DE OMNI.* At the end of the first chapter of the *Prior Analytics,* Aristotle offers an explanation of the semantics of a_X-propositions. This explanation, which has come to be known as the *dictum de omni,* reads as follows:

> We say "predicated of all" when none of those of the subject can be taken of which the other will not be said.[1] (*APr.* 1.1 24b28–30)

First it should be noted that there is a question about the Greek text of this passage. Most manuscripts do not have the phrase "of those," translating the plural article τῶν. But since this article is attested by a major manuscript and is accepted by many commentators, I accept it as well.[2]

Regardless of this textual issue, it seems clear that the *dictum de omni* specifies a necessary and sufficient condition for the truth of a_X-propositions: an a_X-proposition is true just in case none of those of the

1. λέγομεν δὲ τὸ κατὰ παντὸς κατηγορεῖσθαι ὅταν μηδὲν ᾖ λαβεῖν τῶν τοῦ ὑποκειμένου καθ' οὗ θάτερον οὐ λεχθήσεται.

2. The article τῶν is accepted by Bekker (1831: 24b29), Waitz (1844: 147), Maier (1900a: 13n1), Tredennick (1938: 202), Ebert (1995: 230–1n11), Wolff (1998: 162), Drechsler (2005: 286), Ebert & Nortmann (2007: 183), and Smith (2011: 417). The article is also found in Georgius's Syriac translation of the *Prior Analytics,* stemming from the seventh or eighth century (cf. Minio-Paluello 1964: 187). On the other hand, some commentators excise even the phrase τοῦ ὑποκειμένου, although it is found in all manuscripts (Wallies 1917/18: 626–7, Ross 1949: 292, Barnes 2007: 387n34). However, most commentators reject this excision.

subject can be taken of which the predicate is not said. The phrase 'none ... can be taken' may be understood to mean 'there is none ... '.[3] Consequently, the *dictum de omni* states that an a_X-proposition is true just in case there is none of those of the subject of which the predicate is not said; in other words, just in case the predicate is said of everything that is one of those of the subject. Using the resources of modern propositional and quantifier logic, the *dictum de omni* can then be formulated as follows:

Aa_XB if and only if for every Z, if Z is one of those of B, then A is said of Z

PLURALITIES ASSOCIATED WITH TERMS. The phrase 'of those of the subject', governed by the plural article τῶν, introduces the idea of a plurality of items associated with the subject term. Thus the *dictum de omni* states that an a_X-proposition is true just in case the predicate term is said of every member of the plurality associated with the subject term.[4]

Now, Aristotle's phrase 'is said of' seems to express a relation that closely corresponds to that expressed by 'is one of those of'. The *dictum de omni* is generally taken to state that an a_X-proposition is true just in case every item that stands in a certain relation to the subject stands in that same relation to the predicate. If so, then 'is said of' expresses exactly the converse relation of 'is one of those of'. I will indicate this latter relation by the phrase '... is a member of the plurality associated with ...'. This is meant to be a schematic expression that is neutral as to the nature of the relation in question. Thus, the *dictum de omni* states that an a_X-proposition is true just in case every member of the plurality associated with the subject is a member of the plurality associated with the predicate. For the sake of brevity, let us use '... *mpaw* ...' as shorthand for '... is a *m*ember of the *p*lurality *a*ssociated *w*ith ...'. The *dictum de omni* can then be represented by the following schema:

Aa_XB if and only if $\forall Z\ (Z\ \mathit{mpaw}\ B \supset Z\ \mathit{mpaw}\ A)$

---

3. See Barnes (2007: 389).

4. This interpretation of the *dictum de omni* may be adopted even if the article τῶν is rejected.

So far I have said nothing about what the plurality associated with a term is, and what the members of this plurality are. Nor have I specified of what syntactic type the variable 'Z' should be taken to be in the above formulation. I address these questions in Chapters 3 and 4. But for now, let us consider the semantics of the other three kinds of assertoric propositions.

THE ABSTRACT *DICTUM* SEMANTICS. Aristotle's statement of the *dictum de omni* is followed by a brief remark on e_X-propositions: "and likewise for 'predicated of none'" (24b30). This remark indicates the *dictum de nullo,* which characterizes the semantics of e_X-propositions. Aristotle suggests that the *dictum de nullo* should be spelled out in a parallel way to the *dictum de omni.* The critical part of the latter was: 'none of those of the subject can be taken of which the other is *not* said'. Aristotle presumably had in mind that the *dictum de nullo* is obtained from this by omitting the italicized negation *not.*[5] If so, then the *dictum de nullo* states that an e_X-proposition is true just in case the predicate is not said of anything that is one of those of the subject. In other words, it states that an e_X-proposition is true just in case no member of the plurality associated with the subject is a member of the plurality associated with the predicate.

Aristotle does not mention a *dictum de aliquo* characterizing the semantics of i_X-propositions, or a *dictum de aliquo non* characterizing the semantics of o_X-propositions. However, these two *dicta* can easily be supplied with the help of Aristotle's claims concerning the contradictoriness of assertoric propositions. As mentioned above (p. 32), Aristotle states that an o_X-proposition is the contradictory of an a_X-proposition, and an i_X-proposition is the contradictory of an e_X-proposition. In every such pair of contradictories, one proposition is true and the other is false.[6] As we saw above, Aristotle relies on these principles of contradictoriness in his indirect deductions establishing the validity of Baroco and Bocardo.

5. Alexander *in APr.* 25.17–19, 32.20–1, 55.5–7, Ammonius *in APr.* 34.13–15, Maier (1900b: 150), Ebert (1995: 231), Barnes (2007: 390), Ebert & Nortmann (2007: 230), Crivelli (2012: 118).

6. *Int.* 7 17b26–7; cf. Whitaker (1996: 92), Weidemann (2002: 203), Jones (2010: 36), Crivelli (2012: 117–18).

These two principles of contradictoriness allow us to derive the *dictum de aliquo et de aliquo non* from the *dictum de omni et de nullo* in a straightforward way. For instance, an i_X-proposition should be true just in case some member of the plurality associated with the subject is a member of the plurality associated with the predicate. Similarly for o_X-propositions. Thus, the four *dicta* can be represented by the following schemata:[7]

Aa_XB	if and only if	$\forall Z(Z\ mpaw\ B \supset Z\ mpaw\ A)$
Ae_XB	if and only if	$\forall Z(Z\ mpaw\ B \supset \neg\ Z\ mpaw\ A)$
Ai_XB	if and only if	$\exists Z(Z\ mpaw\ B \land Z\ mpaw\ A)$
Ao_XB	if and only if	$\exists Z(Z\ mpaw\ B \land \neg\ Z\ mpaw\ A)$

The formulae on the right-hand side employ the resources of modern propositional and quantifier logic. Of course, these resources were not available to Aristotle. Nevertheless, the four equivalences give, I think, a sufficiently faithful representation of Aristotle's views on the semantics of assertoric propositions.

The four equivalences are abstract in that they do not specify what the plurality associated with a term is. We may therefore call these four equivalences the abstract *dictum* semantics of the assertoric syllogistic. Although the abstract *dictum* semantics does not specify what the relation indicated by '*mpaw*' is, it is strong enough to justify most of Aristotle's claims of validity in the assertoric syllogistic. As I show in the rest of this chapter, it justifies the validity of all of Aristotle's first-figure moods and conversion rules—except the rule of a_X-conversion.

JUSTIFYING THE VALIDITY OF THE PERFECT ASSERTORIC MOODS. As we saw above, the assertoric syllogistic is based on the four perfect moods in the first figure. Aristotle does not establish their validity by direct or indirect deductions within his deductive system of categorical propositions; for their validity is presupposed by the system, as they serve as its deduction rules. Instead, he states that their validity is justified by the *dictum de omni et de nullo*.[8]

---

7. Cf. Crivelli (2012: 118–19).

8. He does so in connection with Barbara, Darii, and Ferio, at *APr.* 1.4 25b39–40, 26a24, 26a27. See Alexander *in APr.* 54.9–11, 55.1–3, 61.3–5, 69.14–20, Wieland (1972: 131n14), Smith (1989: 111), Ebert (1995: 229–30),

Aristotle does not explain exactly how the *dictum de omni et de nullo* is meant to justify the validity of the four perfect moods, but the abstract *dictum* semantics can help us see how such a justification might proceed. The abstract *dictum* semantics states equivalences between categorical propositions on the one hand and formulae of propositional and quantifier logic on the other. The propositional connectives and quantifiers occurring in these formulae are understood to have their usual meaning as determined by twentieth-century classical logic. The equivalences thereby impart certain logical properties to the four assertoric categorical propositions. For example, the abstract *dictum de omni* implies that the relation of a_X-predication is transitive. Given that the abstract *dictum de omni* holds for any A and B, it follows logically that for any A, B, and C: if Aa_XB and Ba_XC, then Aa_XC.[9] Thus, the abstract *dictum de omni* justifies the validity of the perfect mood Barbara. Likewise, the abstract *dictum de omni* and *dictum de nullo* jointly justify the validity of Celarent, as follows:

1. Ae_XB	(major premise)
2. Ba_XC	(minor premise)
3. $\forall Z(Z\ mpaw\ B \supset \neg\ Z\ mpaw\ A)$	(from 1; by abstract *dictum de nullo*)
4. $\forall Z(Z\ mpaw\ C \supset Z\ mpaw\ B)$	(from 2; by abstract *dictum de omni*)
5. $\forall Z(Z\ mpaw\ C \supset \neg\ Z\ mpaw\ A)$	(from 3, 4)
6. Ae_XC	(from 5; by abstract *dictum de nullo*)

The step from lines 3 and 4 to line 5 is justified by rules of classical propositional and quantifier logic. The other steps are underwritten by the abstract *dictum* semantics, so that the validity of Celarent is verified

Byrne (1997: 46), Ebert & Nortmann (2007: 292 and 302), Barnes (2007: 392–4), and Striker (2009: 83–4).

9. This is because the following is logically valid in classical propositional and quantifier logic: if $\forall Z(Z\ mpaw\ B \supset Z\ mpaw\ A)$ and $\forall Z(Z\ mpaw\ C \supset Z\ mpaw\ B)$, then $\forall Z(Z\ mpaw\ C \supset Z\ mpaw\ A)$.

in this semantics. The validity of Darii and Ferio can be justified in a strictly parallel way.

Thus, the abstract *dictum* semantics verifies the validity of all four perfect moods of the assertoric syllogistic, without relying on any assumptions about the nature of the relation indicated by '*mpaw*'.[10] It can thereby explain why Aristotle took their validity to be justified by the *dictum de omni et de nullo* and how this justification proceeds. This does not explain why Aristotle took the four first-figure moods to be perfect (as opposed to merely valid), but it is not my intention here to enter into a discussion of this question.

JUSTIFYING THE VALIDITY OF E_X-CONVERSION. In addition to the perfect moods, the assertoric syllogistic is based on the rules of e_X-, i_X-, and a_X-conversion. Let us begin with e_X-conversion. Aristotle justifies this rule in the second chapter of the *Prior Analytics* as follows:

> If A belongs to none of the Bs, then neither will B belong to any of the As. For if it belongs to some, for instance to C, it will not be true that A belongs to none of the Bs, since C is one of the Bs. (*APr.* 1.2 25a15–17)

Alexander reports that this justification was rejected as circular by some commentators in antiquity. They took Aristotle in the present passage to justify e_X-conversion by means of i_X-conversion, and a few lines later, at 25a20–2, to justify the latter by means of the former.[11] Alexander defends Aristotle's justification of e_X-conversion by arguing that it is based not on i_X-conversion, but on the *dictum de omni et de nullo*.[12] It seems to me that Alexander is on the right track. Further evidence for his proposal comes from the fact that Aristotle's justification contains pluralized phrases such as 'the As' (τῶν A), which introduce the idea of a plurality associated with a given term. These phrases do not occur in Aristotle's standard formulations of categorical propositions, such as 'A belongs to no B' (τὸ A οὐδενὶ τῷ B ὑπάρχει). Their presence

10. Cf. Crivelli (2012: 118–20).

11. Alexander *in APr.* 31.27–32.3; Philoponus *in APr.* 49.6–14.

12. Alexander *in APr.* 32.8–11, 32.28–32. The view that this justification is based on the *dictum de omni et de nullo* is also held by Byrne (1997: 40) and Morison (2008: 213). On the other hand, Barnes (2007: 418) rejects this view.

here may therefore be taken to indicate that Aristotle's justification of e_X-conversion appeals to pluralities associated with terms and hence is based on the abstract *dictum* semantics.[13]

However this may be, the abstract *dictum de nullo* in fact verifies the validity of e_X-conversion. Suppose, for *reductio,* that A is e_X-predicated of B but not vice versa. If B is not e_X-predicated of A, the abstract *dictum de nullo* implies that some member of the plurality associated with A, call it C, is a member of the plurality associated with B. Consequently, some member of the plurality associated with B is a member of the plurality associated with A. Given this, the abstract *dictum de nullo* implies that A is not e_X-predicated of B.[14] Thus, the rule of e_X-conversion is justified without circularity by means of the abstract *dictum de nullo,* in a way that closely corresponds to Aristotle's own presentation of the justification.

The rule of i_X-conversion can be justified in the same way by means of the abstract *dictum de aliquo.* In the case of a_X-conversion, however, it turns out that such a justification is not available.

A PROBLEM WITH A_X-CONVERSION. Aristotle holds that Bi_XA can be validly inferred from Aa_XB. He argues for this as follows:

> If A belongs to all B, then B will belong to some A. For if it belongs to none, then neither will A belong to any B; but it was assumed to belong to all. (*APr.* 1.2 25a17–19)

Although this passage occurs immediately after the justification of e_X-conversion, it does not contain pluralized phrases such as 'the As'.[15] Instead, it contains singular phrases like 'A belongs to all (some, no) B', which are Aristotle's standard expressions for categorical propositions. This may be taken to indicate that the justification is carried out entirely within the language of categorical propositions and does not rely on

---

13. This is argued in more detail in Malink (2008: 521–7).

14. For a more detailed exposition, see p. 89 below. Cf. also Byrne (1997: 40).

15. More precisely, such phrases are not attested in it by the majority of manuscripts; see Smith (1982a: 120; 1989: 237), Williams (1984: 11), and Malink (2008: 526 and 535).

the abstract *dictum* semantics. Thus, Aristotle's justification may be reconstructed as an indirect deduction consisting solely of categorical propositions:

1. $\mathrm{Aa_XB}$ (premise)
2. $\mathrm{Be_XA}$ (assumption for *reductio;* contradictory of $\mathrm{Bi_XA}$)
3. $\mathrm{Ae_XB}$ (from 2; by $\mathrm{e_X}$-conversion)

In line 2 of this deduction, the contradictory of the intended conclusion $\mathrm{Bi_XA}$ is assumed for *reductio.* The step in line 3 is guaranteed by $\mathrm{e_X}$-conversion, whose validity was just established at 25a15–17. In order for the indirect deduction to succeed, the proposition in line 3 must be incompatible with that in line 1. Aristotle holds that $\mathrm{a_X}$-propositions are contrary to the corresponding $\mathrm{e_X}$-propositions and that they cannot be true together.[16] However, he does not explain why this is so.

Within the abstract *dictum* semantics, it is not the case that $\mathrm{a_X}$- and $\mathrm{e_X}$-propositions are contrary or incompatible: they are simultaneously true when the plurality associated with the subject term has no member. For in this case, according to classical propositional and quantifier logic, it is true that *every* and *no* member of the plurality associated with the subject term is a member of the plurality associated with the predicate term.[17] Given this, the abstract *dictum de omni et de nullo* implies that both the $\mathrm{a_X}$-proposition and the corresponding $\mathrm{e_X}$-proposition are true. Accordingly, if the plurality associated with the term B has no member, then $\mathrm{Aa_XB}$ is true, but $\mathrm{Bi_XA}$ is false. Thus, the abstract *dictum* semantics does not validate the rule of $\mathrm{a_X}$-conversion. This problem is known as the problem of existential import.

THREE STRATEGIES TO SOLVE THE PROBLEM OF EXISTENTIAL IMPORT. A number of strategies have been suggested in the secondary literature to solve or avoid the problem of existential import. I here want to mention three of them. These strategies are usually described not with respect

---

16. *APr.* 2.15 63b28–30, *Int.* 7 17b20–3, 10 20a16–18; see Ackrill (1963: 129), Whitaker (1996: 84), Weidemann (2002: 204), and Crivelli (2004: 263n17; 2012: 117).

17. If the plurality associated with the term B has no member, then both $\forall Z(Z\ \mathit{mpaw}\ B \supset Z\ \mathit{mpaw}\ A)$ and $\forall Z(Z\ \mathit{mpaw}\ B \supset \neg Z\ \mathit{mpaw}\ A)$ are true.

to the abstract *dictum* semantics, but with respect to a specific instance of it, namely, the orthodox *dictum* semantics which we will consider in Chapter 3. They can, however, be readily transferred to the more general case of the abstract *dictum* semantics.

Perhaps the most common strategy is to introduce into the assertoric syllogistic a presupposition to the effect that the plurality associated with every term has at least one member. Thus, terms whose associated plurality has no member are not admissible. They are not allowed to serve as argument-terms of categorical propositions. Given this presupposition, the rule of a_X-conversion is valid, without any changes in the four equivalences that constitute the abstract *dictum* semantics.

A second, and related, strategy does require some changes in the equivalences. In particular, it requires that the abstract *dictum de omni* and *dictum de aliquo non* be modified as follows:[18]

Aa_XB	if and only if	(i) $\forall Z$ (Z *mpaw* B $\supset$ Z *mpaw* A) and (ii) $\exists Z$ (Z *mpaw* B)
Ao_XB	if and only if	(i) $\exists Z$ (Z *mpaw* B $\wedge \neg$ Z *mpaw* A) or (ii) not $\exists Z$ (Z *mpaw* B)

The other two *dicta* remain unmodified. On this strategy, unlike on the first, terms whose associated plurality has no member are admissible in the syllogistic. Such terms are allowed to serve as subjects of a_X-propositions; they just cannot serve as subjects of *true* a_X-propositions. Consequently, every a_X-proposition is incompatible with the corresponding e_X-proposition, and the rule of a_X-conversion is validated.

The third strategy, finally, is quite different from the first two. It assumes that Aristotle's *dictum de omni et de nullo* is not governed by classical propositional logic, but by another kind of logic. Specifically, the material implication '$\supset$' that occurs in the abstract *dictum de omni* and *dictum de nullo* is replaced by what is called connexive

18. This strategy is adopted by Moody (1953: 52), Thompson (1953: 256–7), Prior (1962: 169), Wedin (1990: 135), Bäck (2000: 241–3), and Ebert & Nortmann (2007: 333).

implication.[19] Unlike material implication, connexive implication has the property that nothing can imply both a sentence and its negation—a property for which there is also evidence in the *Prior Analytics*.[20] As a result, a_X-propositions are incompatible with e_X-propositions in this kind of connexive *dictum* semantics.[21] The *dictum de aliquo* and *dictum de aliquo non* are modified in such a way that i_X- and o_X-propositions are contradictory to e_X- and a_X-propositions, respectively. Consequently, a_X-conversion is valid in the connexive *dictum* semantics.

Each of the three strategies is viable, but each also has its drawbacks. The first strategy needs to assume that the assertoric syllogistic is not applicable to all terms, but only to terms that have at least one member in the plurality associated with them. The assertoric syllogistic would then rely on a tacit extralogical presupposition about the nature of admissible terms which Aristotle failed to make explicit. It would not be the universally applicable system of formal logic that it is often thought to be.[22] The second strategy too has its problems. It conflicts with some remarks Aristotle makes in *Prior Analytics* 2.15 on o_X-propositions whose predicate term is identical with the subject term, such as Bo_XB. He seems to assert in this chapter that no o_X-proposition of the form Bo_XB can be true.[23] Yet the second strategy implies that such o_X-propositions may be true, if the plurality associated with the term B has no member.[24] Finally, the third strategy seems to be in tension with Aristotle's formulation of the *dictum de omni*, 'none of those of the subject can be

---

19. This strategy is adopted by McCall (1967: 349–56) and Angell (1986: 216–23); cf. also Prior (1962: 170).

20. Aristotle holds that nothing can follow both from something and from its contradictory (*APr.* 2.4 57b3–14). By contraposition, this means that it is also impossible that both something and its contradictory follow from the same thing.

21. This is because if '$\rightarrow$' is connexive implication, the conditional '$\mathrm{Z}\, mpaw\, \mathrm{B} \rightarrow \mathrm{Z}\, mpaw\, \mathrm{A}$' is incompatible with '$\mathrm{Z}\, mpaw\, \mathrm{B} \rightarrow \neg\, \mathrm{Z}\, mpaw\, \mathrm{A}$'.

22. See Rini (2011: 27–8).

23. *APr.* 2.15 64b7–13 (in tandem with 64a4–7 and 64a23–30); see Łukasiewicz (1957: 9), Prior (1962: 169), and Thom (1981: 92). The view that such o_X-propositions cannot be true is also held by Alexander (*in APr.* 34.18–20).

24. This problem with the second strategy has been pointed out by Prior (1962: 169). Moreover, the second strategy has difficulties in accounting for

taken of which the other will not be said'. Since Aristotle does not here use the terminology of implying or following, it seems doubtful that his *dictum de omni* is meant to rely on connexive implication.

Of course, none of these objections is decisive. Nevertheless, they provide some reason not to adopt any of the three strategies just discussed. In what follows, I therefore suggest another solution to the problem of existential import.

ABSTRACT EXISTENTIAL IMPORT. Unlike the second and third strategies, I prefer not to change the logical structure of the abstract *dictum* semantics. Consequently, if a_X-conversion is to be valid, the following requirement of existential import must be satisfied: if A is the subject of an a_X-proposition, then the plurality associated with A has at least one member. This requirement is abstract in the same sense as the abstract *dictum* semantics: it does not specify what the plurality associated with a term is. We may call it the abstract requirement of existential import. Accordingly, the problem of how this requirement can be satisfied is the abstract problem of existential import.

I will not solve this problem, as the first strategy does, by introducing a general presupposition of existential import. Instead, I intend to show that it can be solved merely by specifying what the plurality associated with a term is. In Chapter 3, I consider, and reject, a certain common way to specify this plurality, namely, to identify it with the set of individuals that fall under the term in question. In Chapter 4, I present my own proposal, on which the plurality associated with a term is identified with the set of items of which it is a_X-predicated. This leads to a special instance of the abstract *dictum* semantics in which there is at least one member in the plurality associated with each term, because every term is a_X-predicated of itself.

Aristotle's proof by ecthesis of Bocardo in *Prior Analytics* 1.6 (see p. 96n15 below and Wedin 1990: 147–9).

3

The Orthodox *dictum* Semantics

INTRODUCING THE ORTHODOX *DICTUM* SEMANTICS. The abstract *dictum* semantics appeals to pluralities associated with terms. What are these pluralities? A common answer is that they are sets of individuals, consisting of exactly those individuals that fall under the term in question. On this view, the plurality associated with the term 'man', for example, is the set of individual men such as Socrates, Kallias, and Mikkalos. The plurality associated with the term 'walking' is the set of all walking individuals. Accordingly, the abstract *dictum de omni* states that an a_X-proposition is true just in case every individual that falls under the subject term falls under the predicate term. The abstract *dictum de aliquo* states that an i_X-proposition is true just in case some individual that falls under the subject term falls under the predicate term. Likewise for the other two *dicta.*

The set of individuals that fall under a term is often referred to as the extension of that term. Thus, the plurality associated with a term is, on the present interpretation, identified with the extension of this term. The *dictum de omni* states that an a_X-proposition is true just in case the extension of the subject is a subset of the extension of the predicate. The *dictum de aliquo* states that an i_X-proposition is true just in case the two extensions overlap. Thus, the truth of assertoric categorical propositions depends solely on the extensions of the terms involved.

This has been the prevailing interpretation of the four *dicta* for many centuries. Since the rise of modern logic in the second half of the nineteenth century, it is often rendered in the language of classical first-order logic. This logic is a special instance of classical quantifier logic, the other instances being second-order logic and other kinds of higher-order logic. The language of first-order logic distinguishes between two syntactic types of terms: between first-order predicate terms and zero-order individual terms (or singular terms). The former are usually indicated by uppercase letters, the latter by lowercase letters. On the present interpretation, argument-terms of categorical propositions are taken to be first-order predicate terms, whereas the quantified variable 'Z' that occurs in the abstract *dictum* semantics is taken to be a zero-order individual variable. This leads to the following specification of the abstract *dictum* semantics:

$\mathrm{Aa_XB}$	if and only if	$\forall z(Bz \supset Az)$
$\mathrm{Ae_XB}$	if and only if	$\forall z(Bz \supset \neg Az)$
$\mathrm{Ai_XB}$	if and only if	$\exists z(Bz \wedge Az)$
$\mathrm{Ao_XB}$	if and only if	$\exists z(Bz \wedge \neg Az)$

Frege in effect stated these equivalences in his *Begriffsschrift* (1879), and they have been widely accepted ever since.[1] Following Barnes (2007: 406–9), we may call them the orthodox *dictum* semantics.

TWO ORTHODOX ASSUMPTIONS. The orthodox *dictum* semantics is based on two assumptions. One of them is that the pluralities associated with terms are sets of individuals such as Socrates, Kallias, and so on. The other is that the quantified variable that occurs in the abstract *dictum* semantics is of a different syntactic type than the argument-terms of categorical propositions: the variable is a zero-order individual term, whereas the argument-terms are first-order predicates.

The two assumptions are closely connected. To see this, consider the universal quantification '∀z' in the orthodox *dictum de omni et de nullo*. This is usually understood as an objectual quantification, requiring the formula to which it is applied to be true whatever semantic value is assigned to the variable 'z'. The semantic value assigned to a zero-order individual variable such as 'z' is an individual. The domain of possible semantic values for these variables is a class of individuals. Now, in the orthodox *dictum* semantics, the plurality associated with a term A is

1. See Frege (1879: §12); see also Frege (2004: 69–70).

taken to be the set of those items that make the open formula 'Az' true when assigned as a semantic value to the variable 'z'. Each of these items is an individual. Consequently, the plurality associated with A is a set of individuals. Thus, the first of the above two assumptions follows from the second: given that the quantified variable in the abstract *dictum* semantics is a zero-order individual variable, it follows that the plurality associated with a categorical term is a set of individuals.

AN ALTERNATIVE VIEW. Several commentators, including Mario Mignucci and Michael Frede, reject the orthodox interpretation of the *dictum de omni.* They deny that the quantified variable in it is a zero-order individual variable. Instead, they take it to be a variable of the same syntactic type as the argument-terms of categorical propositions.[2]

This view strongly suggests that the pluralities associated with terms are not sets of individuals, at least not exclusively. It is plausible to take the plurality associated with a term to be a subset of the domain of possible semantic values of the quantified variable. On the alternative view, the quantified variable is of the same syntactic type as the argument-terms of categorical propositions, such as 'man' and 'walking'. It is not immediately clear what the semantic values of these terms are. But it is clear that their domain of possible semantic values does not consist exclusively of individuals such as Socrates or Kallias. For the semantic value of the term 'man' is not an individual man; rather, it might be the universal man, or perhaps the set of all individual men. Whatever the domain of possible semantic values for argument-terms of categorical propositions is, the plurality associated with such a term will be, on the alternative view, a subset of this domain.

In what follows, I argue that the alternative interpretation of the *dictum* semantics is correct and that the orthodox interpretation should be rejected. I adduce four pieces of evidence to undermine the two assumptions on which the orthodox interpretation is based. The first piece of evidence targets the assumption that the quantified variable

2. Maier (1900a: 13n1; 1900b: 150–1), Stekeler-Weithofer (1986: 76), Mignucci (1996: 4–5; 2000: 8–15), and Wolff (1998: 163–4); similarly, Kneale & Kneale (1972: 205). M. Frede did not express this view in print, but his endorsement of it is reported by Barnes (2007: 406n57) and Morison (2008: 212).

is not of the same syntactic type as the argument-terms of categorical propositions. The other three target the assumption that the pluralities associated with terms consist exclusively of individuals. My arguments are in no way unassailable, but since, as far as I can see, there is hardly any positive evidence in favor of the orthodox interpretation in the *Prior Analytics,* the four pieces of evidence constitute a good basis for rejecting it.

THE FIRST PIECE OF EVIDENCE. Consider the distinction between a syntactic type of zero-order singular terms and a syntactic type of first-order predicate terms. However familiar this distinction may be to us post-Fregeans, there is not much evidence for it in Aristotle. Aristotle does distinguish between beings (ὄντα, πράγματα) that are individuals and beings that are universals.[3] Hence some terms such as 'animal' stand for a universal and may therefore be called general terms, while other terms such as 'Mikkalos' stand for an individual and may therefore be called singular terms. However, this does not mean that Aristotle took these terms to be of different syntactic types; at least, there is no indication that he did so in the *Prior Analytics.* Aristotle is happy to use terms such as 'Mikkalos', 'Aristomenes', and 'Pitakkos' as argument-terms of categorical propositions in the *Prior Analytics.*[4] Thus, he seems to take these terms to be of the same syntactic and logical type as other argument-terms of categorical propositions, such as 'animal' and 'walking'.[5] Aristotle holds that syllogistic reasoning is primarily concerned with universals rather than with individuals (1.27 43a42–3). Nevertheless, from a logical point of view, nothing prevents terms such as 'Mikkalos' from being used as argument-terms of categorical propositions—be they universal, particular, or indeterminate.[6]

---

3. See *Cat.* 2 with ὄντων at 1a20; *Int.* 7 17a38–b3 with πραγμάτων at 17a38; and *APr.* 1.27 43a25–43 with ὄντων at 43a25.

4. *APr.* 1.33 47b22, 47b30, 2.27 70a16–18.

5. Smith (1991: 45) writes about Aristotle's treatment of singular and general terms that "[h]e does not treat the latter expressions as fundamentally different in logical form from the former, though he does recognize that they have different features."

6. For this view, see Wieland (1966b: 4–5), Mignucci (1996: 10; 1997: 148; 2000: 11), and Barnes (2007: 158–66).

Aristotle's conversion rules make it clear that if a term can serve as the argument-term of categorical propositions, it can serve both as their subject and as their predicate. Let us call terms that satisfy this condition categorical terms. If I am correct, Aristotle takes terms that stand for individuals, like terms that stand for universals, to be categorical terms. Thus, both singular and general terms are categorical terms. On the other hand, the orthodox *dictum* semantics employs a different concept of singular term, according to which singular terms are not categorical terms but zero-order individual terms. We may call these latter terms noncategorical singular terms.

In his criticism of the orthodox *dictum de omni,* Mignucci writes:

> The tendency to distinguish sharply between singular and general terms depends mainly on the fact that they are thought to be the bearer[s] of different logical relations. (Mignucci 1996: 11)

Aristotle does not seem to think that singular and general terms are bearers of different logical relations. Within the realm of beings, at any rate, he takes the same kind of predication that holds between species and their genera to hold also between individuals and their species. For example, the genus animal is predicated of the species man, and in the same way this species is predicated of the individual Kallias.[7]

These considerations suggest that the quantified variable in the abstract *dictum* semantics should be taken to be a categorical term rather than a noncategorical singular term. They suggest, that is, that one of the two assumptions on which the orthodox *dictum* semantics is based is false. Given this, it might still be the case that the plurality associated with any given categorical term consists exclusively of individuals: the formula 'Z is a member of the plurality associated with A' might be true only if the variable 'Z', while being a categorical term, stands for an individual. The restriction to individuals would then be encoded in the truth conditions of that formula rather than in the syntactic type of the

7. See *APr.* 1.27 43a30–2, *Cat.* 3 1b12–15, 5 3a38–9. Ackrill (1963: 76) comments that Aristotle "does not distinguish between the relation of an individual to its species and that of a species to its genus"; similarly, Barnes (1970: 147–8; 1994: 146), Code (1986: 419), Mignucci (1996: 11–13), Bäck (2000: 178), and Weidemann (2002: 209).

variable 'Z'.[8] The following three pieces of evidence will show that this is not the case, thereby undermining the other assumption on which the orthodox *dictum* semantics is based.

THE SECOND PIECE OF EVIDENCE. At the beginning of the second book of the *Topics,* Aristotle introduces the four standard kinds of assertoric propositions: a_X-, e_X-, i_X-, and o_X-propositions (2.1 108b34–109a1). He goes on to explain several strategies to refute an a_X- or e_X-proposition maintained by one's opponent in a dialectical exchange. One of these strategies is given by the following *topos:*

> Another *topos* is to examine the items to each or to none of which [the predicate] has been said to belong. Investigate species by species, and not in the infinitely many [individuals[9]]; for then the inquiry will proceed more directly and in fewer steps. You should investigate and begin with the first [that is, with the highest species], and then proceed step by step down to those that are not further divisible.[10] (*Top.* 2.2 109b13–16)

The first sentence of this passage makes reference to a plurality of items. This plurality seems to be associated with the subject term of the a_X- or e_X-proposition that is to be refuted. Aristotle evidently takes somebody who states an a_X-proposition to imply that the predicate term belongs to each member of that plurality. In the same way, somebody who states an e_X-proposition implies that the predicate term belongs to no member of that plurality. Aristotle instructs us to consider each member of this plurality and to examine whether or not the predicate term belongs to

8. For an example of such an interpretation, see Crivelli (2004: 261).

9. It seems clear that τοῖς ἀπείροις at 109b14 refers to the "infinitely many" individuals that fall under the term in question; cf. the use of τὰ ἄπειρα at *Met.* B 4 999a26–9. See Primavesi (1996: 121), Wagner & Rapp (2004: 76).

10. ἄλλος τὸ ἐπιβλέπειν οἷς ὑπάρχειν ἢ πᾶσιν ἢ μηδενὶ εἴρηται. σκοπεῖν δὲ κατ' εἴδη καὶ μὴ ἐν τοῖς ἀπείροις· ὁδῷ γὰρ μᾶλλον καὶ ἐν ἐλάττοσιν ἡ σκέψις. δεῖ δὲ σκοπεῖν καὶ ἄρχεσθαι ἀπὸ τῶν πρώτων, εἶτ' ἐφεξῆς ἕως τῶν ἀτόμων. In the first sentence of this passage, πᾶσιν does not indicate that the predicate term is a_X-predicated of each of the items picked out by οἷς. Instead, it indicates that the predicate term belongs (ὑπάρχειν) to *each* of them (for this use of πᾶσιν, cf. *APr.* 1.41 49b29 and Kneale & Kneale 1972: 203). Likewise, μηδενί does not indicate that the predicate term is e_X-predicated of each of those items, but that it belongs to *none* of them.

it.[11] When we find a member to which it does not belong, we have refuted the a_X-proposition; when we find a member to which it belongs, we have refuted the e_X-proposition.

Crucially, the plurality in question is taken to include species. The plurality may or may not include individuals. In any case, it does not consist exclusively of individuals but also has species among its members. Aristotle recommends first examining the highest species that fall under the subject term and then proceeding step by step down to the lowest species. In doing so, he urges, we should not take into account the individuals that fall under the subject term. As soon as we have found a species to which the predicate term belongs or does not belong, we have falsified the e_X- or a_X-proposition, respectively.[12]

The plurality to which Aristotle refers in this *topos* serves a similar function as the plurality associated with the subject term in the *dictum de omni et de nullo*. Like the *dictum*, the *topos* is assessing the truth of categorical propositions by reference to the plurality associated with their subject term. In the *topos*, the plurality is taken to contain not only individuals but also species. Of course, this does not necessarily mean that the same is true for the *dictum de omni et de nullo*, for the *topos* might be viewed as providing merely a dialectical rule of thumb without bearing on the semantics of a_X-propositions. Nevertheless, the similarity between the *topos* and the *dictum* provides some reason to think that both rely on pluralities of the same kind. If so, then the plurality associated with the subject term in the *dictum de omni et de nullo* is not a set exclusively of individuals. It may contain some individuals but does not consist exclusively of them.

THE THIRD PIECE OF EVIDENCE. The third piece of evidence comes from Aristotle's apodeictic syllogistic. It concerns the two moods Barbara NXN and Celarent NXN:

Major premise:	Aa_NB	A necessarily belongs to all B
Minor premise:	Ba_XC	B belongs to all C
Conclusion:	Aa_NC	A necessarily belongs to all C

11. See Alexander *in Top.* 138.6–8.

12. Aristotle illustrates this method by an example at *Top.* 2.2 109b17–24.

Major premise:	Ae_NB	A necessarily belongs to no B
Minor premise:	Ba_XC	B belongs to all C
Conclusion:	Ae_NC	A necessarily belongs to no C

Aristotle takes these moods to be valid and perfect. Although he does not establish their validity by a proof within his deductive system of categorical propositions, he offers a justification of their validity. He does so by assuming the truth of their premises and arguing on this basis for the truth of the conclusion as follows:

> Since A belongs or does not belong by necessity to all B and C is one of the Bs, it is evident that one or the other of these will also apply to C by necessity.[13] (*APr.* 1.9 30a21–3)

As we saw above (pp. 37–39), Aristotle's justification of the perfect moods Barbara and Celarent in the assertoric syllogistic is based on the *dictum de omni et de nullo.* Alexander argues that the present justification of Barbara NXN, too, is based on the *dictum de omni.*[14] More precisely, this justification seems to be based on an apodeictic *dictum de omni* which characterizes the semantics of a_N-propositions.[15] Aristotle mentions such an apodeictic *dictum de omni* at the beginning of the modal syllogistic in chapter 1.8 (30a2–3). He does not spell it out, but merely says that it is similar to the assertoric *dictum de omni.* Thus it seems safe to assume that, like the assertoric *dictum de omni,* the apodeictic one appeals to pluralities associated with terms.[16]

Now, Aristotle's justification relies on the claim that 'C is one of the Bs' (τὸ δὲ Γ τι τῶν B ἐστί). The phrase 'the Bs' seems to refer to the plurality associated with the term B.[17] Crucially, C is said to be a member of this plurality. Aristotle presumably takes C to be one of the Bs on the grounds that, according to the minor premise of Barbara NXN, B is a_X-predicated of C. Thus, he seems to hold that any a_X-proposition

13. ἐπεὶ γὰρ παντὶ τῷ B ἐξ ἀνάγκης ὑπάρχει ἢ οὐχ ὑπάρχει τὸ A, τὸ δὲ Γ τι τῶν B ἐστί, φανερὸν ὅτι καὶ τῷ Γ ἐξ ἀνάγκης ἔσται θάτερον τούτων.

14. Alexander *in APr.* 125.33–126.8.

15. Patterson (1993: 371–2; 1995: 220) and Striker (2009: 116).

16. For further discussion of the apodeictic *dictum de omni* and its role in the justification of Barbara NXN, see pp. 107–113.

17. See pp. 39–40 above and Malink (2008: 521–31).

'B belongs to all C' implies that C is a member of the plurality associated with B. More precisely, we may say that if B is aX-predicated of C, then what the term C stands for, or its semantic value, is a member of the plurality associated with the term B.

This is evidence against the assumption that pluralities associated with terms consist exclusively of individuals. Perhaps there are true aX-propositions whose subject term stands for an individual, like the term 'Mikkalos'. If so, then this individual is a member of the plurality associated with the predicate term. But in most true aX-propositions, the subject does not stand for an individual, but is a general term such as 'man' or 'walking'. As mentioned above, it is not immediately clear what these terms stand for or what their semantic value is. In any case, the items for which such terms stand are not individuals. Nevertheless, they are members of the plurality associated with the predicate term of the aX-proposition. For example, given the truth of the aX-proposition 'Animal belongs to all man', the item for which 'man' stands is a member of the plurality associated with the term 'animal'. Hence the plurality associated with 'animal' is not a set exclusively of individuals.

A defender of the orthodox *dictum* semantics might reply that the phrase 'C is one of the Bs' does not mean that C is a member of the plurality associated with B. Instead, she might take it to mean that B is aX-predicated of C, or that in some sense C is a part of B.[18] I do not claim that this is impossible. Nevertheless, it seems more natural to take the phrase to mean that C is a member of the plurality associated with B. This is confirmed by the fact that virtually the same phrase occurs in Aristotle's justification of the rule of eX-conversion (τὸ γὰρ Γ τῶν Β τί ἐστι, 1.2 25a17). As argued above, this justification is based on the assertoric *dictum de omni et de nullo*.[19] We took the phrase 'C is one of the Bs' there to indicate that C is a member of the plurality

18. For example, Ebert & Nortmann (2007: 31) translate τὸ δὲ Γ τι τῶν Β ἐστί at 30a22 as 'C constitutes a part of the B(-things)'. However, they give a different translation of virtually the same phrase in the justification of eX-conversion (τὸ γὰρ Γ τῶν Β τί ἐστι, 1.2 25a17). In this passage, they take the term C to stand for an individual (2007: 234–5), and translate the phrase as 'C is one of the Bs' (2007: 17).

19. Pp. 39–40; see also p. 89 below.

associated with B. Moreover, we took the term C there to be of the same syntactic type as the quantified variable that occurs in the abstract *dictum* semantics. If this is also true of the phrase 'C is one of the Bs' in the justification of Barbara NXN, then this justification shows that the quantified variable occurring in the abstract *dictum* semantics is a categorical term; for here it is clear that C, being the minor term of Barbara NXN, is a categorical term.

'THE ANIMALS' AND 'THE WHITES'. The term 'animal' is a_X-predicated of terms such as 'man', 'horse', and 'ox'. I have argued that each of these latter terms stands for a member of the plurality associated with the term 'animal'. We may say that each of them stands for a member of the plurality of animals. Since none of them stands for an individual, at least some members of the plurality of animals are not individuals.

On the other hand, there is a strong intuition that the plurality indicated by the English phrase 'the animals' consists only of individuals.[20] Similarly, there is an intuition that phrases like 'the As' (τῶν A) or 'those of the subject' (τῶν τοῦ ὑποκειμένου) refer to a plurality of individuals.[21] If this intuition were correct, it would support the orthodox *dictum* semantics.[22] Now, while such intuitions may be correct with respect to phrases like 'the As' or 'the animals' in English, they are not correct with respect to the corresponding phrases in Aristotle's philosophical Greek. Aristotle is happy to say that the plurality of animals includes species as well as individuals:

> For each one of the animals is either a species or an individual (ἕκαστον γὰρ τῶν ζῴων ἢ εἶδός ἐστιν ἢ ἄτομον). (*Top.* 6.6 144b2–3)

---

20. Thus, Frede (1987: 55) writes: "What makes the concept of an individual so readily available to us is the simple fact that the nouns for kinds, which objects fall under, can also be used in the plural, and that, when they are used in the plural, they apply not to kinds but to individuals."

21. Ross (1949: 193), Smith (1982a: 119–20), Ebert (1995: 230–1n11), Wolff (1998: 159), and Drechsler (2005: 290 and 315); similarly, Buddensiek (1994: 38n67), Bäck (2000: 248), and Ebert & Nortmann (2007: 776–7n1). Smith (1982a: 120) takes Philoponus *in APr.* 49.23–6 to hold the same view.

22. See Ebert (1995: 230–1n11).

Aristotle often uses phrases like 'each of the animals' (ἕκαστον τῶν ζῴων) to refer to infimae species of animals rather than to individual animals.[23] So the plurality indicated by the phrase 'the animals' consists not only of individuals but also of universals. The same would then seem to be true for pluralities indicated by phrases like 'the Bs' in the *Prior Analytics*. Striker writes about Aristotle's use of these phrases:

> The phrase 'one of the Bs' could apply equally well to species or to individuals; both Socrates and man (the species) could be said to be 'one of the animals'. (Striker 2009: 87)

Striker's view is supported by a passage from *Prior Analytics* 1.4, in which Aristotle picks out two members of the plurality indicated by the phrase 'the whites':

> Let swan and snow be taken from among the whites of which man is not predicated. Then, animal is predicated of all of one but of none of the other.[24] (*APr.* 1.4 26b7–9)

The terms 'swan' and 'snow' are general terms; neither of them stands for an individual. Aristotle seems to take each of them to stand for a member of the plurality of 'the whites', and hence of the plurality associated with the term 'white'. If so, then this plurality is not a set exclusively of individuals.

'EVERY' AND 'ALL'. Before turning to the fourth piece of evidence, let me add a remark on the translation into English of Aristotle's a_X-propositions. Aristotle usually expresses them by phrases like τὸ A παντὶ τῷ B ὑπάρχει. Some authors translate the quantifying expression παντὶ as

---

23. See, for example, *GA* 1.20 728b13, 4.10 777a32, *HA* 4.9 536a14, *MA* 11 704a3, *PA* 1.3 644a10–11, 1.5 645a22, 2.1 646a8–9, 2.2 648a16, 3.14 674a13; cf. Balme (1992: 106). At *Met.* Z 12 1038a18, the phrase 'the footed animals' (τὰ ὑπόποδα ζῷα) refers to a collection of infimae species of animal, not of individual animals. The same is presumably true of the phrase 'the particular animals' (τὰ τινὰ ζῷα) at *APost.* 2.14 98a10 (cf. Barnes 1994: 250).

24. εἶτα καὶ ὧν μὴ κατηγορεῖται λευκῶν ὁ ἄνθρωπος, εἰλήφθω κύκνος καὶ χιών· οὐκοῦν τὸ ζῷον τοῦ μὲν παντὸς κατηγορεῖται, τοῦ δὲ οὐδενός; similarly, at 1.4 26b12–14. The occurrence of λευκῶν in these passages is one of the rare examples in *APr.* 1.1–22 of a concrete categorical term used in the plural.

'every': 'A belongs to every B'.[25] Others translate it as 'all': 'A belongs to all B'.[26] The former translation introduces the idea of a plurality of items and suggests that the members of that plurality are particular individuals. For example, a sentence like 'Animal belongs to every man' or 'Every man is an animal' is most naturally understood as meaning that each individual man is an animal. So if the Greek παντὶ meant 'every', this would support the orthodox *dictum* semantics.

It is true that in some contexts παντὶ or other forms of πᾶς mean 'every'. However, these forms can also have other meanings, which are best captured by 'all' or 'the whole', for instance, 'all the water' (τὸ πᾶν ὕδωρ) or 'the whole soul' (πᾶσα ἡ ψυχή).[27] In such contexts, the idea of a plurality of particular individuals is less prominent. Unlike 'every', 'all' does not suggest a preference for the orthodox reading of a_X-propositions. Since 'all' in English best captures the wide range of uses to which πᾶς can be put in Greek, I prefer this translation of Aristotle's a_X-propositions.

THE FOURTH PIECE OF EVIDENCE. The rest of this chapter is devoted to the fourth and last piece of evidence, which is somewhat more complex than the previous ones. It concerns a special notion of conversion described by Aristotle in the second book of the *Prior Analytics:*

> When A belongs to the whole of B and of C and is predicated of nothing else, and B belongs to all C, then it is necessary for A and B to convert. For since A is said only of B and C, and B is predicated both of itself and of C, it is evident that B will be said of everything of which A is said except of A itself. (*APr.* 2.22 68a16–21)

The first sentence contains the phrases 'belongs to the whole of' and 'belongs to all', which both indicate a_X-predications. The second sentence contains the phrase 'A is said only of B and C'. Although this latter phrase lacks quantifying expressions such as 'the whole of' or 'all', it is

25. Owen (1889), Rolfes (1921), Barnes (1984: 39–113), Smith (1989), and Ebert & Nortmann (2007).

26. Jenkinson (1928), Tredennick (1938), and Mueller (1999a, 1999b).

27. For the first example, cf. *Cael.* 3.5 304a28 and *Meteorol.* 1.3 340a13. For the second example, cf. *de An.* 3.9 432a20 and *PA* 1.1 641a28, 641b4. See also *Met.* Δ 26 1024a1–10.

no doubt intended to be equivalent to 'A belongs to the whole of B and of C'. The same is true of all occurrences of 'be said of' and 'be predicated of' in the passage: all of them appear to indicate a_X-predications just as much as 'belong to the whole of' and 'belong to all'.

Aristotle begins by assuming that A is a_X-predicated of both B and C. He also assumes that there is nothing else of which A is a_X-predicated. The latter assumption needs to be qualified, though, for at the end of the passage Aristotle seems to imply that A is a_X-predicated of itself.[28] Thus, A is a_X-predicated of A, B, and C, and of nothing else. In addition, Aristotle assumes that B is a_X-predicated of B and of C.

At the end of the passage, Aristotle denies that B is a_X-predicated of A ('except of A itself'). Thus, the notion of conversion under consideration involves an asymmetry inasmuch as A is a_X-predicated of B, but B not of A. We may call it asymmetric conversion. Asymmetric conversion can be characterized as follows: A is a_X-predicated of everything of which B is a_X-predicated *including* B itself, while B is a_X-predicated of everything of which A is a_X-predicated *except* of A itself.

THE TRADITIONAL INTERPRETATION OF ASYMMETRIC CONVERSION. According to the traditional interpretation of asymmetric conversion, the terms A and B have the same extension.[29] In other words, the set of individuals (such as Socrates, Kallias, and so on) that fall under A is identical to the set of individuals that fall under B. This view is incompatible with the orthodox *dictum* semantics; for if A and B have the same extension, the orthodox *dictum* semantics implies that B is a_X-predicated of A—which Aristotle denies. Thus, the traditional interpretation of asymmetric conversion rejects the orthodox *dictum* semantics. Specifically, it rejects the assumption that an a_X-proposition is true just in case every individual that falls under the subject falls under the predicate. It thereby also rejects the assumption that the plurality associated with a term consists exclusively of individuals.

A defender of the orthodox *dictum* semantics might reply that the traditional interpretation of asymmetric conversion is incorrect. Since

28. Barnes (2007: 494).

29. Waitz (1844: 531), von Kirchmann (1877: 245), Ross (1949: 480), Colli (1955: 887), and Mignucci (1969: 698–9).

B is not a_X-predicated of A, she might insist, there is an individual that falls under A but not under B, so that the two terms do not have the same extension. It seems to me that this is not a viable option. Asymmetric conversion states that B is a_X-predicated of everything of which A is a_X-predicated except of A itself. Intuitively, this seems to imply that every individual that falls under A falls under B—contrary to what the defender of the orthodox *dictum* semantics claims. However, this implication needs to be justified in more detail. I intend to do so in the rest of this chapter. My aim is to show that the orthodox *dictum* semantics cannot give a satisfactory account of asymmetric conversion. Thus, Aristotle's asymmetric conversion will provide another reason to reject the orthodox *dictum* semantics.

THE SET-THEORETIC SEMANTICS. Asymmetric conversion requires that A be a_X-predicated of B and that B be a_X-predicated of everything of which A is a_X-predicated except of A itself. In modern quantifier logic, this amounts to the following:

(i) A is a_X-predicated of B, and
(ii) B is not a_X-predicated of A, and
(iii) for every Z, if A is a_X-predicated of Z and Z is not identical to A, then B is a_X-predicated of Z

What does the quantification 'for every Z' in (iii) mean? As mentioned above, quantifications are usually understood objectually. On this reading, the quantification requires the open formula to which it is applied to be true whatever semantic value is assigned to the free variable 'Z'. Now, the variable 'Z' is a categorical term. So what are the semantic values of categorical terms? The orthodox *dictum* semantics does not, by itself, answer this question. However, it does suggest an answer, as follows. The orthodox *dictum* semantics provides a translation from the language of categorical propositions into the language of classical first-order logic. It thereby allows us to apply the standard model theory of classical first-order logic to the language of categorical propositions: a categorical proposition may be taken to be true in a first-order model just in case the corresponding first-order formula assigned to it by the orthodox *dictum* semantics is true in that model.

As is well known, every first-order model is based on a primitive non-empty set of individuals. The semantic value of a zero-order individual term is an individual from this set. By contrast, the semantic value of a first-order predicate term is a set of individuals: the domain of possible semantic values of first-order predicate terms is the powerset (that is, the set of all subsets) of the primitive set of individuals. In the orthodox *dictum* semantics, categorical terms are regarded as first-order predicate terms. Hence the domain of possible semantic values of categorical terms is the powerset of the primitive set of individuals. The semantic value of a categorical term is a set of individuals. In the orthodox *dictum* semantics, this set is taken to be the plurality associated with that term. Thus, the semantic value of a term is identified with the plurality associated with that term.

We may call the class of all these first-order models for the language of categorical propositions the set-theoretic semantics of the assertoric syllogistic. Given the orthodox *dictum* semantics, the set-theoretic semantics appears to be the most natural class of models for the assertoric syllogistic.

As we saw above, there is a problem of existential import in the abstract *dictum* semantics. This problem carries over to the orthodox *dictum* semantics and to the set-theoretic semantics. A common way to solve it is to introduce a general presupposition to the effect that the plurality associated with any given term has at least one member (p. 42). In the set-theoretic semantics, this amounts to the requirement that the semantic value of every term have at least one member. In other words, the empty set is removed from the domain of possible semantic values of categorical terms. Another way to solve the problem is to modify the *dictum de omni* in such a way that the plurality associated with the subject term of true a_X-propositions is required to have at least one member (p. 42). Within the set-theoretic semantics, this means that the empty set is not removed from the domain of possible semantic values of categorical terms, but cannot serve as the semantic value of the subject term of true a_X-propositions.

Whichever of these solutions is chosen, with the possible exception of the empty set, all sets of individuals are usually admitted as semantic values of categorical terms. Thus, the domain of possible semantic values of categorical terms is the powerset of the primitive set of individuals, perhaps minus the empty set.

ASYMMETRIC CONVERSION IN THE SET-THEORETIC SEMANTICS. Given the set-theoretic semantics and objectual quantification, asymmetric conversion is incoherent. More precisely, the formulae (i), (ii), and (iii) are inconsistent in the set-theoretic semantics unless the term B is empty (that is, unless the semantic value of B is the empty set). This can be seen as follows. Formula (ii) states that B is not a_X-predicated of A. Hence the set-theoretic semantics requires there to be an individual z that falls under A but not under B. Crucially, the singleton set {z} is in the domain of possible semantic values of categorical terms:

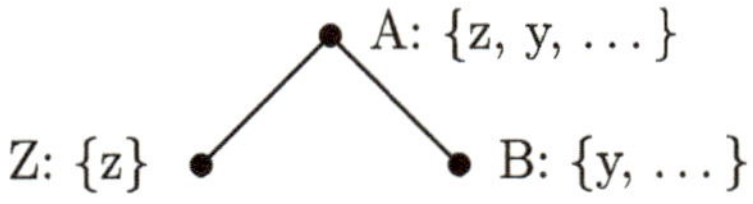

When the singleton {z} is taken to be the semantic value of the variable 'Z', then, given formula (i) and given that B is not empty, the following open formula is false:[30]

> if A is a_X-predicated of Z and Z is not identical to A,
> then B is a_X-predicated of Z

This means that (iii) is false. Hence given that B is not empty, (i), (ii), and (iii) cannot simultaneously be true in the set-theoretic semantics.

The assumption that B is not empty is needed to establish the truth of 'Z is not identical to A' in the above open formula. If B is empty, there are set-theoretic models for asymmetric conversion:

---

30. First, A is a_X-predicated of Z since z falls under A. Second, the truth of 'Z is not identical to A' can be established as follows: given that B is not empty, there is an individual y that falls under B; (i) states that A is a_X-predicated of B; hence y falls under A; y falls under B, but z does not fall under B; so y is not identical to z; therefore two distinct individuals, z and y, fall under A; hence the semantic value of A is distinct from the semantic value of Z (i.e., from the singleton {z}); this implies the truth of 'Z is not identical to A'. Third, B is not a_X-predicated of Z, since z falls under Z but not under B.

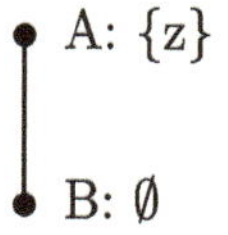

However, in many versions of the set-theoretic semantics, empty terms are not admitted. In other versions, they cannot serve as the subject of a_X-predications (that is, of true a_X-propositions)—but Aristotle assumes that A is a_X-predicated of B. In any case, it is very unlikely that Aristotle would discuss a notion of conversion that implies that one of the terms involved is empty. Thus, the set-theoretic semantics does not give a satisfactory account of asymmetric conversion under objectual quantification.

SUBSTITUTIONAL QUANTIFICATION. As a last resort, the defender of the orthodox *dictum* semantics might interpret the quantification in (iii) not objectually, but substitutionally. On this interpretation, the truth of quantified sentences depends on what terms there are in the language under consideration. Specifically, the quantification 'for every Z' takes into account all terms of the same syntactic type as the variable 'Z': it requires that each such term, when substituted for 'Z' in the open formula to which the quantification is applied, have a true sentence as its result. So substitutional quantification in effect disregards those members of the domain of possible semantic values for which the language has no name. As a result, even if B is not empty, asymmetric conversion is consistent in the set-theoretic semantics. It is satisfiable when the language under consideration does not contain any categorical term D such that the following formula is false: 'if A is a_X-predicated of D and D is not identical to A, then B is a_X-predicated of D'.

This is perhaps one of the best ways to make sense of asymmetric conversion within the set-theoretic semantics. Still, it has its drawbacks. We would have to assume that asymmetric conversion makes sense only inasmuch as the language is unable to designate certain items. Aristotle would be saying 'A belongs to the whole of B and of C and is predicated of nothing else' although there is an individual z that falls under A but not under B or C. Moreover, if substitutional quantification is adopted in formula (iii), it would seem natural to adopt it also in the orthodox *dictum* semantics. If that is done, the language will contain a noncategorical singular term that names the individual z. For, since B is not a_X-predicated of A, the substitutional interpretation of the orthodox *dictum* semantics will require the language to contain a noncategorical singular term 'n', naming the individual z, such that 'An' is true and 'Bn' is false. But then it is difficult to see why the language should not contain a categorical term designating the singleton {z}.

In sum, then, although asymmetric conversion is not unconditionally inconsistent in the set-theoretic semantics, this semantics cannot give a satisfactory account of asymmetric conversion, neither under objectual nor under substitutional quantification. Now, the set-theoretic semantics is the most natural class of models for the assertoric syllogistic if the orthodox *dictum* semantics is accepted. We may therefore conclude that the orthodox *dictum* semantics cannot give a satisfactory account of asymmetric conversion.[31] Instead, the traditional interpretation of asymmetric conversion described above seems to be correct. Thus, Aristotle's discussion of asymmetric conversion shows that he did not endorse the orthodox *dictum* semantics.

SUMMARY. The abstract *dictum* semantics does not specify the nature of the relation indicated by the phrase 'Z is a member of the plurality associated with A'. The orthodox *dictum* semantics provides a way to specify this relation. It is based on two assumptions. The first is that the pluralities associated with terms are sets of individuals such as Socrates, Kallias, and so on. The second is that the quantified variable 'Z' is not a categorical term, but a zero-order individual term. In the present chapter I have argued that these two assumptions, and hence the orthodox *dictum* semantics, should be rejected.

In Chapter 4, I propose an alternative way of specifying the abstract *dictum* semantics, based on the view that the variable 'Z' is a categorical term. This will also yield a satisfactory account of asymmetric conversion (pp. 82–83).

31. Barnes (2007: 494) holds that Aristotle's notion of asymmetric conversion is incoherent. His view is presumably based on acceptance of the orthodox *dictum* semantics.

4

The Heterodox *dictum* Semantics

The heterodox *dictum* semantics is based on the assumption that the plurality associated with a term consists of exactly those items of which the term is a_X-predicated. The relation of a_X-predication is treated as a primitive preorder, in terms of which e_X-, i_X-, and o_X-predication are defined. The aim of the present chapter is to introduce the heterodox *dictum* semantics and to defend it against some objections that have been raised against it. Moreover, we will see how this semantics determines a natural class of first-order models, which may be called the preorder semantics.

INTRODUCING THE HETERODOX *DICTUM* SEMANTICS. I have argued that the quantified variable in the abstract *dictum* semantics is of the syntactic type of categorical terms. Moreover, we have seen that Aristotle takes a_X-predication to be sufficient for being a member of the plurality associated with the predicate term (pp. 52–53). I suggest that it is not only sufficient but also necessary. In other words, C is a member of the plurality associated with A if and only if A is a_X-predicated of C. This leads to the following specification of the abstract *dictum* semantics:

Aa_XB	if and only if	$\forall Z(Ba_XZ \supset Aa_XZ)$
Ae_XB	if and only if	$\forall Z(Ba_XZ \supset \neg Aa_XZ)$
Ai_XB	if and only if	$\exists Z(Ba_XZ \wedge Aa_XZ)$
Ao_XB	if and only if	$\exists Z(Ba_XZ \wedge \neg Aa_XZ)$

Following Barnes, we may call these four equivalences the heterodox *dictum* semantics of the assertoric syllogistic.[1] The first equivalence constitutes the heterodox interpretation of the *dictum de omni.* This interpretation was endorsed by Michael Frede and has been defended by Ben Morison.[2] On the other hand, Jonathan Barnes rejects it, raising two objections to it. Let us consider his objections in turn.

THE FIRST OBJECTION TO THE HETERODOX *DICTUM* SEMANTICS. Barnes's first objection is that it is more natural to read Aristotle's formulation of the *dictum de omni* in the orthodox than in the heterodox way. He claims that "Aristotle's Greek can hardly be construed in the way demanded by the heterodox *dictum*" (2007: 412). Recall Aristotle's formulation: "none of those of the subject can be taken of which the other will not be said" (24b28–30). The orthodox interpretation of this passage may seem more natural because it has been the dominant interpretation for many centuries. But it is difficult to see anything in Aristotle's formulation itself that points to this interpretation.[3]

Although Aristotle's formulation does not explicitly express the heterodox interpretation, it can well be interpreted in the heterodox way. Consider, for instance, the phrase 'none can be taken'. There are only two further occurrences of the construction 'none can be taken' (οὐδὲν ἔστι λαβεῖν) in the *Analytics.* In both of them, the scope of the quantification 'none' is not restricted to individuals or singular terms, but rather encompasses any being whatsoever, or any categorical term.[4]

---

1. Barnes (2007: 406–12). It should be noted that Barnes (2007: 409) has a different version of the heterodox *dictum de nullo:* $\mathrm{Ae_X B}$ if and only if $\forall Z(\mathrm{Ba_X Z} \supset \mathrm{Ae_X Z})$. However, this version conflicts with the view that the *dictum de nullo* is obtained from Aristotle's formulation of the *dictum de omni* by deleting the negation οὐ at 24b30 (cf. p. 36).

2. See Morison (2008: 212–15).

3. Morison (2008: 214).

4. The first passage is concerned with cases in which nothing higher than a certain item can be taken (*APost.* 1.5 74a7–8). For example, if there were no triangles other than isosceles, then nothing higher than 'isosceles' could be taken of which 'having the sum of the interior angles equal to two right angles' is $\mathrm{a_X}$-predicated (cf. *APost.* 1.5 74a16–17; Barnes 1994: 122–3). The second passage states the following thesis of the assertoric syllogistic: if A is $\mathrm{a_X}$-predicated of B, then nothing can be taken that is $\mathrm{a_X}$-predicated of A

We have seen that in *Prior Analytics* 1.9, Aristotle takes C to be one of the Bs on the grounds that B is a_X-predicated of C (pp. 52–53). Similarly, the phrase 'none of those of the subject' in Aristotle's formulation of the *dictum de omni* may refer to items of which the subject is a_X-predicated. Also, we have seen that the phrase 'be said of' is used to indicate a_X-predication in *Prior Analytics* 2.22.[5] Similarly, the phrase 'of which the other will not be said' in Aristotle's formulation of the *dictum de omni* may refer to items of which the predicate term is not a_X-predicated.

It is true that, if he wanted, Aristotle could have explicitly stated the heterodox *dictum de omni.* For instance, he might have used a phrase such as 'what B is said of all of, A is said of all of it' (1.41 49b24–5). But there may be a number of reasons why he did not do so. For example, he might have wished to make a terminological distinction between the language of categorical propositions on the one hand and their *dictum* semantics on the other. Nevertheless, it is not implausible that what he meant in his formulation of the *dictum* is adequately captured by the heterodox interpretation.

THE SECOND OBJECTION TO THE HETERODOX *DICTUM* SEMANTICS. Barnes's second objection is that the heterodox *dictum de omni* is of no use as an explanation of the meaning of a_X-propositions, on the grounds that the explanans itself contains a_X-propositions (2007: 412). Thus, a_X-predication is explained or defined in terms of a_X-predication, so that the explanation becomes circular.

Consequently, the heterodox *dictum de omni* is less informative about the nature of a_X-predication than the orthodox one. The latter provides an explicit definition of a_X-predication in terms of a more primitive notion, namely, that of an individual's falling under a term. This definition allows us to determine the truth of a_X-propositions when we know the extension of the terms involved. The heterodox *dictum de omni* gives no such explicit definition of a_X-predication. Rather, a_X-predication is treated as a primitive, undefined relation. The question of how to determine the truth of a_X-propositions is not answered. Thus,

---

and e_X-predicated of B or vice versa (*APost.* 1.16 80a28–30; cf. Barnes 1994: 165).

5. P. 57. Cf. also the use of 'be predicated of' and 'be said of' at *APr.* 1.27 43a30–2 and 43a41–2.

unlike the orthodox *dictum de omni,* the heterodox one does not determine the intuitive meaning of a_X-propositions; it does not define what a_X-predication is.

REPLY TO THE SECOND OBJECTION. Did Aristotle mean the *dictum de omni* as an explicit definition of a_X-predication in terms of a more primitive notion? Morison suggests that the answer is no:

> There is nothing in the language at *APr.* 24b25–30 to suggest that we must construe the *dictum* as a definition. But if it isn't a definition, what is it? The obvious thought is that it is a characterisation of the relations of 'being predicated of every' and 'being predicated of no' in which we are told precisely those facts about the relations which are needed to explain the perfect syllogisms and the conversion rules. (Morison 2008: 214)

On this view, Aristotle's *dictum de omni et de nullo* is not intended as a definition of what a_X-predication is.[6] Instead, it specifies logical properties of a_X- and e_X-predication that account for the validity of his perfect moods and conversion rules. As such, the heterodox *dictum de omni et de nullo* is informative. It states that, for any A and B, A is a_X-predicated of B if and only if A is a_X-predicated of everything of which B is a_X-predicated. Given classical propositional and quantifier logic, this implies that the relation of a_X-predication is both reflexive and transitive. In other words, it implies that the following holds for any A, B, and C:[7]

Reflexivity: Aa_XA
Transitivity: if Aa_XB and Ba_XC, then Aa_XC

Relations that are reflexive and transitive are called preorders. Thus, the heterodox *dictum de omni* implies that a_X-predication is a preorder. This is not a specific feature of the heterodox *dictum* semantics. That a_X-predication is a preorder is already implied by the abstract *dictum de*

6. Cf. also Crivelli (2012: 143–4n47).

7. For any A, $\forall Z(Aa_XZ \supset Aa_XZ)$ is logically valid in classical propositional and quantifier logic. Hence the heterodox *dictum de omni* implies that Aa_XA holds for any A. In the same way, the heterodox *dictum de omni* implies the transitivity of a_X-predication; cf. p. 38 above.

omni, and hence also by the orthodox one. However, unlike the abstract and the orthodox *dictum de omni,* the heterodox one is equivalent to the statement that a_X-predication is a preorder.[8] Thus the heterodox *dictum de omni* is just another way of saying that a_X-predication is a preorder.

In the heterodox *dictum* semantics, the three relations of e_X-, i_X-, and o_X-predication are defined in terms of the primitive a_X-preorder. These definitions imply certain logical properties of the three relations, for example, that i_X-predication is symmetric. Thus, the heterodox *dictum* semantics is informative and useful inasmuch as it determines logical properties of, and logical relations between, a_X-, e_X-, i_X-, and o_X-predication.

These logical properties and relations suffice to establish the validity of the seven deduction rules on which Aristotle's deductive system is based: the three conversion rules and the four perfect moods. As we saw above, the abstract *dictum* semantics validates all seven of these rules except a_X-conversion. Clearly, every rule that is valid in the abstract *dictum* semantics is also valid in the heterodox one, since the latter is a special instance of the former. In addition, the heterodox *dictum* semantics, unlike the abstract and orthodox ones, also validates the rule of a_X-conversion (that is, the inference from Aa_XB to Bi_XA). This can be seen as follows:

1. Aa_XB	(premise)
2. Ba_XB	(by reflexivity of a_X-predication)
3. $Aa_XB \wedge Ba_XB$	(from 1, 2)
4. $\exists Z(Aa_XZ \wedge Ba_XZ)$	(from 3)
5. Bi_XA	(from 4; by heterodox *dictum de aliquo*)

The step in line 2 relies on the reflexivity of a_X-predication, which is guaranteed by the heterodox *dictum de omni.*

8. The heterodox *dictum de omni* follows from the statement that a_X-predication is a preorder: the implication from left to right in the heterodox *dictum de omni* follows from the transitivity of a_X-predication, and the converse follows from the reflexivity of a_X-predication. By contrast, the abstract and the orthodox *dictum de omni* do not follow from the statement that a_X-predication is a preorder.

In the abstract *dictum* semantics, the rule of a_X-conversion gives rise to the abstract problem of existential import. This rule is invalid in the abstract *dictum* semantics, unless there is a guarantee that in every a_X-proposition the plurality associated with the subject term has at least one member (see pp. 41–44). In the heterodox *dictum* semantics, the plurality associated with a term is the set of those items of which the term is a_X-predicated. Since a_X-predication is reflexive, the plurality associated with any term A has at least one member, namely, A itself. (More precisely, we may say that the semantic value of the term A is a member of the plurality associated with the term A.) Thus, the abstract problem of existential import is solved, and a_X-conversion is valid in the heterodox *dictum* semantics.

When the abstract *dictum* semantics is specified in the orthodox way, the problem of existential import remains. Like in the abstract *dictum* semantics, a_X-predication is a preorder in the orthodox one. But this does not help solve the abstract problem of existential import in the orthodox *dictum* semantics. So the problem must be solved in another way, typically by one of the three strategies we saw above. It is an advantage of the heterodox *dictum* semantics over the orthodox one that it provides a solution to the problem of existential import.

In sum, the heterodox *dictum* semantics validates all seven rules of Aristotle's deductive system. Consequently, it also validates all moods and conversion rules held to be valid by Aristotle in the assertoric syllogistic. As such, the heterodox *dictum* semantics is informative and useful, even though it does not provide a definition of what a_X-predication is. Thus, Barnes's second objection can be dismissed.

REFLEXIVITY OF A_X-PREDICATION. The above solution to the problem of existential import relies crucially on the reflexivity of a_X-predication. Its reflexivity is expressed by means of a_X-propositions whose subject term is identical to the predicate term, such as Aa_XA. Now, some commentators think that such propositions are not admissible in Aristotle's syllogistic but that the predicate and the subject must be two distinct terms in every categorical proposition.[9] If this were correct, they could

9. Corcoran (1972: 696; 1973: 201–2; 1974: 99; 2003: 279), Smith (1983: 225), Boger (2004: 131 and 238).

argue that the *dictum de omni* does not imply the reflexivity of a_X-predication, on the grounds that the *dictum de omni* 'Aa_XB if and only if . . . ' has no admissible instances of the form 'Aa_XA if and only if . . . '.

As Jonathan Barnes has pointed out, however, there is hardly any evidence for such a view in the *Prior Analytics*.[10] On the contrary, Aristotle repeatedly acknowledges categorical propositions whose subject is identical with the predicate. As we saw above, he assumes the truth of a_X-propositions such as Ba_XB in *Prior Analytics* 2.22.[11] Moreover, he accepts o_X- and e_X-propositions of the form Bo_XB and Be_XB throughout chapter 2.15 of the *Prior Analytics*. This suggests that categorical propositions whose subject is identical to the predicate are well-formed and perfectly acceptable expressions of Aristotle's syllogistic language. If so, then the abstract *dictum de omni* and hence also its orthodox and heterodox instances imply that a_X-predication is reflexive. If the abstract *dictum de omni* or one of its instances correctly represents Aristotle's views on the semantics of a_X-propositions, then he is committed to the reflexivity of a_X-predication.

This is confirmed by Aristotle's assertion in *Prior Analytics* 2.15 to the effect that o_X-propositions of the form Bo_XB cannot be true (see p. 43 above). Since o_X-propositions are the contradictories of a_X-propositions, this implies that all propositions of the form Ba_XB are true. Thus, Aristotle seems to endorse the reflexivity of a_X-predication, although he does not explicitly assert it.[12]

10. Barnes (2007: 387–8). Both Corcoran (1972) and Smith (1983) are interested in showing that a certain deductive system for Aristotle's syllogistic is complete with respect to (i.e., strong enough to prove everything valid in) a certain semantics. The proposition Aa_XA is not provable in their deductive systems, but is valid in their semantics. Hence the proof of completeness fails when such propositions are admitted (cf. Martin 1997: 15 and 4–6).

11. In his discussion of asymmetric conversion, Aristotle assumes the truth of Aa_XA and Ba_XB; see p. 57 above. Some commentators take Aristotle's remark 'B is predicated of itself' at *APr.* 2.22, 68a19–20, to assert the reflexivity of a_X-predication (Łukasiewicz 1957: 149, van Rijen 1989: 209, Barnes 2007: 494; similarly, Bocheński 1956: 70 12.18).

12. Accordingly, the reflexivity of a_X-predication is accepted by Łukasiewicz (1957: 88), McCall (1963: 37), and Barnes (2007: 142 and 408).

INTRODUCING THE PREORDER SEMANTICS. As we saw above, the orthodox *dictum* semantics determines a class of models, namely, the class of standard first-order models that satisfy the four equivalences of the orthodox *dictum* semantics (pp. 58–59). We called this class the set-theoretic semantics.

In the same way, the heterodox *dictum* semantics too determines a class of models. This can be seen as follows. First of all, the heterodox *dictum* semantics employs the language of classical propositional and quantifier logic. It is convenient to regard this as the language of first-order logic (as opposed to second-order logic or other kinds of higher-order logic). In first-order logic, quantifications are applied only to zero-order individual variables. Now, the variables to which quantifications are applied in the heterodox *dictum* semantics are of the syntactic type of categorical terms. Hence categorical terms are, on this view, regarded as zero-order individual terms. Copulae such as 'a_X' or 'e_X', on the other hand, are regarded as binary first-order predicates. For example, the heterodox *dictum de nullo*

$$\mathrm{Ae_XB} \qquad \text{if and only if} \qquad \forall \mathrm{Z}(\mathrm{Ba_XZ} \supset \neg \mathrm{Aa_XZ})$$

is taken to be of the form

$$\mathrm{R_1ab} \qquad \text{if and only if} \qquad \forall \mathrm{z}(\mathrm{R_2bz} \supset \neg \mathrm{R_2az}),$$

where 'R_1' and 'R_2' are binary first-order predicates representing the copulae 'e_X' and 'a_X'. Thus, categorical propositions are viewed as atomic formulae of a first-order language.[13] Given this, the heterodox *dictum* semantics determines the class of first-order models that satisfy the four heterodox equivalences. More precisely, it determines the class of models that satisfy the universal closure of the four equivalences, in which both 'A' and 'B' are bound by universal quantifiers.

These models are also models for the language of categorical propositions. They are based on a primitive set of items, each of which can

13. In the orthodox *dictum* semantics, by contrast, categorical propositions are not atomic formulae of the first-order language.

serve as the semantic value of a categorical term. (I will discuss the nature of these items in Chapter 5.) The set of these items is the domain of quantification in such a model. The semantic value of a categorical copula is a binary relation in that domain. The value of the copula 'a_X' is a primitive preorder relation. The values of the other three copulae are definable in terms of this preorder. Thus, the first-order models that satisfy the universal closure of the four heterodox equivalences are based on a primitive preorder relation. We may call the class of all these models the preorder semantics of the assertoric syllogistic.

SUMMARY. The preorder semantics is the class of first-order models determined by the heterodox *dictum* semantics. Whereas the former

abstract *dictum* semantics

Aa_XB iff $\forall Z(Z\ mpaw\ B \supset Z\ mpaw\ A)$
Ae_XB iff $\forall Z(Z\ mpaw\ B \supset \neg Z\ mpaw\ A)$
Ai_XB iff $\exists Z(Z\ mpaw\ B \wedge Z\ mpaw\ A)$
Ao_XB iff $\exists Z(Z\ mpaw\ B \wedge \neg Z\ mpaw\ A)$

instances of the abstract *dictum* semantics

orthodox *dictum* semantics

Aa_XB iff $\forall z(Bz \supset Az)$
Ae_XB iff $\forall z(Bz \supset \neg Az)$
Ai_XB iff $\exists z(Bz \wedge Az)$
Ao_XB iff $\exists z(Bz \wedge \neg Az)$

heterodox *dictum* semantics

Aa_XB iff $\forall Z(Ba_XZ \supset Aa_XZ)$
Ae_XB iff $\forall Z(Ba_XZ \supset \neg Aa_XZ)$
Ai_XB iff $\exists Z(Ba_XZ \wedge Aa_XZ)$
Ao_XB iff $\exists Z(Ba_XZ \wedge \neg Aa_XZ)$

class of first-order models determined by the orthodox and heterodox *dictum* semantics, respectively

set-theoretic semantics

preorder semantics

Figure 1

semantics is a class of models, the latter is the set of the four heterodox equivalences introduced above. Because of the completeness of classical first-order logic, these two semantics are equivalent: everything valid in one of them is valid in the other and vice versa (see pp. 75–76 below).

In exactly the same way, the set-theoretic semantics is the class of first-order models determined by the orthodox *dictum* semantics (pp. 58–59). Again, the two semantics are equivalent. The orthodox and heterodox *dictum* semantics, in turn, are special instances of the abstract *dictum* semantics. Each of these three semantics is a set of four equivalences characterizing the meaning of the four kinds of assertoric propositions. Their interrelations can be represented by the diagram in Figure 1. In Chapter 5, I discuss the preorder semantics in more detail and compare it with the set-theoretic semantics. Chapter 6 will then deal with Aristotle's proofs by ecthesis, showing how these can be reconstructed within the abstract and heterodox *dictum* semantics. These two chapters provide additional explanation of and justification for the heterodox account of the assertoric syllogistic. They are not strictly presupposed by the subsequent discussion and can be omitted without loss of continuity. Readers primarily interested in the modal syllogistic may skip to Chapter 7.

5

The Preorder Semantics

This chapter provides a closer look at the preorder semantics and some of its logical properties. We will consider how the preorder semantics relates to Aristotle's claims of validity and invalidity in the assertoric syllogistic and how it compares to the set-theoretic semantics. The discussion will be less exegetical and somewhat more technical than it has been so far.

WHAT ARE THE SEMANTIC VALUES OF CATEGORICAL TERMS? The preorder semantics treats categorical terms as zero-order individual terms. In the standard models of first-order logic, the semantic value of these terms is a single primitive item, or at least it is considered as such. The semantic value assigned to a categorical term like 'horse' or 'walking' in the preorder semantics is such a primitive item. This is a major difference from the set-theoretic semantics, in which the semantic value of these terms is a complex item, namely, a set of individuals. Another difference is that in the set-theoretic semantics, the semantic value of a categorical term is identified with the plurality associated with this term in the abstract *dictum* semantics. In the preorder semantics, by contrast, the plurality associated with a term is not the term's semantic value, but a subset of the domain of possible semantic values of categorical terms.

Unlike the set-theoretic semantics, the preorder semantics does not specify the nature of the semantic values of categorical terms. As a result, the distinction between a term and its semantic value is not as important in the preorder semantics as it is in the set-theoretic semantics. We may therefore adopt the view that the semantic value of a given term is this term itself.[1] For example, the semantic value of the linguistic expression 'horse' is this expression itself. Of course, I do not want to attribute this view to Aristotle. He presumably did not have the notion of a semantic value as it is employed in modern model theory. More important, he does not discuss the meaning or semantic status of categorical terms in the *Prior Analytics*. But if the techniques of model theory are to be applied to Aristotle's syllogistic, we need to assign semantic values to categorical terms. Taking each term to be its own semantic value is a convenient way to do so without substantive commitments about the meaning of categorical terms. I will therefore adopt this strategy, although it is not essential to the preorder semantics. This is not meant to be an interpretive statement about Aristotle's views; rather, it is a simplifying assumption of a predominantly technical nature.

Given this assumption, the domain of possible semantic values of categorical terms is the (nonempty) set of categorical terms available in the language under consideration. Thus, I will say that the plurality associated with a categorical term consists of categorical terms—instead of possible semantic values of categorical terms. Specifically, the plurality associated with a given term consists of those terms of which it is a_X-predicated. Similarly, I will say, for example, that i_X-predication requires there to be a categorical term—instead of a possible semantic value of categorical terms—of which both the subject and the predicate are a_X-predicated.

REMARK ON ANTISYMMETRY. Given that every categorical term is its own semantic value, no two distinct categorical terms have the same semantic value.[2] At the same time, two distinct categorical terms may be a_X-predicated of each other, for example, 'man' and 'biped rational animal'. In other words, A and B may be a_X-predicated of each other while A is not identical to B. This means that the relation of a_X-predication is not antisymmetric in the preorder semantics. However,

1. An example of a semantics in which every categorical term is its own semantic value is given by Smith (1982b: 123–4).

2. Again, this is not an essential feature of the preorder semantics. It would be perfectly possible to consider models in which the semantic value of a categorical term is a categorical term distinct from it, so that two terms can have the same semantic value.

nothing will depend on this failure of antisymmetry in the preorder semantics. Rather, the issue of antisymmetry is not relevant in Aristotle's syllogistic. In *Prior Analytics* 1.1–22, Aristotle does not discuss identity statements like 'A is identical with B'. Accordingly, the first-order language of the heterodox *dictum* semantics does not contain a symbol of identity, so that the condition of antisymmetry is not even expressible in it.

FOUR KINDS OF VALIDITY. Next I want to consider how the preorder semantics relates to Aristotle's claims about the validity and invalidity of moods and conversion rules. These moods and conversion rules are special cases of what we may call a categorical inference. A categorical inference consists of a set of categorical propositions that serve as the premises, and a categorical proposition that serves as the conclusion. The set of premises may also be empty. Thus, a categorical inference is a pair consisting of a set of categorical propositions and a single categorical proposition. For present purposes, we only need to consider assertoric categorical inferences (that is, inferences that consist solely of assertoric propositions).

It will be helpful to distinguish four kinds of validity of categorical inferences. First are Aristotle's claims of validity. In this sense, a categorical inference is valid just in case its validity is asserted by Aristotle in *Prior Analytics* 1.1–22. Second, there is validity in Aristotle's deductive system of categorical propositions based on the four perfect moods and the three conversion rules (see pp. 31–33). In this sense, a categorical inference is valid just in case it is deducible in the system. In other words, it is valid just in case there is a direct or indirect deduction in the deductive system from its premises to the conclusion. Third, there is validity in the heterodox *dictum* semantics. In this sense, a categorical inference is valid just in case its conclusion follows logically, in classical first-order logic, from the premises and the four equivalences of the heterodox *dictum* semantics. Fourth, there is validity in the preorder semantics. In this sense, a categorical inference is valid just in case its conclusion is true in every model of the preorder semantics in which all of its premises are true.

Because of the completeness of classical first-order logic, validity in the heterodox *dictum* semantics is equivalent to validity in the preorder semantics: a categorical inference is valid in one of them just in case it is valid in the other. Hence the third kind of validity may be identified with the fourth, and it remains to consider only three kinds of validity. For assertoric categorical inferences, these three kinds are related as follows:

validity asserted by Aristotle in *APr.* 1.1–22	$\longrightarrow$ $\not\longleftarrow$	validity in the deductive system	$\longrightarrow$ $\not\longleftarrow$	validity in the heterodox *dictum* semantics and the preorder semantics

Every assertoric categorical inference whose validity Aristotle asserts in *Prior Analytics* 1.1–22 is valid in the deductive system. The converse is not true: for example, the inference from $\mathrm{Ae_XB}$ to $\mathrm{Bo_XA}$ is valid in the deductive system but is not mentioned by Aristotle.[3] Moreover, validity in the deductive system implies validity in the preorder semantics, for the deduction rules of the system are all valid in the preorder semantics. In other words, the deductive system is sound with respect to the preorder semantics. Again, the converse is not true: the deductive system is not complete with respect to the preorder semantics. Consider, for instance, the reflexivity of $\mathrm{a_X}$-predication. The categorical inference with the empty set of premises and the conclusion $\mathrm{Aa_XA}$ is valid in the preorder semantics. But it is not valid in the deductive system, for it is not deducible by means of the four perfect moods and the three conversion rules.[4]

3. See Ebert & Nortmann (2007: 238–9).

4. There are classes of models in which all deduction rules of the deductive system are valid but $\mathrm{a_X}$-predication is not reflexive. For example, the preorder semantics may be modified in such a way that $\mathrm{a_X}$-predication is no longer required to be reflexive but only serial, where seriality is the condition that every term be $\mathrm{a_X}$-predicated of some term. (An example of an irreflexive serial relation is the proper part relation in a domain of infinitely divisible magnitudes.) Everything else remains as it is. In the original preorder semantics, reflexivity is only needed to account for the validity of $\mathrm{a_X}$-conversion. This is also accounted for by seriality. The proof of $\mathrm{a_X}$-conversion given on p. 67 can easily be modified so that it relies on seriality instead of reflexivity. Hence everything valid in the deductive system is valid in the modified serial semantics, but $\mathrm{a_X}$-predication is not reflexive in it.

FOUR KINDS OF INVALIDITY. Corresponding to the four kinds of validity, there are four kinds of invalidity. First are Aristotle's claims of invalidity. In this sense, a categorical inference is invalid just in case its invalidity is asserted in *Prior Analytics* 1.1–22. Of course, not every categorical inference is claimed to be valid or invalid by Aristotle; many of them are not mentioned by him. Consequently, some categorical inferences are neither valid nor invalid in the first sense. As to the other three kinds of invalidity, a categorical inference is invalid in one of these senses if and only if it is not valid in that sense. The four kinds of invalidity are related as follows:

invalidity asserted by Aristotle in *APr.* 1.1–22	$\longrightarrow$ $\not\longleftarrow$	invalidity in the heterodox *dictum* semantics and the preorder semantics	$\longrightarrow$ $\not\longleftarrow$	invalidity in the deductive system

Invalidity asserted by Aristotle in *Prior Analytics* 1.1–22 implies invalidity in the preorder semantics. This can be proved by suitable models of the preorder semantics in which the premises of the categorical inference in question are true and the conclusion is false. These models are given in Appendix B (pp. 313–319). The converse is not true: invalidity in the preorder semantics does not imply invalidity asserted by Aristotle, simply because a large number of obviously invalid inferences are not mentioned by Aristotle.

In sum, as far as assertoric categorical inferences are concerned, validity asserted by Aristotle in *Prior Analytics* 1.1–22 implies the other three kinds of validity, and invalidity asserted by Aristotle in these chapters implies the other three kinds of invalidity. Thus, the deductive system, the heterodox *dictum* semantics, and the preorder semantics are in accordance with Aristotle's claims of validity and invalidity in chapters 1.1–22. In this sense, they are adequate with respect to Aristotle's assertoric syllogistic.

THE *EX FALSO QUODLIBET* AND THE *VERUM EX QUOLIBET*. All assertoric categorical inferences whose invalidity Aristotle asserts in *Prior Analytics* 1.1–22 are invalid in the preorder semantics. However, a note of caution is in order. Some assertoric categorical inferences whose invalidity Aristotle asserts elsewhere in the *Prior Analytics* are valid in the preorder semantics. Consider, for example, the following inference, in which $\mathrm{Aa_XA}$ is inferred from its contradictory:

Premise:	$\mathrm{Ao_XA}$
Conclusion:	$\mathrm{Aa_XA}$

Aristotle does not discuss such inferences in *Prior Analytics* 1.1–22. But he denies their validity in chapter 2.4, where he states that nothing can follow from its own contradictory.[5] In the preorder semantics, by contrast, the inference is valid on the grounds that its premise cannot be true. For the preorder semantics is governed by modern classical logic, which accepts the principle of *ex falso quodlibet* (according to which anything follows from logically false premises).

Here are two more examples:

Premise:	$\mathrm{Aa_XB}$	$\mathrm{Aa_XB}$
Premise:	$\mathrm{Ao_XB}$	$\mathrm{Ba_XC}$
Conclusion:	$\mathrm{Ca_XD}$	$\mathrm{Da_XD}$

Both inferences are valid in the preorder semantics as a result of the validity in classical logic of the *ex falso quodlibet* and of the *verum ex quolibet* (according to which a logically true conclusion follows from any premises). On the other hand, there is reason to think that Aristotle would deny their validity. In *Prior Analytics* 2.15, Aristotle asserts that $\mathrm{Aa_XA}$ cannot be deduced from the premise pair $\mathrm{Aa_XB}$, $\mathrm{Ao_XB}$ (2.15 64a20–2). This suggests that $\mathrm{Ca_XD}$ cannot be deduced from this premise pair either. In chapter 1.23, he states that every deduction (συλλογισμός) comes about through one of his three figures (1.23 40b20–22). The two inferences given above are not in the three figures, nor can they be proved to be valid by means of inferences in the three figures, because the terms that occur in their conclusions do not occur in the premises. Aristotle seems to hold, then, that the conclusion of the two inferences cannot be

5. *APr.* 2.4 57b13–14; see Łukasiewicz (1957: 49–50), Patzig (1959: 191), and McCall (1966: 415).

deduced from the premises and hence that the inferences are invalid.[6] In this respect, Aristotle's account of deduction differs from that of modern classical logic but bears some similarity to modern systems of relevance logic and paraconsistent logic.

None of the three categorical inferences given above are moods, for moods are, by definition, in one of the three figures (p. 30). It is therefore still true that every assertoric mood whose invalidity Aristotle asserts somewhere in the *Prior Analytics* is invalid in the preorder semantics. Also, it should be noted that the three inferences in question are already valid in the abstract *dictum* semantics, and hence also in the orthodox *dictum* semantics and the set-theoretic semantics. All these semantics validate categorical inferences that Aristotle regards as invalid. This is one of the costs of applying tools of modern classical logic to Aristotle's syllogistic, both on the orthodox and on the heterodox interpretation. I hope this study will show that the cost is outweighed by the insights gained through the application of these tools.

COMPARING THE PREORDER SEMANTICS AND THE SET-THEORETIC SEMANTICS. In the remainder of this chapter, I point out some more differences between the preorder semantics and the set-theoretic semantics. Let me begin by explaining what the latter semantics looks like from the perspective of the former. Recall that the set-theoretic semantics is based on a primitive nonempty set of individuals and that the domain of possible semantic values of categorical terms is the powerset of this primitive set (p. 59). The semantic value of the a_X-copula is the subset relation in that powerset. The subset relation is a preorder. As such, it is comparable to the primitive preorder in the preorder semantics. In a powerset, the subset relation constitutes a very special kind of preordered structure, namely, a complete atomic Boolean algebra. These algebras have a number of special properties. For instance, they possess a zero element, that is, an element that is a subset of all elements. In the powerset algebra, this is the empty set. Moreover, they satisfy the condition of atomicity, according to which every nonzero element includes at least one atom as a subset (where an atom is a nonzero element that includes only itself and the zero element as subsets). In the powerset algebra, the atoms are

6. See Priest (2006: 12n16; 2007: 132).

the singleton sets. By contrast, the primitive preorder employed in the preorder semantics need not satisfy these properties; it need not even constitute a Boolean algebra, but is only required to be a preorder.

In the set-theoretic semantics, the semantic value of a categorical term is a set of individuals. Sets are not single primitive items, but have a complex structure, determined by the relation of set-theoretic membership. This relation is critical for the interpretation of categorical propositions in the set-theoretic semantics. For example, a_X-propositions are interpreted by the condition that every member of the semantic value of the subject term be a member of the semantic value of the predicate term. This condition can be straightforwardly expressed in terms of the subset preorder: the semantic value of the subject term is a subset of the semantic value of the predicate term. This corresponds neatly to the interpretation of a_X-propositions in the preorder semantics.

However, the set-theoretic interpretation of i_X-propositions does not correspond neatly to that in the preorder semantics. The set-theoretic interpretation requires there to be an individual that is a member of both the semantic value of the predicate term and that of the subject term. This condition cannot straightforwardly be expressed in terms of the subset preorder. More precisely, when trying to do so, we obtain a condition of greater logical complexity than the original version: we obtain the condition that there be an *atomic* element in the powerset that is a subset of both semantic values. In an atomic Boolean algebra, this is equivalent to the condition that there be a *nonzero* element in the powerset that is a subset of both semantic values.[7] In this latter condition, the existential quantification is restricted to a proper subset of the domain of possible semantic values of categorical terms, namely, to nonzero elements. No such restriction is present in the original version formulated in terms of membership. Thus, the reformulation in terms of the preorder is logically more complex than the original version, since it introduces a restriction to nonzero elements.

The interpretation in the set-theoretic semantics of i_X-propositions looks natural when formulated in terms of membership. But it looks less natural when formulated in terms of the subset preorder, because it introduces an additional stipulation of nonzeroness.

---

7. A similar order-theoretic description of a semantics for the assertoric syllogistic is given by Martin (1997: 5).

THE NONEMPTY SET SEMANTICS. As we saw above, the set-theoretic semantics has a problem of existential import. A common way to solve it is to exclude empty terms, that is, to remove the empty set from the domain of possible semantic values of categorical terms (p. 59). The resulting domain is a complete atomic Boolean algebra with the zero element removed. Let us call this the nonempty set semantics of the assertoric syllogistic.

In this semantics, the set-theoretic interpretation of i_X-propositions can be straightforwardly expressed in terms of the subset preorder, as follows: there is some member in the domain of possible semantic values of categorical terms that is a subset of both the semantic value of the subject term and that of the predicate term. In this condition, the quantification 'there is some' is not restricted to a proper subset of the domain. The reformulation in terms of the subset preorder has the same logical structure as the interpretation of i_X-propositions in the preorder semantics. Thus, the nonempty set semantics can be viewed as a special instance of the preorder semantics. From the perspective of the resulting preorder, however, the appeal to a primitive set of individuals and to the relation of set-theoretic membership appears as an unnecessary complication.

The complication may be described as follows. First a complete atomic Boolean powerset is generated from a primitive set of individuals. Then the zero element is removed because of the problem of existential import. As a result, the set-theoretic interpretation can be naturally reformulated solely in terms of the preorder, so that we obtain a special instance of the preorder semantics. So why not forget membership and start with the preorder right away? After all, there is no evidence in the *Prior Analytics* that Aristotle generated a domain of semantic values of categorical terms from a set of items that cannot serve as semantic values of categorical terms.

EMPTY TERMS. A term is empty when no individual falls under it, that is, when its extension is the empty set. Examples of empty terms are 'goat-stag' and 'nonbeing'. Although they are empty, Aristotle accepts these two terms as argument-terms of categorical propositions in the *Prior Analytics*.[8] This is evidence against the nonempty set semantics,

8. *APr.* 1.38 49a24 in connection with 49a11–19, cf. Thom (1981: 79), Bäck (2000: 243), and Crivelli (2004: 162n39).

in which empty terms are not allowed. Aristotle does not seem to exclude empty terms in the syllogistic. Rather, the question of whether or not an individual falls under a term seems to be irrelevant in *Prior Analytics* 1.1–22. Accordingly, the question is not relevant in the preorder semantics; in fact, it cannot even be properly expressed in the preorder semantics. Nothing prevents 'nonbeing' from being a_X-predicated of 'goat-stag' in the preorder semantics, and nothing prevents 'goat-stag' from being a_X-predicated of itself. On the assumption that every term is its own semantic value, the semantic value of the term 'goat-stag' is the term 'goat-stag'.

The models of the nonempty set semantics are obtained from those of the set-theoretic semantics by removing the zero element (that is, the empty set). The models of the preorder semantics, on the other hand, may have a zero element (that is, an element of which every term is a_X-predicated). Zero elements do not play a significant role in the preorder semantics: some models have a zero element; others do not. The rule of a_X-conversion from Aa_XB to Bi_XA is valid in the preorder semantics even when the semantic value of B is a zero element. In fact, if a model of the preorder semantics has a zero element, then every i_X-proposition is true in it.

ASYMMETRIC CONVERSION REVISITED. The nonempty set semantics is a special instance of the preorder semantics. The preorder semantics therefore does not exclude set-theoretic models, but denies only that these are the only admissible ones. It is more general than the nonempty set semantics, including also a large number of non-set-theoretic models. Such non-set-theoretic models are not necessarily needed to account for Aristotle's claims of invalidity in the assertoric syllogistic in chapters 1.1–7 of the *Prior Analytics;* these claims can also be accounted for by the nonempty set semantics. But non-set-theoretic models are needed to give a satisfactory account of Aristotle's notion of asymmetric conversion. For, as we saw above, the nonempty set semantics and the set-theoretic semantics cannot achieve this (pp. 56–62).

Asymmetric conversion concerns two terms, A and B, which are both a_X-predicated of themselves. It requires that A be a_X-predicated of everything of which B is a_X-predicated including B itself, while B is a_X-predicated of everything of which A is a_X-predicated except of A itself. In the preorder semantics, a model for asymmetric conversion can be constructed as follows (a_X-predication is indicated by downward lines):

This model may be constructed in such a way that every downward path of a_X-predications terminates in a term that is not a_X-predicated of another term. These terms may be thought of as categorical singular terms standing for individuals such as Socrates and Kallias.[9] If so, then A and B are a_X-predicated of exactly the same categorical singular terms. We may therefore say that A and B have the same extension, that is, the same set of individuals that fall under them. This matches the traditional interpretation of asymmetric conversion, according to which B is not a_X-predicated of A although both terms have the same extension (see p. 57). The fact that B is not a_X-predicated of A does not depend on the extension of the two terms, but rather on some of their nonextensional features.[10] Thus, the preorder semantics gives a satisfactory account of asymmetric conversion in line with the traditional interpretation.

SUPPLEMENTATION. Asymmetric conversion implies that B is not a_X-predicated of A. Moreover, it implies that A is not a_X-predicated of anything of which B is e_X-predicated. Thus, asymmetric conversion violates the following statement:

If not Ba_XA, then there is a Z such that
(i) Aa_XZ, and
(ii) for every Y, if Za_XY then not Ba_XY

This statement is often called the mereological principle of (strong) supplementation. The qualification "mereological" indicates that the relation of a_X-predication in it is regarded as a part-whole relation. More

9. Cf. pp. 48–49 above. Aristotle characterizes individuals as those items that are "predicated of nothing else truly universally" (*APr.* 1.27 43a25–6).

10. For an account of these nonextensional features, see Malink (2009a: 125–39).

precisely, it is regarded as a reflexive part-whole relation, according to which every item is a part of itself. The principle states that if A is not a part of B, then some part of A is entirely disjoint from B. Such a part of A is called a supplement. In order to see more clearly what role the principle of supplementation plays in the assertoric syllogistic, it will be helpful to consider the preorder semantics from a mereological perspective.

A MEREOLOGICAL PERSPECTIVE ON THE PREORDER SEMANTICS. Alexander and Philoponus regard a_X-predication as a kind of part-whole relation.[11] This seems to be in accordance with Aristotle's views. Aristotle describes the relation between terms such as 'science' and 'medicine' as that of 'a whole to a part'.[12] Similarly, he tends to think of universals as wholes including as parts their species.[13] Aristotle sometimes expresses a_X-propositions by the phrase 'being in a whole'.[14] Also, his terminology for particular and universal propositions is derived from part-whole terminology: ἐν μέρει ('in part') versus καθόλου (literally: 'of the whole').

In modern formalized systems of mereology, two basic requirements are imposed on the part-whole relation: that it be a preorder and that it be antisymmetric.[15] This basic system of mereology is often strengthened by additional principles, which assert the existence of certain items given the existence of other items. For example, they assert the existence of complements, atoms, sums, or products. Another example is the principle of supplementation, asserting the existence of supplements. When the principle of supplementation is added to the basic system of mereology, the result is called extensional mereology.[16] It is so called because the addition of the principle implies that any two items that have the

11. Alexander *in APr.* 25.2–4, Philoponus *in APr.* 47.23–48.2, 73.22–3, 104.11–16, 164.4–7; similarly, Maier (1900b: 151–4).

12. *APr.* 2.15 64a17 and 64b12–13; cf. 64a4–7 and Smith (1989: 203).

13. *Met.* Δ 25 1023b18–19, 1023b24–5, Δ 26 1023b29–32, *Phys.* 1.1 184a25–6.

14. *APr.* 1.1 24a13, 24b26–7, 1.4 25b33, 1.8 30a2–3, 2.1 53a21–4, *APost.* 1.15 79a37–b20.

15. See, for example, Simons (1987: 25–41) and Varzi (1996: 260–7).

16. Varzi (1996: 262).

same nonempty set of proper parts are parts of each other.[17] Given antisymmetry, this means that any two such items are identical. A classical example of an extensional mereology is any powerset (of a nonempty set) with the empty set removed—that is, the underlying structure of the nonempty set semantics.

In the preorder semantics, a_X-predication is regarded as a very weak part-whole relation. This relation need not satisfy antisymmetry, nor need it satisfy the additional principles that assert the existence of supplements, complements, sums, products, and so on. Aristotle's syllogistic, I suggest, does not rely on any such assumptions of existence, but only on the requirement that a_X-predication be a preorder.

SUPPLEMENTATION AND ECTHESIS. The principle of supplementation is not valid in the preorder semantics. This might give rise to an objection to the preorder semantics. The objection would be that Aristotle implicitly relies on this principle in his proofs by ecthesis in the assertoric syllogistic. It is often thought that Aristotle's proofs by ecthesis are based on the following implication:[18]

If Bo_XA, then there is a Z such that
(i) Aa_XZ, and
(ii) Be_XZ

Let us call this the strong principle of o_X-ecthesis. In the preorder semantics, this principle is equivalent to the mereological principle of supplementation. So if Aristotle relied on this principle in his proofs by ecthesis, there would be good reason to reject the preorder semantics. In Chapter 6, I argue that, contrary to what is often thought, Aristotle does not rely on the strong principle of o_X-ecthesis.

17. Simons (1987: 28–9). Proper parts of an item are those parts of which the item is not a part.

18. Authors who hold this view include Galen *Inst. Log.* X 8, Alexander *in APr.* 104.3–7, Łukasiewicz (1957: 65), Patzig (1968: 161), Rescher & Parks (1971: 685), Rescher (1974: 11), Smith (1983: 227; 1989: xxiii), Detel (1993: 164), Lagerlund (2000: 8), Martin (2004: 19), and Nortmann (2005: 164). The implication is also accepted as valid by Barnes (2007: 404–5).

6

Ecthesis

In order to establish the validity of imperfect syllogistic moods, Aristotle employs direct or indirect deductions based on conversion rules and perfect moods. In the assertoric syllogistic, all valid imperfect moods are proved to be valid by such deductions. For some of these moods, however, Aristotle indicates alternative proofs based on the method of what he calls ecthesis. These proofs by ecthesis have often been interpreted in a way that conflicts with the heterodox *dictum* semantics and the preorder semantics. For example, they have been taken to rely on the strong principle of o_X-ecthesis mentioned above.

My aim in this chapter is to give an account of Aristotle's proofs by ecthesis in the assertoric syllogistic and to show that they do not conflict with the heterodox *dictum* semantics and the preorder semantics. In particular, I argue that they do not rely on the strong principle of o_X-ecthesis. Thus the two semantics will be defended against potential objections.

There are two places in the assertoric syllogistic where Aristotle employs ecthesis: to establish the validity of e_X-conversion in chapter 1.2, and to prove certain assertoric third-figure moods in chapter 1.6. There is one more occurrence of ecthesis in the modal syllogistic, in proofs of Baroco NNN and Bocardo NNN in chapter 1.8. I will postpone discussion of these proofs (pp. 181–185) and will focus here on those in chapters 1.2 and 1.6.

OF WHAT SYNTACTIC TYPE IS THE TERM SET OUT? The most characteristic feature of ecthesis is the introduction, the setting out, of a new term that did not occur previously in the proof. Aristotle describes this step as follows:

One of the Ss is taken, for instance N (ληφθῇ τι τῶν Σ οἷον τὸ N, *APr.* 1.6 28a24–5)

One of the Ss is taken (ληφθῇ τι τῶν Σ, *APr.* 1.6 28b21)

In both passages, the term S is a categorical term, that is, an argument-term of a categorical proposition. The new term set out, N, stands for a member of the plurality of the Ss.

There is a controversy in the secondary literature over the syntactic type of the term set out. Some commentators take it to be a categorical term.[1] Others take the term set out to be a noncategorical singular term, at least in the context of the assertoric syllogistic.[2] Aristotle's treatment of ecthesis in the assertoric syllogistic does not settle the controversy. But in the modal syllogistic, in chapter 1.8, he states that the ecthetic proof involves a "deduction about" the term set out (30a10). By this he seems to mean that the term set out serves as the argument-term of a categorical proposition that is used as the premise of a syllogistic mood.[3] If so, then the term set out is a categorical term, at least in the modal syllogistic. In view of this, some commentators assume that the term set out is a noncategorical singular term in the assertoric syllogistic, but a categorical term in the modal syllogistic.[4]

---

1. Maier (1900a: 101n2, 105, and 106n1), Ross (1949: 317–18), Łukasiewicz (1957: 59–67), Patzig (1968: 161–8), Smith (1983: 226–7; 1989: xxiii–xxv), and Wolff (1998: 141–54).

2. Smiley (1962: 63), Thom (1981: 166–74), Smith (1982a: 117–21), Mignucci (1991: 21–8), Patterson (1995: 71–2), Drechsler (2005: 206–14), and Ebert & Nortmann (2007: 234–5, 327–8, and 335–7). This view goes back to Alexander, who thinks that the item set out in proofs by ecthesis in the assertoric syllogistic is a perceptible individual; cf. Alexander *in APr.* 33.2–5, 99.31–100.14, 104.1–3, 112.33–113.1, 122.17–21.

3. Alexander *in APr.* 121.15–122.29.

4. Patterson (1995: 71–3), Drechsler (2005: 215–17), and Ebert & Nortmann (2007: 234–5, 327–8, 335–7, and 376–9). Again, this view goes back to Alexander *in APr.* 122.17–29; he thinks that in the modal syllogistic the item set out is not a perceptible individual, but 'some part or species' (122.28) of a universal. Others assume, despite the evidence to the contrary, that the term set out is a noncategorical singular term even in the modal syllogistic (Mignucci 1991: 24–8; Thom 1993: 196–7; Johnson 1993: 179; 2004: 280).

ECTHESIS AND THE *DICTUM DE OMNI ET DE NULLO.* The controversy over the syntactic type of the term set out is connected to that over the orthodox or heterodox interpretation of Aristotle's *dictum de omni et de nullo*. In the two passages just quoted, the term set out stands for a member of the plurality of the Ss. That is, it stands for a member of the plurality associated with the categorical term S.[5] Pluralities associated with categorical terms figure crucially in the *dictum de omni et de nullo*. Aristotle's formulation of the *dictum de omni* is: 'none of those of the subject can be taken of which the other will not be said' (24b28–30). The formulation is parallel to the phrase 'one of the Ss is taken' in the ecthetic proofs in chapter 1.6. As Smith points out, this suggests that Aristotle's proofs by ecthesis are based on the *dictum de omni et de nullo*.[6]

If this is correct, proofs by ecthesis should be closely connected to the abstract *dictum* semantics. Specifically, the term set out in these proofs should be of the same syntactic type as the quantified variable used in the abstract *dictum* semantics; for both the term set out and the quantified variable are of such a type as to stand for a member of the plurality associated with a categorical term. On the orthodox interpretation, the quantified variable is a noncategorical singular term, whereas on the heterodox interpretation it is a categorical term. Now, I have argued in detail for the latter interpretation. Given this interpretation, the natural view is that the term set out in proofs by ecthesis is a categorical term—in the assertoric as well as in the modal syllogistic.

My aim here is not so much to prove this view of ecthesis as to show that there is no evidence against it. To this end, I will show how all of Aristotle's proofs by ecthesis in the assertoric syllogistic can be adequately reconstructed within the abstract *dictum* semantics. So let us recall the four equivalences of the abstract *dictum* semantics (p. 37):

$$\begin{array}{llll}
\mathrm{Aa_XB} & \text{if and only if} & \forall \mathrm{Z}\,(\mathrm{Z}\ mpaw\ \mathrm{B} \supset & \mathrm{Z}\ mpaw\ \mathrm{A}) \\
\mathrm{Ae_XB} & \text{if and only if} & \forall \mathrm{Z}\,(\mathrm{Z}\ mpaw\ \mathrm{B} \supset & \neg\ \mathrm{Z}\ mpaw\ \mathrm{A}) \\
\mathrm{Ai_XB} & \text{if and only if} & \exists \mathrm{Z}\,(\mathrm{Z}\ mpaw\ \mathrm{B} \wedge & \mathrm{Z}\ mpaw\ \mathrm{A}) \\
\mathrm{Ao_XB} & \text{if and only if} & \exists \mathrm{Z}\,(\mathrm{Z}\ mpaw\ \mathrm{B} \wedge & \neg\ \mathrm{Z}\ mpaw\ \mathrm{A})
\end{array}$$

5. See pp. 39–40 above. Cf. also the phrase 'C is one of the Bs' (τὸ γὰρ Γ τῶν Β τί ἐστιν, *APr.* 1.2 25a17) in the ecthetic proof of $\mathrm{e_X}$-conversion, where C is the new term set out.

6. Smith (1989: xxv).

These equivalences, in combination with the abstract requirement of existential import, will suffice to account for Aristotle's ecthetic proofs in the assertoric syllogistic.

THE ECTHETIC PROOF OF $\mathrm{E_X}$-CONVERSION. As we saw above (pp. 39–40), Aristotle justifies the conversion from $\mathrm{Ae_XB}$ to $\mathrm{Be_XA}$ by means of an indirect argument relying on the *dictum de nullo* (1.2 25a15–17). Although he does not explicitly say so, the justification employs the method of ecthesis.[7] It can be reconstructed as a proof by ecthesis relying on the abstract *dictum de nullo*. The step of ecthesis, setting out a new term C, occurs in line 4:

1. $\mathrm{Ae_XB}$	(premise)
2. not $\mathrm{Be_XA}$	(assumption for *reductio*)
3. not $\forall \mathrm{Z}(\mathrm{Z}\ mpaw\ \mathrm{A} \supset \neg \mathrm{Z}\ mpaw\ \mathrm{B})$	(from 2; by abstract *dictum de nullo*)
4. $\mathrm{C}\ mpaw\ \mathrm{A} \wedge \mathrm{C}\ mpaw\ \mathrm{B}$	(from 3; by existential instantiation)
5. not $\forall \mathrm{Z}(\mathrm{Z}\ mpaw\ \mathrm{B} \supset \neg \mathrm{Z}\ mpaw\ \mathrm{A})$	(from 4)
6. not $\mathrm{Ae_XB}$	(from 5; by abstract *dictum de nullo*)

Since the statement in line 6 contradicts the premise in line 1, the indirect proof of $\mathrm{e_X}$-conversion is complete.[8] The steps in lines 4 and 5 are justified by rules of classical propositional and quantifier logic. In particular, the introduction in line 4 of the new term C is justified by the rule of existential instantiation in modern quantifier logic. Of course, I do not want to suggest that Aristotle had a clear grasp of all these rules.

7. Alexander *in APr.* 32.28–33.3, Philoponus *in APr.* 49.20–31, Maier (1900a: 20), Ross (1949: 293), Smiley (1962: 63), Patzig (1968: 138 and 162), Lear (1980: 4), Patterson (1995: 71), Hintikka (2004: 146), Drechsler (2005: 155 and 315–16), Ebert & Nortmann (2007: 234 and 327), and Striker (2009: 87).

8. Alternatively, the statements in lines 2 and 6 could be replaced by $\mathrm{Bi_XA}$ and $\mathrm{Ai_XB}$, respectively. In this case, the proof would rely on the abstract *dictum de aliquo* instead of the abstract *dictum de nullo*.

Nevertheless, the above proof gives, I think, an adequate reconstruction of his justification of e_X-conversion.

THE ECTHETIC PROOF OF DARAPTI. In *Prior Analytics* 1.6, Aristotle gives ecthetic proofs of four third-figure moods, namely, Darapti, Disamis, Datisi, and Bocardo (28a22–6, b14–15, b20–1). Each of these moods has already been proved to be valid without ecthesis, either by a direct or by an indirect deduction. Ecthesis is only used as an alternative method to establish their validity.

Let us begin with the case of Darapti:

Major premise:	Pa_XS	P belongs to all S
Minor premise:	Ra_XS	R belongs to all S
Conclusion:	Pi_XR	P belongs to some R

The ecthetic proof of this mood appears as follows:

> The demonstration can also be carried out ... by setting out (τῷ ἐκθέσθαι). For if both terms belong to all S, and if one of the Ss is taken, for instance N, then both P and R will belong (ὑπάρξει) to this; consequently, P will belong to some R. (*APr.* 1.6 28a22–6)

Aristotle sets out a member of the plurality associated with the middle term S, and calls it N. This step cannot be justified by the abstract *dictum* semantics alone, because this semantics does not guarantee that there is a member of the plurality associated with S. However, the step can be justified by means of the abstract requirement of existential import, according to which the plurality associated with any term must have at least one member (p. 44). This is the only passage where that requirement is needed to account for Aristotle's ecthetic proofs within the abstract *dictum* semantics.

Aristotle goes on to infer that both P and R belong (ὑπάρξει) to N. What does it mean to say, in this context, that a term belongs to another? Remarkably, the verb 'belong' is not accompanied here by a quantifying expression such as 'all', 'some', or 'none'. In the whole assertoric syllogistic (*Prior Analytics* 1.1–2, 4–7), there are only two passages where 'belong' is used without a quantifying expression: the present passage in the ecthetic proof of Darapti, and a passage in the ecthetic proof of Bocardo (which we will consider shortly).[9] In both passages, the term to which something is said to belong is the term set out by ecthesis. This suggests that the two unquantified occurrences of 'belong' have some significance in connection with ecthesis. Alexander takes them to indicate that the item set out is an individual, on the grounds that quantifying expressions cannot occur in propositions whose subject term stands for an individual.[10] This is in line with the orthodox, but not with the heterodox *dictum* semantics.

However, there may well be another explanation of the unquantified occurrences of 'belong' in proofs by ecthesis. In order to see this, let us briefly reconsider Aristotle's ecthetic justification of e_X-conversion:

> If A belongs to none of the Bs, then neither will B belong to any of the As. For if it belongs to some, for instance to C, it will not be true that A belongs to none of the Bs, since C is one of the Bs. (*APr.* 1.2 25a15–17)

The phrase 'if it belongs to some, for instance to C' in this passage implies that B belongs to C. So we have another, albeit implicit, occurrence of 'belong' used without a quantifying expression. What does it mean that B belongs to C? At the end of the passage, C is said to be one of the Bs. That is, C is said to be a member of the plurality associated with B (see pp. 52–54). Hence the unquantified occurrence of 'belong' in the passage indicates a relation between B and a member of the plurality associated with B. It is then natural to assume that, in this context, the verb 'belong' is used to indicate that C is a member of the plurality associated with B. Thus, I want to suggest that, in proofs by ecthesis, Aristotle uses unquantified phrases like 'B belongs to C' to express that C is a member of the plurality associated with B, without indicating the presence of individuals or noncategorical singular terms.[11]

If this is correct, then Aristotle's statement in the ecthetic proof of Darapti that P and R belong to N means that N is a member of the

9. Smith (1982a: 119).

10. Alexander *in APr.* 100.9–14; similarly, Smith (1982a: 119–20).

11. For further justification of this suggestion, see Malink (2008: 524–30). As argued above (p. 35), this is also the meaning of the unquantified occurrence of 'be said' in Aristotle's formulation of the *dictum de omni* at 24b30.

pluralities associated with P and R. In the following reconstruction, this corresponds to lines 7 and 8:

1. $\mathrm{Pa_XS}$	(major premise)
2. $\mathrm{Ra_XS}$	(minor premise)
3. $\forall \mathrm{Z}(\mathrm{Z}\ \mathit{mpaw}\ \mathrm{S} \supset \mathrm{Z}\ \mathit{mpaw}\ \mathrm{P})$	(from 1; by abstract *dictum de omni*)
4. $\forall \mathrm{Z}(\mathrm{Z}\ \mathit{mpaw}\ \mathrm{S} \supset \mathrm{Z}\ \mathit{mpaw}\ \mathrm{R})$	(from 2; by abstract *dictum de omni*)
5. $\exists \mathrm{Z}(\mathrm{Z}\ \mathit{mpaw}\ \mathrm{S})$	(abstract requirement of existential import)
6. $\mathrm{N}\ \mathit{mpaw}\ \mathrm{S}$	(from 5; by existential instantiation)
7. $\mathrm{N}\ \mathit{mpaw}\ \mathrm{P}$	(from 3, 6)
8. $\mathrm{N}\ \mathit{mpaw}\ \mathrm{R}$	(from 4, 6)
9. $\exists \mathrm{Z}(\mathrm{Z}\ \mathit{mpaw}\ \mathrm{R} \wedge \mathrm{Z}\ \mathit{mpaw}\ \mathrm{P})$	(from 7, 8)
10. $\mathrm{Pi_XR}$	(from 9; by abstract *dictum de aliquo*)

As before, this reconstruction relies on the abstract *dictum* semantics (lines 3, 4, and 10), and on several rules of classical propositional and quantifier logic (lines 6–9). In addition, it relies on the abstract requirement of existential import (line 5). The step of ecthesis, setting out the new term N, occurs in line 6.

THE ECTHETIC PROOF OF DARAPTI IN THE HETERODOX *DICTUM* SEMANTICS. Given the above reconstruction, every specification of the abstract *dictum* semantics that satisfies the abstract requirement of existential import can account for Aristotle's ecthetic proof of Darapti. The heterodox *dictum* semantics satisfies this requirement by means of the reflexivity of $\mathrm{a_X}$-predication, which is guaranteed by the *dictum de omni* (see p. 68). It is obtained from the abstract *dictum* semantics by identifying the relation indicated by '*mpaw*' with $\mathrm{a_X}$-predication. This leads to the following reconstruction of the ecthetic proof of Darapti:

1. $\mathrm{Pa_XS}$	(major premise)
2. $\mathrm{Ra_XS}$	(minor premise)

3. $\forall Z(Sa_X Z \supset Pa_X Z)$	(from 1; by heterodox *dictum de omni*)
4. $\forall Z(Sa_X Z \supset Ra_X Z)$	(from 2; by heterodox *dictum de omni*)
5. $\exists Z(Sa_X Z)$	(by reflexivity of a_X-predication)
6. $Sa_X N$	(from 5; by existential instantiation)
7. $Pa_X N$	(from 3, 6)
8. $Ra_X N$	(from 4, 6)
9. $\exists Z(Ra_X Z \wedge Pa_X Z)$	(from 7, 8)
10. $Pi_X R$	(from 9; by heterodox *dictum de aliquo*)

Note that in this reconstruction, line 5 is justified by the reflexivity of a_X-predication instead of the abstract requirement of existential import.

A number of commentators have expressed dissatisfaction with this reconstruction of the proof. In particular, they have raised two objections concerning the inference from $Pa_X N$ and $Ra_X N$ in lines 7 and 8 to the conclusion in line 10. The first objection is that $Pa_X N$ and $Ra_X N$ differ from the two original premises in lines 1 and 2 only in that the middle term S is replaced by the new term N. Hence, the objection goes, the conclusion could also be directly inferred from the two original premises, so that the step of ecthesis seems to be superfluous. But why then should Aristotle have made use of ecthesis in the proof, setting out the new term N? The second objection is that the inference from lines 7 and 8 to line 10 is, in effect, equivalent to an application of Darapti. Thus, the ecthetic proof of Darapti seems to be circular, on the grounds that Darapti is proved valid by means of an application of Darapti. Alexander takes these difficulties to show that the above reconstruction is incorrect, and that the item set out is an individual.[12] Likewise, later commentators take the difficulties to show that the term set out is a singular term, which stands for an individual.[13] This view is incompatible with the above reconstruction within the heterodox *dictum* semantics, on which the term set out is a categorical term which may or may not stand for an individual.

Let us begin with the second objection, the charge of circularity. It is true that the reconstruction involves an inference from $Pa_X N$ and

12. Alexander *in APr.* 99.27–100.9.

13. Wieland (1966b: 24–5), Smith (1982a: 118–20), Mignucci (1991: 16 and 22), Striker (1996: 208; 2009: 87 and 104); similarly, Drechsler (2005: 314–15).

Ra_XN to Pi_XR. Crucially, however, this inference does not rely on the mood Darapti. Rather, it is a complex inference relying on the heterodox *dictum de aliquo* (line 10) and some elementary rules of classical propositional and quantifier logic (line 9). This complex inference has the same premises and the same conclusion as Darapti, but it must not be identified with Darapti. The latter is a mood whose validity can be established in Aristotle's deductive system of categorical propositions by means of conversion rules and first-figure moods. The former is based on the heterodox *dictum* semantics and on rules of propositional and quantifier logic. The two inferences differ essentially in their status and the role they play in Aristotle's syllogistic. The reconstruction of the ecthetic proof of Darapti within the heterodox *dictum* semantics is no more circular than the earlier reconstruction within the abstract *dictum* semantics. For neither reconstruction contains an application of Darapti.

Now for the first objection, the charge of superfluity. It is true that in the heterodox reconstruction the step of setting out the new term N is, strictly speaking, superfluous. Unlike in the abstract reconstruction, lines 9 and 10 could, in principle, follow immediately after the original premises in lines 1 and 2. Lines 3–8 could simply be deleted. However, there may be several reasons why Aristotle did not wish to present the proof in this way. First, the distinction between the deductive system of categorical propositions and the *dictum* semantics would be less clear in such a proof. Moreover, such a proof could not serve, as Aristotle's proof does, as a paradigm for the ecthetic proofs of Disamis and Datisi later in chapter 1.6. In these proofs, a new term needs to be set out. In the case of Disamis and Datisi, the introduction of the new term is justified by an application of the *dictum de aliquo* to the i_X-premise, without appealing to the abstract requirement of existential import or to the reflexivity of a_X-predication. Apart from that, the ecthetic proofs of Disamis and Datisi are entirely parallel to the above proof of Darapti. The paradigmatic character of the ecthetic proof of Darapti would be lost if Aristotle did not set out a new term in it. Thus, there are good reasons for Aristotle to make use of ecthesis in the proof, even if this is unnecessary from a purely logical perspective.

As we have seen, Aristotle's formulations of ecthetic proofs typically contain phrases like 'the As', referring to pluralities associated with categorical terms. These phrases point to the abstract *dictum* semantics, suggesting that Aristotle thought of himself as performing his ecthetic proofs within this semantics rather than within the heterodox *dictum* semantics. Nevertheless, his formulation of the ecthetic proof of Darapti is entirely compatible with the heterodox interpretation and can be adequately reconstructed within it.

THE ECTHETIC PROOF OF BOCARDO. The last ecthetic proof mentioned by Aristotle in *Prior Analytics* 1.6 is that of the mood Bocardo:

Major premise:	$\mathrm{Po_XS}$	P does not belong to some S
Minor premise:	$\mathrm{Ra_XS}$	R belongs to all S
Conclusion:	$\mathrm{Po_XR}$	P does not belong to some R

Aristotle indicates the ecthetic proof of Bocardo only very briefly, as follows:

> This can also be proved without *reductio,* if one of the Ss is taken to which P does not belong. (*APr.* 1.6 28b20–1)

Among Aristotle's ecthetic proofs in the *Prior Analytics,* this is the only one that involves $\mathrm{o_X}$-propositions. Based on the major premise $\mathrm{Po_XS}$, Aristotle sets out a member of the plurality associated with the term S and states that P 'does not belong' to it. This is the second of the two occurrences of the verb 'belong' without a quantifying expression in the assertoric syllogistic. What does it mean that P does not belong to the term set out? According to the strong principle of $\mathrm{o_X}$-ecthesis (p. 85), it should mean that P is $\mathrm{e_X}$-predicated of the term set out:

> If $\mathrm{Po_XS}$, then there is a Z such that
> (i) $\mathrm{Sa_XZ}$, and
> (ii) $\mathrm{Pe_XZ}$

However, there is a passage from *de Interpretatione* 7 that casts doubt on this view. The passage concerns the meaning of negative sentences that lack a quantifying expression, such as 'Man is not white'. Aristotle's point there is that such sentences might seem to be equivalent to $\mathrm{e_X}$-propositions such as 'No man is white', but that in fact they are not (17b34–7).[14] So it is questionable whether 'does not belong' in the ecthetic proof of Bocardo indicates an e_X-predication.

Fortunately, there is no need to take the ecthetic proof of Bocardo to involve e_X-predications. The proof can be reconstructed within the abstract *dictum* semantics in a similar fashion to the proof of Darapti. There, we took the unquantified phrase 'P belongs to N' to mean that N is a member of the plurality associated with P. Accordingly, 'P does not belong to N' should be taken to mean that N is not a member of the plurality associated with P. In the following reconstruction, this corresponds to line 5:

1. $\mathrm{Po_XS}$	(major premise)
2. $\mathrm{Ra_XS}$	(minor premise)
3. $\exists Z(Z\ \mathit{mpaw}\ S \wedge \neg\, Z\ \mathit{mpaw}\ P)$	(from 1; by abstract *dictum de aliquo non*)
4. $\forall Z(Z\ \mathit{mpaw}\ S \supset Z\ \mathit{mpaw}\ R)$	(from 2; by abstract *dictum de omni*)
5. $N\ \mathit{mpaw}\ S \wedge \neg\, N\ \mathit{mpaw}\ P$	(from 3; by existential instantiation)
6. $N\ \mathit{mpaw}\ R$	(from 4, 5)
7. $\exists Z(Z\ \mathit{mpaw}\ R \wedge \neg\, Z\ \mathit{mpaw}\ P)$	(from 5, 6)
8. $\mathrm{Po_XR}$	(from 7; by abstract *dictum de aliquo non*)

As before, the reconstruction relies on the abstract *dictum* semantics (lines 3, 4, and 8) and on rules of propositional and quantifier logic (lines 5–7). Unlike the proof of Darapti, it does not rely on the abstract requirement of existential import.[15]

---

14. Cf. Whitaker (1996: 85), Crivelli (2004: 244n18), and Jones (2010: 38–45).

15. As mentioned above (p. 42), some authors suggest an alternative interpretation of o_X-propositions: $\mathrm{Ao_XB}$ if and only if (i) $\exists Z\ (Z\ \mathit{mpaw}\ B \wedge \neg\ Z\ \mathit{mpaw}\ A)$ or (ii) not $\exists Z\ (Z\ \mathit{mpaw}\ B)$. (More precisely, they adopt the orthodox versions of (i) and (ii).) On this interpretation, the inference from line 1 to lines 3 and 5 in the above reconstruction is invalid.

The reconstruction does not rely on specific assumptions about the meaning of the formula '$\neg$ N *mpaw* P' in line 5. In particular, the formula is not taken to mean that P is e_X-predicated of N. Thus, Aristotle's ecthetic proof of Bocardo provides no evidence for the strong principle of o_X-ecthesis. Since this is his only ecthetic proof involving o_X-propositions, there is no evidence that Aristotle endorsed that principle.[16] Instead, given the heterodox *dictum* semantics, the ecthetic proof of Bocardo in effect relies on the following weak principle of o_X-ecthesis:

If Po_XS, then there is a Z such that
(i) Sa_XZ, and
(ii) Po_XZ

When the above reconstruction of the ecthetic proof of Bocardo is transferred from the abstract into the heterodox *dictum* semantics, it relies on the weak principle of o_X-ecthesis. The resulting reconstruction is open to objections similar to those that we saw in connection with the heterodox interpretation of the ecthetic proof of Darapti, but these objections can be answered in the same way as in the case of Darapti.

COMPLEMENTS OF TERMS. The strong principle of o_X-ecthesis is not valid in the heterodox *dictum* semantics. I have argued that, contrary to what is often thought, Aristotle does not endorse this principle, but only the weak principle of o_X-ecthesis. In the rest of this chapter, I offer an explanation of why Aristotle did not endorse the former principle and why he would have been reluctant to accept it.

Suppose that 'man' is o_X-predicated of 'white'. In this case, it is easy to set out a term like 'swan', of which 'white' is a_X-predicated and of which 'man' is e_X-predicated.[17] However, this does not mean that such a term can be set out in all o_X-predications. For instance, suppose that 'moving' is o_X-predicated of 'man'. Which term should Aristotle set out in this case to verify the strong principle of o_X-ecthesis? He needs to set out a term of which 'man' is a_X-predicated and of which 'moving' is

16. Patzig (1968: 162) argues that Aristotle asserts the strong principle of o_X-ecthesis at *APr.* 1.28 44a9–11, but Patzig's argument is not convincing (see Malink 2009a: 124–5).

17. See *APr.* 1.4 26b7–8 and 26b12–13.

e_X-predicated. Perhaps 'not-moving man' is a suitable candidate. The question is, however, whether such a term exists in the language under consideration. Correspondingly, there is a question whether a suitable semantic value for such a term exists in the domain of possible semantic values of categorical terms.

'Not-moving man' is the conjunction of 'not-moving' and 'man', with 'not-moving' being the complement (or negation) of 'man'. In the set-theoretic semantics and the nonempty set semantics, such conjunctions and complements exist: if 'moving' is o_X-predicated of 'man', then there is a nonempty set that is the semantic value of the term 'not-moving man'. This is true regardless of whether the language contains the term 'not-moving man'. So given the objectual interpretation of quantification, the strong principle of o_X-ecthesis is valid in the set-theoretic semantics and the nonempty set semantics.[18] Many authors who endorse the strong principle of o_X-ecthesis seem to presuppose one of these two semantics.[19]

In the set-theoretic semantics, the domain of possible semantic values of categorical terms is a Boolean algebra (namely, the powerset of the primitive nonempty set of individuals). Every item in this algebra has a complement, and any two items have a conjunction (namely, their set-theoretic intersection). In the nonempty set semantics, most items have complements, and many pairs of items have a conjunction. More precisely, every item except the top item (that is, the universal set) has a complement, and any two items that have a nonempty intersection have a conjunction. In the preorder semantics, on the other hand, there is no guarantee that such complements or conjunctions exist. As a result, the strong principle of o_X-ecthesis is not valid in the preorder semantics. If 'moving' is o_X-predicated of 'man', there need not be a semantic value corresponding to the term 'not-moving man'.

Earlier, I suggested that, in the preorder semantics, every term can be taken to be its own semantic value (pp. 73–74). Given this, the question whether the domain of possible semantic values contains a semantic value for the term 'not-moving man' comes down to the question whether the

---

18. Patzig (1968: 161) and Smith (1983: 228).

19. See, for example, Łukasiewicz (1957: 65), Smith (1989: xxiv), Wolff (1998: 142), Nortmann (2005: 164), and Barnes (2007: 404–5).

language under consideration contains this term. Even if the language under consideration contains the two terms ‘man’ and ‘moving’, there is no guarantee that it also contains the term ‘not-moving man’.

In the *de Interpretatione,* Aristotle states that terms such as ‘not-man’ are not names in the proper sense, but merely indefinite names.[20] Aristotle does not use such terms in *Prior Analytics* 1.1–22.[21] Of course, this does not mean that they cannot be used in the syllogistic. But in any case, it should not be presupposed that every term possesses a complement in Aristotle’s language of categorical propositions. Complements are usually taken to satisfy what are known as the traditional laws of obversion, for example, that A is a_X-predicated of B just in case the complement of A is e_X-predicated of B. In the preorder semantics, the existence of such complements for every term implies the mereological principle of (strong) supplementation.[22] However, as we saw above (p. 83), Aristotle rejects this principle in *Prior Analytics* 2.22. Thus he is committed to denying the universal existence of complements satisfying the traditional laws of obversion: some terms may have such complements, but not all.[23]

CONJUNCTIONS OF TERMS. Aristotle maintains a skeptical attitude not only toward complements but also toward conjunctions of terms. There are only two examples of conjunctive terms in *Prior Analytics* 1.1–22, namely, ‘sleeping horse’ and ‘waking horse’.[24] However, these do not play an essential role in the syllogistic and can easily be replaced by simple nonconjunctive terms.[25] In any case, the fact that Aristotle occasionally

---

20. *Int.* 2 16a29–32, 10 19b8–10; similarly, 3 16b11–14.

21. For his reluctance to use complements of terms in the syllogistic, see Patzig (1968: 144), Flannery (1987: 468–9), and Bäck (2000: 252); similarly, Nortmann (1996: 131–2 and 194) and Schmidt (2000: 60 and 174).

22. Cf. Fact 139, p. 323 below.

23. Moreover, it can be shown that Aristotle’s apodeictic syllogistic is incompatible with the traditional laws of obversion for X- and N-propositions (Brenner 2000: 342).

24. *APr.* 1.22 40a37–8 and 40b10–12.

25. The two conjunctive terms occur in chapter 1.22 in Aristotle’s discussion of the premise pairs Aa_QB, Ce_NB and Ai_QB, Ce_NB. We may label these premise pairs ae-3-QN and ie-3-QN, where the number 3 indicates that they are in the third figure. Aristotle uses the two conjunctive terms in counter-

uses conjunctive terms does not imply that there is a conjunctive term for every pair of terms in Aristotle's language of categorical propositions.

In the *de Interpretatione,* Aristotle states that the three terms 'walking', 'white', and 'man' do not constitute a unity.[26] Thus, a term such as 'walking white man' does not stand for a single unity. Likewise, terms such as 'walking man' do not stand for a single unity either.[27] At the same time, Aristotle states in the *de Interpretatione* that the subject and predicate of simple declarative sentences must stand for a single unity. Even if 'horse' and 'man' were grouped together under a single name 'cloak', he argues, the sentence 'Cloak is white' would not be a simple declarative sentence; the reason for this is that 'cloak' does not stand for a single unity because 'man' and 'horse' do not constitute a unity.[28] Similarly, the sentence 'Walking man is white' would not be a simple declarative sentence, because 'walking man' does not stand for a unity.

Now, the categorical propositions employed in the *Prior Analytics* are simple declarative sentences.[29] Given the above considerations, one might therefore expect that conjunctive terms such as 'walking man' cannot serve as argument-terms of categorical propositions. Obviously, this does not prevent Aristotle from using conjunctive terms such as 'sleeping horse' in the syllogistic. Still, the above considerations suggest

examples designed to establish that these two premise pairs are inconcludent (i.e., that they do not yield any conclusion in their figure). Now, Aristotle has already established the inconcludence of the premise pair ae-1-QN (i.e., Aa_QB, Be_NC) by counterexamples that do not involve conjunctive terms (1.16 36a30). These examples are also used to establish the inconcludence of ae-1-QX and ao-1-QX (1.15 35a24 and 35b10). The same examples can be used to establish the inconcludence of ae-3-QN and ie-3-QN. Thus, Aristotle's counterexamples involving conjunctive terms in chapter 1.22 can easily be replaced by counterexamples not involving any conjunctive terms.

26. *Int.* 11 20b15–22 and 21a7–14; cf. Whitaker (1996: 150–1), Weidemann (2002: 370), and Crivelli (2004: 156 and 177).

27. Ackrill (1963: 147).

28. *Int.* 8 18a18–21; cf. Ackrill (1963: 130–1), Whitaker (1996: 95–8), and Crivelli (2004: 155–80).

29. See *APr.* 1.1 24a16–17 and *Int.* 5–6 17a20–6; cf. Alexander *in APr.* 11.6–9.

that Aristotle would not want to accept proofs whose validity depends on the universal existence of conjunctions of terms.

If I am correct, Aristotle is not willing to assume in the syllogistic that every term has a complement or that there is a conjunctive term for every pair of terms. As a result, he lacks the resources to set out suitable terms to verify the strong principle of o_X-ecthesis for all o_X-predications. If ‘moving’ is o_X-predicated of ‘man’, he cannot take it for granted that a term such as ‘not-moving man’ is available to be set out. Thus, his reluctance to accept the strong principle of o_X-ecthesis can be explained by his skeptical attitude toward complements and conjunctions of terms.

SUMMARY. The purpose of this chapter was to show that all of Aristotle’s proofs by ecthesis in the assertoric syllogistic can be reconstructed within the abstract *dictum* semantics (given the abstract requirement of existential import). Moreover, I have defended the heterodox interpretation of these proofs by ecthesis. In particular, I have argued that the term set out in proofs by ecthesis is not a noncategorical singular term and that Aristotle does not employ the strong principle of o_X-ecthesis. Finally, I have offered an explanation of why Aristotle would not be willing to accept this principle.

This completes our discussion of ecthesis, and indeed of the whole assertoric syllogistic. It is now time to move on to the modal syllogistic.

II

The Apodeictic Syllogistic

AN OVERVIEW OF THE APODEICTIC SYLLOGISTIC. Aristotle's modal syllogistic begins with what is known as the apodeictic syllogistic (*Prior Analytics* 1.3 and 1.8–12). The apodeictic syllogistic is concerned with necessity propositions, that is, with propositions whose copula contains a modal qualifier such as 'necessarily'. Aristotle focuses on four kinds of necessity propositions:

Aa_NB	A necessarily belongs to all B
Ae_NB	A necessarily belongs to no B
Ai_NB	A necessarily belongs to some B
Ao_NB	A necessarily does not belong to some B

The apodeictic syllogistic consists of three parts. In the first part, Aristotle states the conversion rules for necessity propositions. These rules exactly mirror those for assertoric propositions: e_N- and i_N-propositions are convertible, a_N-propositions can be converted to i_N-propositions, and o_N-propositions are not convertible (chapter 1.3).

The second part deals with syllogistic moods of the form NNN, in which both premises and the conclusion are necessity propositions (chapter 1.8). Again, Aristotle's treatment of these moods exactly mirrors the assertoric syllogistic: he takes an NNN-mood to be valid just in case the corresponding purely assertoric mood is valid. Thus, Barbara, Celarent, Darii, and Ferio NNN are valid and perfect moods. The valid NNN-moods of the second and third figures are mostly proved by direct deductions relying on conversion rules and the perfect NNN-moods. There are only two exceptions whose validity cannot be proved in this way, namely, Baroco NNN and Bocardo NNN. Aristotle establishes the validity of these two moods by means of the method of ecthesis.

In the third part of the apodeictic syllogistic, Aristotle discusses moods with a mixed premise pair consisting of a necessity proposition and an assertoric proposition (chapters 1.9–12). He focuses on moods of the form NXN and XNN, declaring some of them valid and others invalid. His main claim is that Barbara, Celarent, Darii, and Ferio are valid and perfect as NXN-moods but invalid as XNN-moods (chapter 1.9). For example, Barbara NXN is valid according to Aristotle:

Major premise:	Aa_NB	A necessarily belongs to all B
Minor premise:	Ba_XC	B belongs to all C
Conclusion:	Aa_NC	A necessarily belongs to all C

whereas Barbara XNN is invalid:

Major premise:	Aa_XB	A belongs to all B
Minor premise:	Ba_NC	B necessarily belongs to all C
Conclusion:	Aa_NC	A necessarily belongs to all C

In the second and third figures, there are valid moods of the form XNN as well as NXN. These are all proved by direct deductions relying on conversion rules and the four perfect NXN-moods (chapters 1.10–11).

The validity of the four perfect NXN-moods is the cornerstone of the apodeictic syllogistic. Given the principle of N-X-subordination, according to which every necessity proposition implies the corresponding assertoric proposition, the validity of these four moods entails the validity of the four perfect NNN-moods. In fact, given N-X-subordination and Aristotle's conversion rules, the four perfect NXN-moods suffice to establish the validity of all moods held to be valid by Aristotle in the apodeictic syllogistic, with the exception only of Baroco NNN and Bocardo NNN.

DOUBTS ABOUT ARISTOTLE'S PERFECT NXN-MOODS. Despite their central role in the apodeictic syllogistic, the validity of the four perfect NXN-moods has been disputed since antiquity. Theophrastus and Eudemus, two of Aristotle's pupils, denied the validity of these moods. They gave several counterexamples to Barbara NXN and argued that no necessity proposition can follow from a premise pair that contains an assertoric proposition.[1] Their rejection of Barbara NXN was widely influential and was embraced by Platonists such as Themistius, Syrianus, and Proclus.[2] On the other hand, some Peripatetics undertook to defend Aristotle against these objections.[3] The arguments put forward on both sides of the dispute were discussed by Alexander of Aphrodisias in a lost treatise entitled "On the Difference between Aristotle and His Associates Concerning Mixtures of Premises."[4]

The validity of Aristotle's NXN-moods is also disputed in the more recent literature. Since Albrecht Becker (1933), it is often thought that these moods are incompatible with Aristotle's conversion rules for necessity propositions. More specifically, it is thought that the validity of the perfect NXN-moods requires a *de re* reading of necessity propositions, whereas the validity of Aristotle's conversion rules requires a *de dicto* reading (see pp. 9–10 above). On this view, Aristotle's apodeictic syllogistic would employ two different readings of necessity propositions and would therefore be incoherent.

PLAN OF THE PART. In view of the controversy over Aristotle's perfect NXN-moods and given their fundamental role in the modal syllogistic, my discussion of the apodeictic syllogistic focuses on these four moods. On what grounds did Aristotle take them to be valid, and how is their validity justified? Is their validity consistent with Aristotle's conversion rules for necessity propositions? These are the questions with which we will be mainly concerned in this part of the book.

I begin with the case of Barbara NXN (Chapters 7–10). Aristotle seems to justify the validity of this mood by means of a *dictum de omni* for a_N-propositions, but, as I argue, this *dictum* does not satisfactorily

---

1. See Alexander *in APr.* 124.8–30, Philoponus *in APr.* 124.9–125.18.

2. See Ammonius *in APr.* 38.38–39.2, Philoponus *in APr.* 123.15–17.

3. These Peripatetics include Alexander and two of his teachers, Sosigenes and Herminus (for Alexander, see Ammonius *in APr.* 39.1; for Sosigenes, see Ammonius *in APr.* 39.24–6 and Philoponus *in APr.* 126.20–3; for Herminus, see Ammonius *in APr.* 39.31–4 and also Alexander *in APr.* 125.3–29).

4. See Alexander *in APr.* 125.30–1 and Moraux & Wiesner (2001: 94–105). Some of the arguments in defense of Aristotle's NXN-moods are summarized by Alexander in his commentary, *in APr.* 126.9–127.15.

explain why he took Barbara NXN to be valid (Chapter 7). His endorsement of Barbara NXN can instead be explained, I suggest, by means of the account of predication that Aristotle develops in his *Topics.* In particular, it can be explained by the *Topics*' theory of predicables and categories. Both a_N- and a_X-predication can be taken to be governed by this theory in such a way that Barbara NXN is valid (Chapters 8 and 9). This also allows us to reject Theophrastus's and Eudemus's arguments against Barbara NXN (Chapter 10).

Next, we will consider the other three perfect NXN-moods and Aristotle's conversion rules for necessity propositions (Chapters 11 and 12). I show how the relations of e_N- and i_N-predication can be defined solely in terms of a_N- and a_X-predication. They are defined in such a way that the perfect NXN-moods and the conversion rules are all simultaneously valid, without attributing to Aristotle an ambiguity between two readings of necessity propositions. Thus it will be shown that, contrary to what is often thought, the apodeictic syllogistic is not inconsistent. Finally, I discuss Aristotle's proofs by ecthesis of Baroco NNN and Bocardo NNN, and the problems associated with his claim that Baroco XNN and Bocardo NXN are invalid (Chapter 12).

7

The Apodeictic *dictum de omni*

THE ABSTRACT APODEICTIC *DICTUM DE OMNI.* Aristotle justifies the validity of Barbara NXN in *Prior Analytics* 1.9, in a passage I discussed in connection with the *dictum de omni* (p. 52). Alexander and others take his justification of Barbara NXN to be based on a version of the *dictum de omni* that characterizes the semantics of a_N-propositions.[1] As we saw above, Aristotle justified the validity of assertoric Barbara by means of the assertoric *dictum de omni,* which characterizes the semantics of a_X-propositions.[2] It is therefore plausible that he would justify the validity of Barbara NXN by means of a corresponding apodeictic *dictum de omni* for a_N-propositions.

Aristotle mentions an apodeictic *dictum de omni* in *Prior Analytics* 1.8. He does not spell it out, but merely states that it is similar to the assertoric *dictum de omni.*[3] Aristotle's assertoric *dictum de omni* reads: "We say 'predicated of all' when none of those of the subject can be taken of which the other will not be said" (24b28–30). Given this, a natural way to formulate the corresponding apodeictic *dictum de omni* is: "We say 'predicated necessarily of all' when none of those of the subject can be taken of which the other will not be said by necessity."[4]

---

1. Alexander 125.33–126.8, Patterson (1993: 371–2; 1995: 220), and Striker (2009: 116).

2. *APr.* 1.4 25b39–40; cf. p. 37n8 above.

3. Aristotle writes that for a_N-propositions "we will explain 'being in something as a whole' and 'predicated of all' in the same way" as for a_X-propositions (*APr.* 1.8 30a2–3).

4. Patterson (1993: 371; 1995: 220).

As argued above, the assertoric *dictum de omni* states that an a_X-proposition is true just in case every member of the plurality associated with the subject is a member of the plurality associated with the predicate (pp. 34–37). Using '*mpaw*' as shorthand for being a member of the plurality associated with a given term, we have:

Aa_XB if and only if $\forall Z$ (Z *mpaw* B ⊃ Z *mpaw* A)

This is the abstract assertoric *dictum de omni*. It is abstract in that it does not specify what the relation indicated by '*mpaw*' is. The corresponding abstract apodeictic *dictum de omni* should then be taken to state that an a_N-proposition is true just in case every member of the plurality associated with the subject is something of which the predicate is said by necessity. In other words,

Aa_NB if and only if $\forall Z$ (Z *mpaw* B ⊃ A is said of Z by necessity)

A WAY TO JUSTIFY THE VALIDITY OF BARBARA NXN. Although the abstract apodeictic *dictum de omni* does not specify what the relation of being 'said of by necessity' is, it helps justify the validity of Barbara NXN, as follows:

1. Aa_NB	(major premise)
2. Ba_XC	(minor premise)
3. $\forall Z$(Z *mpaw* B ⊃ A is said of Z by necessity)	(from 1; by abstract apodeictic *dictum de omni*)
4. $\forall Z$(Z *mpaw* C ⊃ Z *mpaw* B)	(from 2; by abstract assertoric *dictum de omni*)
5. $\forall Z$(Z *mpaw* C ⊃ A is said of Z by necessity)	(from 3, 4)
6. Aa_NC	(from 5; by abstract apodeictic *dictum de omni*)

This argument is parallel to the justification of the perfect assertoric moods we saw above (p. 38). As before, lines 1, 2, and 6 belong to the language of categorical propositions, whereas lines 3–5 belong to the language of the abstract *dictum* semantics. The argument contains two applications of the abstract apodeictic *dictum de omni,* and one of the abstract assertoric *dictum*. The step in line 5 relies on standard rules of propositional and quantifier logic. Thus, the two abstract *dicta* entail the validity of Barbara NXN.

ARISTOTLE'S OWN JUSTIFICATION OF BARBARA NXN. The above argument shows how the abstract apodeictic and assertoric *dictum de omni* can be used to justify the validity of Barbara NXN. However, this argument seems to differ from the argument Aristotle himself gives for the validity of this mood in *Prior Analytics* 1.9. Since his argument provides a justification of the validity of both Barbara and Celarent NXN, the major premise is either Aa_NB or Ae_NB, and the minor premise is Ba_XC. On the basis of these premises, Aristotle argues as follows:

> Since A belongs or does not belong by necessity to all B and C is one of the Bs, it is evident that one or the other of these will also apply to C by necessity. (*APr.* 1.9 30a21–3)

Aristotle's argument relies on the statement that 'C is one of the Bs', which he seems to have inferred from the minor premise Ba_XC. As argued above (pp. 52–53), the phrase 'C is one of the Bs' indicates that C is a member of the plurality associated with B. This is not in accordance with the above justification of Barbara NXN by means of the two abstract *dicta de omni,* for this justification does not assume that C is a member of the plurality associated with B. Instead, it states, in line 4, that every member of the plurality associated with C is a member of the plurality associated with B. By contrast, Aristotle's own justification does not mention any members of the plurality associated with C.

Aristotle's statement that C is one of the Bs cannot be explained by the abstract assertoric *dictum de omni.* But it can be explained by the heterodox assertoric *dictum de omni,* in which the relation of being a member of the plurality associated with a term is identified with a_X-predication. On the heterodox interpretation, the minor premise of Barbara NXN implies that C is a member of the plurality associated with B. Aristotle's justification of Barbara NXN seems to be based on this heterodox interpretation (see pp. 51–54).

The major premise of Barbara NXN, along with the abstract apodeictic *dictum de omni,* implies that every member of the plurality associated with B is something of which A is said by necessity. Since C is a member of that plurality, it follows that A is said of C by necessity. How does this help establish the conclusion of Barbara NXN, that A is a_N-predicated of C? Aristotle seems to assume here that the relation of being 'said of by necessity' used in the apodeictic *dictum de omni* implies a_N-predication. Given this assumption, Aristotle's brief justification of Barbara NXN can be reconstructed as follows:

1. Aa_NB	(major premise)
2. Ba_XC	(minor premise)
3. $\forall Z(Z\ \textit{mpaw}\ B \supset$ A is said of Z by necessity)	(from 1; by abstract apodeictic *dictum de omni*)
4. C *mpaw* B	(from 2)
5. A is said of C by necessity	(from 3, 4)
6. Aa_NC	(from 5)

Line 4 in this argument corresponds to Aristotle's statement that 'C is one of the Bs'. Unlike the earlier argument, the present argument contains no application of the assertoric *dictum de omni* and only one application of the abstract apodeictic *dictum de omni* (in line 3). The step in line 4 is justified by the heterodox *dictum* semantics, in which the relation indicated by '*mpaw*' is identified with a_X-predication. The step in line 6 relies on the assumption that being 'said of by necessity' implies a_N-predication.

Conversely, it seems obvious that the relation of a_N-predication should imply the relation of being 'said of by necessity'. Thus, Aristotle's justification in effect relies on the assumption that these two relations coincide. So given that the relation indicated by '*mpaw*' is identified with a_X-predication, the abstract apodeictic *dictum de omni* amounts to the claim that an a_N-proposition is true just in case the predicate is a_N-predicated of everything of which the subject is a_X-predicated:

$$Aa_NB \quad \text{if and only if} \quad \forall Z(Ba_XZ \supset Aa_NZ)$$

We may call this the heterodox apodeictic *dictum de omni.* If I am correct, Aristotle's justification of Barbara NXN implicitly relies on the heterodox apodeictic *dictum de omni.* This is not to say that Aristotle directly applies this heterodox *dictum;* according to the above reconstruction, he applies the abstract, not the heterodox, apodeictic *dictum de omni* in deriving line 3. Rather, the assumptions that underlie the derivation of lines 3, 4, and 6 commit Aristotle to the heterodox apodeictic *dictum de omni*.[5] In this connection it is also worth noting that the heterodox apodeictic *dictum de omni* was used by some commentators in antiquity to justify the validity of Barbara NXN.[6]

REMARK ON THE ORTHODOX APODEICTIC *DICTUM DE OMNI*. Before having a closer look at the heterodox apodeictic *dictum de omni*, let us briefly consider its orthodox variant. On the orthodox interpretation, the quantification '∀Z' in the abstract *dictum* is taken to mean 'for every individual z'. Thus, the orthodox apodeictic *dictum de omni* reads as follows:

Aa_NB	if and only if	for every individual z, if z falls under B, then z falls under A by necessity

This is known as the *de re* reading of a_N-propositions. It is called "*de re*" because it quantifies over individuals, that is, over *res*, and because the predicate term A is viewed as expressing a necessary property of these individuals. Many commentators attribute to Aristotle the *de re* reading of a_N-propositions and assume that he takes the validity of Barbara NXN to be justified by this reading.[7] It is true that Barbara NXN is valid on the *de re* reading of a_N-propositions, but this does not mean that Aristotle had this reading in mind when he asserted the validity of Barbara NXN. The *de re* reading and the orthodox apodeictic *dictum de omni* cannot properly explain Aristotle's statement that 'C is one of the Bs' in his justification of Barbara NXN; for this statement indicates that

5. If Aristotle directly applied the heterodox apodeictic *dictum de omni*, his justification could be shortened by deleting lines 4 and 5 and by replacing the statement in line 3 with $\forall Z(Ba_XZ \supset Aa_NZ)$; the conclusion in line 6 could then be inferred immediately from lines 2 and 3.

6. As reported by Alexander *in APr.* 126.23–8; see Mueller (1999a: 120n44).

7. For example, Becker (1933: 38–42), Hintikka (1973: 139–40), Sorabji (1980: 201–2), Johnson (1989: 274; 2004: 272), Kosman (1990: 351–2), Thom (1993: 206), Thomason (1993: 116), Mignucci (1998: 50–2 and 61–2), Brenner (2000: 336), and Rini (2011: 72–4).

C is a member of the plurality associated with B, whereas on the orthodox interpretation this plurality consists exclusively of the individuals that fall under B (see pp. 51–54). Accordingly, the above reconstruction of Aristotle's justification is not compatible with the orthodox apodeictic *dictum de omni.* Moreover, as we saw in Chapter 3 above, there are a number of objections to the orthodox *dictum* semantics in general. All of this suggests that Aristotle's justification of Barbara NXN is not based on the orthodox apodeictic *dictum de omni* and hence not on the *de re* reading of a_N-propositions.

THE HETERODOX APODEICTIC *DICTUM DE OMNI.* If I am correct, Aristotle's justification of Barbara NXN relies, in effect, on the heterodox apodeictic *dictum de omni,* according to which A is a_N-predicated of B if and only if A is a_N-predicated of everything of which B is a_X-predicated. In this 'if and only if'-claim, the implication from right to left can be taken for granted, because it follows from the reflexivity of a_X-predication (which is guaranteed by the assertoric *dictum de omni*). The converse implication is more substantive, stating that if A is a_N-predicated of B, then A is a_N-predicated of everything of which B is a_X-predicated. Asserting this implication for any A and B is just another way of asserting the validity of Barbara NXN. As a result, given that a_X-predication is reflexive, the heterodox apodeictic *dictum de omni* is simply equivalent to the statement that Barbara NXN is valid. Thus, if Aristotle's justification of Barbara NXN relies on the heterodox apodeictic *dictum de omni,* the justification appears to be circular and therefore cannot explain on what grounds he took Barbara NXN to be valid.

Had Aristotle employed the orthodox apodeictic *dictum de omni,* his justification would not be circular. In the orthodox apodeictic *dictum de omni,* the implication from left to right is not equivalent to the validity of Barbara NXN. Instead, this *dictum* provides an explicit definition of a_N-predication by means of independent notions, namely, by means of the notions of an individual's falling under a term and of its falling under a term by necessity. The heterodox apodeictic *dictum de omni,* on the other hand, does not provide such a definition. It determines logical properties of a_N-predication, but it does not define what the relation of a_N-predication is or when it holds between two terms.

This is similar to what we saw in connection with the heterodox assertoric *dictum de omni* and a_X-predication (pp. 65–68). In the heterodox assertoric *dictum de omni,* the implication from left to right is equivalent to the validity of assertoric Barbara. Consequently, the heterodox assertoric *dictum de omni* entails the validity of assertoric Barbara, but it cannot explain on what grounds Aristotle took this mood to be valid. Now, the validity of assertoric Barbara is obvious and generally accepted. But the validity of Barbara NXN is not immediately obvious, and not everyone accepts it. So the question why Aristotle took Barbara NXN to be valid is more pressing than the question why he took assertoric Barbara to be valid. Although Aristotle does not give a satisfactory answer to the former question, a proper interpretation of the modal syllogistic should, I think, attempt to give one. This will be our main task in Chapters 8–10. As mentioned above, I will argue that the validity of Barbara NXN can be explained by the theory of predication Aristotle developed in the *Topics.*

8

Barbara NXN and the Four Predicables

INTRODUCING THE PREDICABLES. Aristotle's modalized propositions have a tripartite syntax, consisting of a predicate term, a subject term, and a copula. For example, the a_N-proposition 'A necessarily belongs to all B' consists of the predicate term A, the subject term B, and the a_N-copula 'necessarily belongs to all' (pp. 23–28). In order to specify the semantics of Aristotle's modalized propositions, it is vital to give a semantic interpretation of the modal copulae occurring in them. Each of these copulae stands for a relation between terms; for example, the a_N-copula stands for the relation of a_N-predication. Thus, giving a semantic interpretation of the modal syllogistic requires an account of the relations of a_N-predication, e_N-predication, and so on.

Since Aristotle does not offer an account of these relations in the *Prior Analytics,* any attempt to do so on his behalf is bound to be conjectural to some extent. However, Aristotle offers several classifications of relations between terms in his *Organon,* and some of them may be helpful in interpreting the modal syllogistic. A prominent example of such a classification is found in the *Topics*' theory of what are called the four predicables: genus (with differentia), definition, proprium, and accident. Richard Patterson has suggested that the predicables can be used to interpret Aristotle's modal syllogistic:

> What are we to do ... about determining which syllogisms or conversion principles are rightly taken as valid and primary? The answer is that we can consult the underlying relations among genus, species, accident and proprium for which the modal system is supposed to provide a logical calculus. (Patterson 1995: 48–9)

In what follows, I pursue and develop Patterson's idea, albeit in a different way than he does.

Aristotle does not mention the predicables in the syllogistic in *Prior Analytics* 1.1–22. But he does mention them in later chapters of the *Prior Analytics*. For example, he recommends that we should distinguish between definitions, propria, and accidents when searching for suitable premises to deduce a given conclusion (1.27 43b1–8). He also makes use of the predicables to describe states of affairs in which given assertoric propositions are true, as for example in the following passage:[1]

> It is possible that A should belong neither to any B nor to any C, and that B should not belong to any C, e.g. a genus to species of another genus—for animal belongs neither to music nor to medicine, nor does music belong to medicine. (*APr.* 2.2 54a36–b1)

Although this passage is concerned with assertoric propositions, Aristotle's strategy is also applicable to modalized propositions. The predicables, I suggest, can be used to describe states of affairs in which modalized propositions are true. But before turning to this task, let us have a look at Aristotle's treatment of the predicables in the *Topics*.

THE PREDICABLES IN *TOPICS* 1.8. Aristotle introduces the predicables by claiming that every premise indicates one of the four predicables (*Topics* 1.4 101b17–25). He does not explain exactly what this claim means, but he undertakes to prove it in *Topics* 1.8 (103b6–19), and this proof contains a helpful account of the predicables.

The proof proceeds by division. Aristotle considers all cases in which something is predicated of something (πᾶν τὸ περί τινος κατηγορούμενον, 103b7–8). He divides the class of these predications by two cuts. The first cut is based on the criterion of counterpredication (ἀντικατηγορεῖσθαι), by which Aristotle means that the predicate is true of everything of which the subject is true and vice versa. For example, 'capable of learning grammar' counterpredicates with 'man', whereas 'animal' does not. In any predication the predicate either counterpredicates with the subject

1. Further examples are *APr.* 2.2 54a31, 54b5–6, 54b12, 55a13–14, 55a22–3, 55a31–2, 2.3 55b18–19, 56a27–8.

counterpredication \ *essential predication*	+	–
+	definition	proprium
–	genus / differentia	accident

Figure 2

or it does not. If it does, it is a definition or a proprium of the subject; otherwise, it is a genus, differentia, or accident of the subject.

The second cut is based on the criterion of whether the predicate, as Aristotle puts it, signifies the essence of the subject or is part of the definition of the subject (103b10 and b13). In other words, it is based on the criterion of essential predication. For example, 'animal' is predicated essentially of 'man', whereas 'capable of learning grammar' is not. In any predication the predicate is either predicated essentially of the subject or it is not. If it is, it is a definition, genus, or differentia of the subject; otherwise it is a proprium or accident of the subject.

These two cuts yield an exhaustive division of the class of all predications into four subclasses, each of which corresponds to one of the predicables, as shown in Figure 2.[2] Aristotle does not explain what the predications are that he considers in his proof. But the proof makes it clear that something is predicated of something just in case it stands in the relation of one of the predicables to it. In other words, the following holds for any A and B:

S1: A is predicated of B if and only if A is a genus, differentia, definition, proprium, or accident of B

The label "S1" is shorthand for "Statement 1". In what follows, we will encounter more such statements about the predicables.

Aristotle seems to take S1 to justify his claim that every premise indicates one of the predicables. In addition to S1, Aristotle's proof

2. See Barnes (2003: 303–4) and Wagner & Rapp (2004: 282).

makes it clear that something is predicated essentially of something just in case it is a genus, differentia, or definition of it:

S2: A is predicated essentially of B if and only if A is a genus, differentia, or definition of B

Choosing 'man' as the subject term, typical examples of the predicables are

'biped terrestrial animal'	is a definition of	'man'
'capable of learning grammar'	is a proprium of	'man'
'animal'	is a genus of	'man'
'biped'	is a differentia of	'man'
'just'	is an accident of	'man'

In this list, the term 'just' is regarded as an accident of the term 'man', and the term 'animal' is regarded as a genus of 'man'. Since terms are linguistic items, one might want to say that the terms 'just' and 'animal' stand for—rather than are—an accident or genus of what the term 'man' stands for. For present purposes, however, we need not take into account this distinction, and can simply say that a term is an accident, definition, proprium, genus, or differentia of another term.

PREDICABLES AND CATEGORICAL PROPOSITIONS. As we have seen, Aristotle claims in the *Topics* that every premise indicates one of the four predicables. However, this is not true for the premises that he discusses in the *Prior Analytics.*[3] Consider, for example, the categorical e_X-proposition 'Man belongs to no horse'. Although this proposition can serve as a premise of a syllogistic mood, it does not indicate any of the predicables: neither is 'man' a genus, differentia, definition, proprium, or accident of 'horse', nor does the e_X-proposition indicate that it

3. The premises considered in the *Prior Analytics* differ from those considered in the *Topics.* In the latter work, premises are yes-no questions such as 'Is animal the genus of man?', whereas in the *Prior Analytics*, premises are declarative sentences capable of being true or false; see *Top.* 1.4 101b29–32 and *APr.* 1.1 24a22–b15.

is. Consequently, 'man' is also not predicated of 'horse', since the predicables give an exhaustive classification of the class of all predications (S1).

In the *Prior Analytics*, by contrast, the predicate term of any categorical proposition is regarded as something that is predicated (κατηγορούμενον) of the subject term; for example, 'man' is predicated of 'horse' in the e_X-proposition 'Man belongs to no horse'.[4] Accordingly, since this proposition is true, we say that 'man' is e_X-predicated of 'horse'; in fact, it is even e_N-predicated of it. This notion of predication is different from that employed in S1 and in the *Topics*' discussion of the predicables. Although 'man' is e_X- and e_N-predicated of 'horse', it is not predicated of 'horse' in the *Topics*' sense of predication. In what follows, unprefixed occurrences of the words 'predicated' and 'predication' will always indicate predications in the sense of the *Topics*. The prefixed variants, such as 'e_X-predicated' and 'i_N-predicated', indicate the various kinds of predications discussed in the syllogistic in the *Prior Analytics*.

Although the predicables are not applicable to all categorical propositions discussed in the *Prior Analytics*, they still seem to be applicable to some of them. In particular, one may think that the predicables are applicable to all true a_X- and a_N-propositions, and that in every such proposition the predicate term is a genus, differentia, definition, proprium, or accident of the subject term.[5] Thus, I argue that the predicables can help specify the semantics of a_X- and a_N-propositions, and thereby also explain on what grounds Aristotle took Barbara NXN to be valid. To this end, I consider Aristotle's theory of predicables in more detail, focusing on his treatment of genera in book 4 of the *Topics*.

EVERY GENUS IS A GENUS OF EVERYTHING OF WHICH IT IS PREDICATED. The fourth book of the *Topics* is devoted to genera. One of Aristotle's claims there is that 'being' is not a genus of anything (4.6 127a26–34). Aristotle proves this claim by *reductio*. He argues that if 'being' were a genus of anything, it would be a genus of everything, and then shows that certain

---

4. *APr.* 1.1 24b16–17, 1.5 26b36, 1.6 28a12–13; see Alexander *in APr.* 11.22–6.

5. Barnes (1970: 146–7). For further discussion of this point, see pp. 127–129 and 162 below.

absurd consequences would follow from this. His argument that if 'being' were a genus of anything it would be a genus of everything is based on the premise that 'being' is predicated of everything. The argument is as follows:

> If he has rendered being as a genus, clearly it will be a genus of everything, since it is predicated of everything; for the genus is never predicated of anything except of its species. (*Top.* 4.6 127a28–30)

If somebody renders 'being' as a genus, then, since 'being' is predicated of everything, she will be committed to the view that 'being' is a genus of everything. More generally, if somebody renders something as a genus, she is committed to it being a genus of everything of which it is predicated.

It is not immediately clear what it means to render something as a genus. To see what this means, it is useful to recall that the predicables are relational: something is a genus, definition, proprium, or accident *of something.* Thus, saying of something that it is a genus presumably means saying that that thing is a genus *of something:* A is a genus if and only if there is a B such that A is a genus of B. Likewise, rendering something as a genus presumably means stating that it is a genus of something. Now, Aristotle's argument implies that if somebody renders something as a genus, she is committed to it being a genus of everything of which it is predicated. Thus, Aristotle seems to hold that if something is a genus of something, it is a genus of everything of which it is predicated. In other words, he takes the following to be true for every A and every B:

S3: If A is a genus of B, then A is a genus of everything of which A is predicated

The term B occurs in the antecedent of S3, but not in the consequent. Thus, asserting that S3 is true for every A and every B is equivalent to asserting the following: for every A, if there is a B such that A is a genus of B, then A is a genus of everything of which A is predicated. This is to say, if A is a genus (of something), then A is a genus of everything of which it is predicated.

Aristotle justifies S3 by saying that "the genus is never predicated of anything except of its species." Given that the predicables are relational, the phrase "the genus" in this remark seems to pick out any item that is a genus of something. Thus, Aristotle states that if something is a genus of something, then it is predicated of its species and of nothing else. In the fourth book of the *Topics,* Aristotle appears to treat being a species of something as the converse relation of being a genus of something: B is a species of A if and only if A is a genus of B.[6] So Aristotle's remark justifies S3. As we will see in a moment, Aristotle will qualify his remark in book 6 of the *Topics.* There, he will state that genera are predicated not only of their species but also of the individuals that fall under these species. This qualified version of the remark, too, will suffice to justify S3.

FURTHER EVIDENCE FOR S3. In book 6 of the *Topics,* Aristotle argues that no genus of a species is predicated of a differentia of that species. For example, if 'animal' is a genus of 'man', and 'biped' is a differentia of 'man', then 'animal' is not predicated of 'biped'. In order to justify this claim, Aristotle invokes a premise to the effect that a genus is predicated exclusively of its species and of the individuals that fall under these species:

> The general view is that the genus is predicated, not of the differentia, but of that of which the differentia is predicated. For instance, animal is predicated of man and ox and other footed animals, not of the differentia itself which is said of the species. For if animal is to be predicated of each of the differentiae, then ... the differentiae will be all either species or individuals, if they are animals; for each of the animals is either a species or an individual. (*Top.* 6.6 144a32–b3)

In the earlier passage from *Topics* 4.6, Aristotle implied that a genus is predicated exclusively of its species. In the present passage, he allows that a genus is also predicated of individuals.[7] It seems clear that the genus in question is a genus not only of its species but also of the

6. See *Top.* 4.1 121a30–b4; cf. Slomkowski (1997: 86) and Smith (1997: 63).

7. In chapter 4.2 of the *Topics,* Aristotle makes a similar claim for 'partaking' instead of 'being predicated': he states that only species and individuals partake of a genus, and that therefore no differentia partakes of a genus (122b20–3).

individuals that fall under these species.[8] Consequently, the present passage implies that a genus is predicated only of items of which it is a genus. As before, this should be taken to mean that if something is a genus of something, it is a genus of everything of which it is predicated. Thus, the present passage confirms S3.

IF A IS A GENUS OF B, THEN A IS A GENUS OF EVERYTHING OF WHICH B IS PREDICATED. Aristotle holds, very plausibly, that

> of those of which the species is predicated, the genus must be predicated as well. (*Top.* 4.1 121a25–6)

This means

S4: If A is a genus of B, and B is predicated of C, then A is predicated of C

Now, S3 and S4 entail[9]

S5: If A is a genus of B, and B is predicated of C, then A is a genus of C

This last statement is similar in structure to Barbara NXN. It could be used to justify the validity of Barbara NXN if a_N-predication were identified with the relation of being a genus of something. However, this would be too narrow an interpretation of a_N-predication, since Aristotle often assumes a_N-predications in which the predicate is not a genus of the subject, for example, that 'white' is a_N-predicated of 'swan' and of 'snow' (see p. 330 below).

8. For the notion of a genus of an individual, see *Cat.* 5 3a2. If individuals have genera, the claim that B is a species of A just in case A is a genus of B (p. 120) should be modified as follows: B is a species of A just in case A is a genus of B and B is not an individual.

9. If A is a genus of B, and B is predicated of C, then A is predicated of C (S4), and hence A is a genus of C (S3). S5 is confirmed by *Top.* 6.6 144b4–8, where Aristotle states the following: if B is a species of A, and B is predicated of C, then C is a species of A. Given that B is a species of A just in case A is a genus of B (p. 120), this is equivalent to S5.

In order to justify the validity of Barbara NXN, S5 needs to be modified in such a way that the relation of being a genus of something is replaced by a_N-predication. This will be done in S6–13 by appealing to some more of Aristotle's claims about the predicables. I argue that, for Aristotle, every subject of an essential predication is predicated essentially of everything of which it is predicated. In other words, if there is an A that is predicated essentially of B, then B is predicated essentially of everything of which it is predicated (S9). Moreover, I argue that every subject of an a_N-predication is also the subject of an essential predication (S11). This implies that if A is a_N-predicated of B, then B is predicated essentially—and hence is a_N-predicated—of everything of which it is predicated. So given the transitivity of a_N-predication, it will follow that if A is a_N-predicated of B, then A is also a_N-predicated of everything of which B is predicated (S13).

IF A IS A GENUS OF B, THEN B IS PREDICATED ESSENTIALLY OF EVERYTHING OF WHICH IT IS PREDICATED. To begin, let us consider a passage that is slightly more complex than the previous ones. It describes a procedure for establishing that something is a genus of something. Suppose we want to establish that B is a genus of C. Aristotle's procedure starts from the assumption that B belongs to C, which means that B is predicated of C.[10] In order to establish that B is a genus of C, it suffices to find some A that is a genus of B and is predicated essentially of C:

> If that which has been called a genus (B) is admitted to belong to the species (C), but it is disputed whether it belongs as a genus, then it is enough to show that one of the higher genera (A) is predicated of the species (C) essentially. For if one of them is predicated essentially, all of them, both higher and lower than this one (A), if predicated at all of the species (C), will be predicated of it essentially; so that which has been given as a genus (B) is also predicated essentially. (*Top.* 4.2 122a10–17)

Aristotle refers to A as one of the genera that are higher than B. As Alexander points out, this means that A is a genus of B.[11] Thus,

10. The assumption is first formulated in terms of belonging (*Top.* 4.2 122a11), and later in terms of predication (122a15).

11. Alexander *in Top.* 310.11–13.

Aristotle holds that if B is predicated of C and some genus of B is predicated essentially of C, then B is a genus of C, and is therefore predicated essentially of C.[12] In other words,

S6: If A is a genus of B, and B is predicated of C, and A is predicated essentially of C, then B is predicated essentially of C

This statement takes the form of a conditional whose antecedent consists of three conjuncts. The third of these conjuncts is superfluous because it follows from the first two; for if A is a genus of B and B is predicated of C, then A is a genus of C (S5) and hence A is predicated essentially of C (S2). Thus, the third conjunct can be omitted, which leads to the following statement:

S7: If A is a genus of B, and B is predicated of C, then B is predicated essentially of C

EVERY SUBJECT OF AN ESSENTIAL PREDICATION HAS A GENUS. If A is predicated essentially of B, then A is a definition, or a genus, or a differentia of B (S2). Aristotle holds that every definition consists of a genus and one or more differentiae.[13] So if A is a definition of B, then there is a genus of B. Also, every differentia is presumably part of a definition.[14] This means that if A is a differentia of B, then there is a genus of B, namely, the genus in combination with which the differentia constitutes a definition.[15] In sum, then, the following is true for every B:

S8: If there is an A that is predicated essentially of B, then there is a C that is a genus of B

12. See Alexander *in Top.* 310.9–17.

13. *Top.* 1.8 103b14–16, 6.1 139a28–9, 6.4 141b25–7, 6.6 143b19–20, 7.3 153b14–15.

14. See *Top.* 6.6 143b6–9.

15. This definition will usually be a definition of B, except possibly when B is an individual that has no definition. In this case, the definition will be a definition of a genus or species of B.

In other words, every subject of an essential predication has a genus. Now, S7 and S8 entail that every subject of an essential predication is predicated essentially of everything of which it is predicated:

S9: If there is an A that is predicated essentially of B, and B is predicated of C, then B is predicated essentially of C

Consequently, no term that is predicated nonessentially as an accident or proprium of something is the subject of an essential predication.[16] These terms do not have a definition, and hence do not have an essence. We will consider these terms in detail in Chapter 9.

ESSENTIAL PREDICATION IMPLIES A_N-PREDICATION. In order for the theory of predicables to explain the validity of Barbara NXN, the relations of a_N- and a_X-predication need to be connected to the predicables in some way. In particular, a_N-predication should be connected to the predicables' relation of essential predication. This will be done in S10–11. First, it is safe to assume that essential predication implies a_N-predication:[17]

---

16. For example, if 'capable of learning grammar' is a proprium of 'man', then it is not the subject of an essential predication. Now, 'man' is the subject of an essential predication, given that 'animal' is its genus. Consequently, the relation of counterpredication (ἀντικατηγορεῖσθαι) that holds between 'capable of learning grammar' and 'man' cannot be analyzed as mutual predication (i.e., as the condition that the one term be predicated of the other and vice versa). For otherwise, given S9, 'man' would be predicated essentially of 'capable of learning grammar', so that the latter term would be the subject of an essential predication. Instead of mutual predication, counterpredication may then be analyzed as coextensiveness: two terms counterpredicate just in case every individual that falls under one of them also falls under the other and vice versa. This is in accordance with Aristotle's characterization of counterpredication at *Top.* 5 102a18–22.

17. This is confirmed by Aristotle's discussion of per se predication in the *Posterior Analytics.* The most prominent case of per se predication obtains when the predicate belongs to the subject essentially (ἐν τῷ τί ἐστιν, *APost.* 1.4 73a34–7). Aristotle states that all these per se predications are necessary (*APost.* 1.4 73b16–19 and 1.6 74b6–8; cf. McKirahan 1992: 83–4 and Barnes 1994: 117 and 126).

S10: If A is predicated essentially of B, then Aa_NB

This means that if A is a definition, genus, or differentia of B, then it is a_N-predicated of it. While this implication is generally accepted, the converse implication does not seem to be valid. Two of Aristotle's standard examples of a_N-predications in the modal syllogistic are that 'white' is a_N-predicated of 'swan' and of 'snow'. But in the *Topics* Aristotle denies that 'white' is predicated essentially of 'snow', thereby also denying that it is a differentia of 'snow' (4.1 120b21–9). Thus, he seems to accept a_N-predications whose predicate is not predicated essentially of the subject.[18] Despite this, I suggest that a_N-predication is grounded in essential predications in a certain way, as follows.

EVERY SUBJECT OF AN A_N-PREDICATION IS THE SUBJECT OF AN ESSENTIAL PREDICATION. Alexander regards 'white' as what he calls an inseparable accident of 'swan' and 'snow', by which he means an accident that is predicated of its subject necessarily but not essentially.[19] He holds that although inseparable accidents are not predicated essentially of their subject, they follow from the essence of the subject:

> These predicates are accidents because they are not part of the essence of the things to which they belong, for the essence of snow does not consist in being white . . . ; but they follow their essence, and they are inseparable accidents. (Alexander *in Top.* 50.8–11)

Alexander goes on to explain that being white is a necessary consequence of certain material features of snow that belong to the essence of snow (50.18–21 and 50.29–31). On this view, the essence of snow is the cause that explains why 'white' is predicated necessarily of 'snow'. More generally, the necessity with which an inseparable accident is predicated of its subject is grounded in the essence of the subject. Such a view of necessity can also be attributed to Aristotle, even though he does not

18. Aristotle also denies that 'having the sum of the interior angles equal to two right angles' is predicated essentially of 'triangle' (*Met.* Δ 30 1025a30–2, *PA* 1.3 643a27–31), although he presumably accepts that it is a_N-predicated of it.

19. Alexander *in Top.* 50.6–51.5.

explicitly state it. Thus, David Charles holds that, for Aristotle, necessary properties of kinds are grounded in the kind's essence:[20]

> the essence is the one cause of all the kind's derived necessary properties. (Charles 2000: 203)

I suggest that this is also true for Aristotle's notion of a_N-predication in the modal syllogistic, and that the necessity with which the predicate of an a_N-predication is predicated of its subject derives from the essence of the subject. This presupposes that the subject has an essence and hence that it is the subject of an essential predication. So on this view, every subject of an a_N-predication is the subject of an essential predication:

S11: If there is an A such that Aa_NB, then there is a C such that C is predicated essentially of B

This is a substantive statement, and it is vital for the account of the validity of Barbara NXN pursued here. There is no direct evidence for it in the *Prior Analytics*, as Aristotle does not explain the nature of a_N-predication in his modal syllogistic. Nevertheless, it seems plausible that Aristotle employed a notion of a_N-predication that makes reference to the essence of the subject, and thereby requires the subject to have an essence. Thus, S11 can be plausibly attributed to Aristotle and can help explain why he took Barbara NXN to be valid.[21]

---

20. Similarly, T. Irwin attributes to Aristotle the view that "an essential property will explain the necessary properties of a natural kind" (Irwin 1980: 38–9), and J. Barnes takes Aristotle in *Posterior Analytics* 1.4 to argue that "necessity is ultimately grounded in essential or definitional connections" (Barnes 1994: 120); see also Loux (1991: 73).

21. It is worth noting that S10 and S11 are, mutatis mutandis, in line with the account of essence and necessity given by Kit Fine (1994). S10 corresponds to Fine's statement: "I accept that if an object essentially has a certain property then it is necessary that it has the property . . . ; but I reject the converse" (1994: 4; *pace* Zalta 2006: 661 and 681–3). S11 corresponds to Fine's conception of necessity as based on the essence of certain objects: "the necessity has its source in those objects which are the subject of the underlying essentialist claim . . . Socrates being a man having its source in the identity of Socrates, 2 being a number having its source in the identity of 2" (1994: 8–9).

IF A IS A_N-PREDICATED OF B, THEN A IS A_N-PREDICATED OF EVERYTHING OF WHICH B IS PREDICATED. While the validity of Barbara NXN has been disputed since antiquity, the validity of Barbara NNN is undisputed and widely accepted:

S12: If Aa_NB, and Ba_NC, then Aa_NC

Now, S9–12 entail a statement that comes close to asserting the validity of Barbara NXN, namely,

S13: If Aa_NB, and B is predicated of C, then Aa_NC

In order to see how this follows from S9–12, suppose that A is a_N-predicated of B and B is predicated of C. Since B is the subject of an a_N-predication, it is also the subject of an essential predication (S11). Consequently, since B is predicated of C, B is predicated essentially of it (S9) and hence also a_N-predicated of C (S10). Given that A is a_N-predicated of B, it follows that A is also a_N-predicated of C (S12).

S13 would justify the validity of Barbara NXN if the relation of a_X-predication were identified with the *Topics*' relation of being predicated of something. However, as we will see, there is reason to think that these two relations cannot be identified. In order to justify the validity of Barbara NXN, S13 thus needs to be modified in such a way that predication is replaced by a_X-predication. This will be done in S14–15.

A_X-PREDICATION AND THE PREDICABLES. According to the *Topics*' theory of predication, a term is predicated of another just in case it is a genus, differentia, definition, proprium, or accident of it (S1). Aristotle holds that, unlike the other predicables, accidents can belong nonuniversally to their subjects (*Topics* 2.1 109a11–25). Thus, a term can be an accident of another without being a_X-predicated of it; for example, 'just' is an accident of 'man' but fails to be a_X-predicated of 'man' because not all men are just.[22] Moreover, a term might be a_X-predicated of another without being predicated of it. As we saw above (p. 120), Aristotle denies that a genus of a species can be predicated of a differentia of this species. For example, 'animal' cannot be predicated of 'biped'. Yet it

22. See Alexander *in Top.* 128.24–30.

is not unreasonable to think that 'animal' is a_X-predicated of 'biped'.[23] Given this, a_X-predication cannot be identified with the *Topics*' relation of predication. However, we can still define a restricted version of a_X-predication that is sufficiently close, for our purposes, to predication. This can be done as follows.

I have argued that the assertoric syllogistic should be interpreted by the preorder semantics, based on a primitive relation of a_X-predication (pp. 66–71). The preorder semantics does not specify what this primitive relation is. The relation may or may not include cases in which the predicate fails to be predicated, in the sense of the *Topics*' predicables, of the subject. If it does, we can now confine ourselves to those a_X-predications in which the predicate is a definition, genus, differentia, proprium, or accident of the subject. Thus, the primitive relation of a_X-predication is restricted to predications in the sense of the predicables, eliminating cases like 'Animal belongs to all biped'.

The resulting restricted relation should again be able to serve as the relation of a_X-predication in the preorder semantics. To this end, it needs to be a preorder, that is, it needs to be reflexive and transitive. It is not immediately clear whether the *Topics*' relation of predication is reflexive and transitive and whether restricting a_X-predication to predication yields a preorder. If not, then we can consider the reflexive and transitive closure of the restricted relation, that is, the smallest reflexive and transitive relation that contains the restricted relation. The resulting preorder can serve as the relation of a_X-predication in the modal syllogistic. This relation of a_X-predication may include cases in which the predicate is not predicated of the subject, but these cases must have resulted from generating the reflexive or transitive closure of the restricted relation of predication. Hence if A is a_X-predicated of B but is not predicated of it, then either A is identical with B (reflexive closure) or A is connected to B by a finite chain of predications (transitive closure). Thus, the following is true for the resulting relation of a_X-predication:

S14: If Aa_XB, then (i) A is predicated of B, or (ii) A is identical with B, or (iii) there are $D_1, \ldots, D_n$ such that A

23. At *APr.* 2.2 54b4–7 and 2.3 56a26–9, Aristotle assumes that 'animal' is a_X-predicated of 'footed', referring to the former term as a genus and to the latter as a differentia; see p. 162 below.

is predicated of D_1, D_1 is predicated of D_2, ..., and D_n is predicated of B

S14 encodes the assumption that a_X-predication is restricted by the predicables in the way just described. It is an immediate consequence of taking a_X-predication to be the reflexive and transitive closure of a subset of the *Topics*' relation of predication. In combination with S13, this suffices to justify the validity of Barbara NXN.

THE VALIDITY OF BARBARA NXN. Given that a_N- and a_X-predication have the properties just described, Barbara NXN is valid. S13 and S14 entail the following:[24]

S15: If Aa_NB, and Ba_XC, then Aa_NC

This, then, is the way I think the *Topics*' theory of predicables can explain why Barbara NXN is valid, and on what grounds Aristotle took it to be valid. The crucial contribution of the theory of predicables is the statement S9, that every subject of an essential predication is predicated essentially of everything of which it is predicated. The modal syllogistic's relations of a_N- and a_X-predication are connected to the predicables in S10–11 and S14. S11 states that every subject of an a_N-predication is the subject of an essential predication; the underlying assumption is that the necessity with which the predicate of an a_N-predication is predicated of its subject is grounded in the essence of the subject. In S14, a_X-predication is restricted by the predicables, in such a way that it is taken to be the reflexive and transitive closure of a subclass of the relation of predication.

As argued above, Aristotle's justification of Barbara XXX and NXN appeals to the heterodox assertoric and apodeictic *dictum de omni*,

24. Suppose Aa_NB and Ba_XC. Given S14, Ba_XC implies that (i) B is predicated of C, or (ii) B is identical with C, or (iii) there are $D_1, \ldots, D_n$ such that B is predicated of D_1, D_1 is predicated of D_2, ..., and D_n is predicated of C. In case (i), Aa_NC follows by S13. In case (ii), Aa_NC follows immediately from Aa_NB, since B is identical to C. In case (iii), Aa_NC follows by finitely many applications of S13: since Aa_NB and B is predicated of D_1, S13 implies Aa_ND_1; since D_1 is predicated of D_2, S13 implies Aa_ND_2, and so on; finally, since D_n is predicated of C, S13 implies Aa_NC.

respectively. These two heterodox *dicta* determine logical properties of the relations of a_X- and a_N-predication, but they do not define what these relations are or when they hold between two terms. Following Aristotle's lead, I will not undertake to give an explicit definition of them, but will treat them as primitive relations that have the logical properties stated in S10–12 and S14. These properties connect a_X- and a_N-predication to the theory of predicables, and thereby explain the validity of Barbara NXN. I submit that Aristotle's decision to take Barbara NXN to be valid was implicitly motivated by these properties, even though he did not spell them out.

A_X-PREDICATION IN THE ASSERTORIC AND MODAL SYLLOGISTIC. If I am correct, Aristotle's modal syllogistic employs a specific notion of a_X-predication which is obtained from the notion of a_X-predication used in the assertoric syllogistic in two steps. First, the latter relation is restricted to predications in the sense of the predicables, and then the restricted relation is extended to its reflexive and transitive closure.

This way of constructing a_X-predication is somewhat technical and artificial, especially the second step. I do not want to suggest that Aristotle had in mind exactly this construction. Rather, I suggest that the notion of a_X-predication employed by Aristotle in the modal syllogistic is governed by two different kinds of principle. On the one hand, it is governed by the *Prior Analytics*' assertoric *dictum de omni,* which requires a_X-predication to be reflexive and transitive; at the same time, it is governed by the *Topics*' theory of predicables. The two steps of the above construction are meant to represent these two kinds of principle, respectively.

Of course, the relation of a_X-predication that results from this construction can also be used in the assertoric syllogistic; any reflexive and transitive relation can serve as the primitive relation in the preorder semantics. But unlike the assertoric syllogistic, I suggest, the modal syllogistic requires the more specific and restricted version of a_X-predication which is determined by the theory of predicables in the way described by S14. Thus, the assertoric syllogistic is applicable to a wider class of a_X-predications than the modal one.

N-X-SUBORDINATION. Let me at this point add a remark on the relation between a_N- and a_X-predication. It is natural to assume that the former implies the latter:

S16: If Aa_NB, then Aa_XB

This is an instance of the more general principle of N-X-subordination, according to which every N-proposition implies the corresponding X-proposition of the same quantity and quality.

Curiously, Aristotle does not assert the principle of N-X-subordination in the modal syllogistic.[25] He also does not use it in his proofs—or at least there is no clear evidence of his doing so. This is all the more striking since an appeal to N-X-subordination would have been straightforward and helpful in many cases. Aristotle repeatedly asserts the validity of moods with an N-premise after he has already asserted the validity of the same mood with the corresponding X-premise instead of the N-premise. It would be straightforward simply to infer the validity of these moods from the validity of the moods with the corresponding X-premise by means of N-X-subordination. But this is not what Aristotle does. Instead, he gives an independent justification for the moods with the N-premise.[26] Thus, one may say that he conspicuously avoids using N-X-subordination.

The modal syllogistic might therefore, in principle, be compatible with interpretations in which N-X-subordination is not valid. However, Aristotle never rejects N-X-subordination or implies its invalidity.[27] So the modal syllogistic is equally compatible with interpretations in which

---

25. Wieland (1966a: 52–5) and Thom (1996: 34).

26. Here are three examples. First, the validity of first-figure QNQ-moods (1.16 36a2–7, 36a17–21) is not inferred from that of the corresponding QXQ-moods (1.15 33b33–40. cf. Thom 1996: 41). Instead, the QNQ-moods are taken to be perfect in the same way as the QXQ-moods. Second, the validity of Barbara and Darii NQM (1.16 35b39–36a2, 36b1–2) is not inferred from that of Barbara and Darii XQM (1.15 34a34–b6, 35a35–40). Instead, Aristotle states that their validity should be established by means of the same kind of (fairly complicated) proof he gave for the latter two moods. Third, the validity of Darapti NQM and Disamis NQQ (1.22 40a11–16, 40a39–b8) is not inferred from that of Darapti XQM and Disamis XQQ (1.21 39b10–16, 39b26–31). Instead, the former two moods are reduced by conversion to Darii NQM and Darii QNQ in the same way as the latter two are reduced to Darii XQM and Darii QXQ.

27. *Pace* Wieland (1966a: 40–55; 1972: 136), who argues that Aristotle denies N-X-subordination and that he takes N-propositions to be incompatible with the corresponding X-propositions. Most commentators reject Wieland's

N-X-subordination is valid. Most commentators prefer an interpretation of this latter kind, and I follow them, accepting S16 as valid.

A PROBLEM. I have argued that if a term is the subject of an essential predication, it is predicated essentially of everything of which it is predicated (S9). Clearly, some terms are predicated as accidents or propria of a subject without being predicated essentially of it. If I am correct, these terms cannot be the subject of an essential predication. In other words, nothing can be predicated essentially of them. This leads to the question as to which terms are the subject of an essential predication, and which not.

A standard example of an essential predication is that 'animal' is a genus of 'man'. But Aristotle also holds that 'color' is a genus of 'white' and that 'virtue' is a genus of 'justice', and so on.[28] Thus, 'white' is the subject of an essential predication. At the same time, however, Aristotle holds that 'white' is predicated of some subjects as an accident and hence is predicated nonessentially of them.[29] For example, 'white' is predicated as an accident of 'snow'. This appears to contradict S9, according to which every subject of an essential predication should be predicated essentially of everything of which it is predicated. How can this apparent contradiction be resolved? Moreover, if 'man', 'white', and 'justice' are subjects of essential predications, which terms are not?

These questions cannot properly be addressed within the theory of predicables alone. Rather, they seem to be connected with Aristotle's ten categories of substance, quantity, quality, and so on. The purpose of Chapters 9 and 10 is to explain how those questions can be answered by means of Aristotle's theory of the ten categories. I argue that his account of the categories in the *Topics* relies on the distinction between essence terms and nonessence terms mentioned on p. 7 above:

---

view (van Rijen 1989: 217, Schmidt 1989: 83n16, Thom 1996: 34–5, Buddensiek 1994: 98–101, Patterson 1995: 106–15, Nortmann 1996: 142–8, Ebert & Nortmann 2007: 441–3, Striker 2009: 116).

28. For the first example, see *Top.* 2.2 109a37–8, 4.1 121a7–9, 4.3 123b25–7, *Cat.* 11 14a20–3. 'Virtue' and 'good' are regarded as genera of 'justice' at *Cat.* 11 14a22–3 and *Top.* 3.1 116a24. Also, 'evil' is a genus of both 'defect' and 'excess', *Top.* 4.3 123b28–9.

29. *Top.* 1.5 102b8, 4.1 120b21–2.

Essence terms:	'animal', 'man', 'color', 'redness', 'motion', ...
Nonessence terms:	'colored', 'red', 'walking', ...

As I argue in detail, essence terms are subjects of essential predications but cannot be predicated nonessentially of anything. Conversely, nonessence terms can be predicated nonessentially but are not the subjects of essential predications. All substance terms are essence terms, examples being 'man' and 'animal'. On the other hand, some nonsubstance terms are essence terms, whereas others are nonessence terms. For example, 'whiteness' is an essence term, whereas 'white' is a nonessence term. Hence when Aristotle says that 'color' is a genus of 'white', this is not entirely precise, and he should have said instead that 'color' is a genus of 'whiteness'.

I first show how this distinction between essence and nonessence terms is grounded in the *Topics*' account of the categories (Chapter 9) and then explain how the distinction can be put to use in the modal syllogistic (Chapter 10).

9

Categories in the *Topics*

CATEGORIES AS A CLASSIFICATION OF TERMS. In a well-known passage from chapter 4 of the *Categories*, Aristotle introduces the ten categories as follows:

> Of things said without any combination, each signifies either substance or quantity or quality or a relative or where or when or being-in-a-position or having or doing or being-affected. To give a rough idea, examples of substance are man, horse; of quantity: four-foot, five-foot; of quality: white, grammatical; of a relative: double, half, larger; of where: in the Lyceum, in the market-place; of when: yesterday, last-year; of being-in-a-position: is-lying, is-sitting; of having: has-shoes-on, has-armor-on; of doing: cutting, burning; of being-affected: being-cut, being-burned. (*Cat.* 4 1b25–2a4)

The first category is that of substance (οὐσία). The other ones are those of quantity, quality, relation, and so on. The categories provide a classification of items that Aristotle calls 'things said without any combination'. These can be taken to be terms, that is, linguistic items such as 'man' and 'white'. At the same time, the categories can also be viewed as providing a classification of the nonlinguistic items signified by terms. For present purposes, however, I am mainly interested in them as a classification of terms. For example, the terms 'man' and 'horse' belong to the category of substance, and 'four-foot' and 'white' belong to one of the nine nonsubstance categories.

EVERY GENUS IS IN THE SAME CATEGORY AS THAT OF WHICH IT IS A GENUS. The classification of terms provided by the ten categories plays a prominent role in the *Topics*. Aristotle gives a detailed account of the categories in *Topics* 1.9 and makes extensive use of them later on in the treatise. At the beginning of chapter 1.9, Aristotle states that every genus, differentia, definition, proprium, and accident is in one of the ten categories (103b20–5). Thus, there is a close connection between the categories and the four predicables. For example, Aristotle holds that every genus is in the same category as its species:

> Moreover, see whether the genus and the species are not found in the same division, but the one is a substance while the other is a quality, or the one is a relative while the other is a quality, as, e.g., snow and swan are each a substance, while white is not a substance but quality, so that white is not a genus either of snow or of swan.... To speak generally, the genus ought to fall under the same division as the species; for if the species is a substance, so too should be the genus, and if the species is a quality, so too the genus should be a quality; e.g. if white is a quality, so too should color be. Likewise, also, in the other cases. (*Top.* 4.1 120b36–121a9)

In this passage, Aristotle states that if A is a genus of B, and B is a species of A, then A and B are in the same category. As mentioned above, book 4 of the *Topics* is governed by the assumption that B is a species of A just in case A is a genus of B (p. 120). Thus, Aristotle states,

S17: If A is a genus of B, then A and B are in the same category

NO SUBJECT OF AN ESSENTIAL PREDICATION IS THE PREDICATE OF A CROSS-CATEGORIAL PREDICATION. Let us say that a predication is cross-categorial just in case the predicate and the subject are not in the same category. S17 states that there are no cross-categorial genus-predications. This, in combination with Aristotle's theory of predicables, implies that no subject of an essential predication is the predicate of a cross-categorial predication. To see this, suppose that some A is predicated essentially of B, and that B is predicated of C. Since B is the subject of an essential predication, there is a D that is a genus of B (S8). D and B are in the same category (S17). Since D is a genus of B, and B is predicated of C, D is a genus of C (S5). So D and C are in the same category (S17). Consequently, B and C are in the same category.

In sum, the following is true for any B and C:

S18: If there is an A that is predicated essentially of B, and B is predicated of C, then B and C are in the same category

PARONYMOUS PAIRS OF NONSUBSTANCE TERMS. S18 leads to a problem concerning nonsubstance terms similar to the problem we saw above in connection with S9. Aristotle accepts that nonsubstance terms are subjects of essential predications. For example, 'justice' has 'virtue' as its genus (p. 132n28). On the other hand, he also accepts that nonsubstance terms are predicates of cross-categorial predications, especially when they are predicated of substance terms. For example, it might be thought that 'justice' is predicated as an accident of 'man'.[1] So it seems that one and the same nonsubstance term can be both the subject of an essential predication and the predicate of a cross-categorial predication—which would contradict S18.

The apparent contradiction can be resolved by means of a distinction that Aristotle draws between two kinds of nonsubstance terms. In the *Categories,* Aristotle distinguishes between nouns such as 'justice', 'blindness', and 'whiteness' on the one hand and corresponding terms such as 'just', 'blind', and 'white' on the other.[2] In Aristotle's terminology, terms of the latter kind are called paronymous or paronyms.[3] The distinction between nouns and corresponding paronyms is also

1. See, for example, *Top.* 2.1 109a21–6.

2. *Cat.* 1 1a12–15, 8 10a27–32; see also *EE* 3.1 1228a35–6.

3. In the *Categories,* Aristotle takes paronyms (παρώνυμα) to be not linguistic expressions such as 'just' and 'blind', but the nonlinguistic items to which these expressions apply (*Cat.* 1 1a12–15; see Ackrill 1963: 72–3). For example, a just thing is called a paronym on the grounds (i) that the term 'just' applies to it, and (ii) that this term stands in a certain linguistic relation to the noun 'justice'. Elsewhere, however, Aristotle takes paronyms to be linguistic expressions: for example, he states that the terms 'two' and 'three' are paronymous names (τὸ γὰρ τρία καὶ δύο παρώνυμα ὀνόματά ἐστιν, *Phys.* 3.7 207b8–9; see Ross 1936: 559–60). Ross characterizes Aristotle's use of "paronym" in this passage as follows: "A word is called a παρώνυμον of another when (1) there is a linguistic connexion between the words (*Cat.* 10a32–b9) and (2) the latter stands for something metaphysically simpler and more fundamental than what the former stands for. λευκός is παρώνυμον from

prominent in the *Topics*. Many of Aristotle's *topoi* there rely on this distinction.[4] Now, Aristotle holds that whereas paronymous terms can be predicated of substance terms, the corresponding nouns cannot. For example, 'justice' and 'blindness' cannot be predicated of substance terms, on the grounds that a particular substance can be said to be just or blind, but not justice or blindness:[5]

> If blindness were the same as being blind, both would be predicated (κατηγορεῖτο) of the same thing. But though a man is said to be blind, a man is certainly not said to be blindness. (*Cat.* 10 12a39–b1)

Given that 'justice' and 'blindness' are not predicated of substance terms, it also seems unlikely that they are predicated, in the sense of the predicables, of terms in a nonsubstance category other than their own. Thus, there is reason to think that unlike the corresponding paronymous terms, nonsubstance nouns cannot be predicates of cross-categorial predications. If so, they can be subjects of essential predications without violating S18.

The passage just quoted implies that paronymous terms, such as 'just' and 'blind', are predicated of substance terms. Hence given that Aristotle is committed to S18, he is also committed to denying that paronymous terms are subjects of essential predications.

This result is confirmed by another consideration, as follows. If nonsubstance paronymous terms were subjects of essential predications, they would have a genus (S8). This genus would be either a substance term or a nonsubstance term. Since no substance term can be the genus of a nonsubstance term (S17), the genus would be a nonsubstance term. But it is presumably not a nonsubstance noun such as 'virtue'; for although 'virtue' is a genus of 'justice', it is unlikely that it is a genus of the paronymous term 'just'.[6] Instead, it might be thought that the

---

λευκότης because it means 'characterized by λευκότης'." I follow this latter usage, referring to terms such as 'just', 'blind', and 'white' as paronyms.

4. See, for example, *Top.* 2.2 109a34–b12, 2.4 111a33–b4, 2.8 114a26–b5, 3.1 116a23–8, 3.3 118a34–9, 4.3 124a10–14, 7.1 151b28–33, 7.3 153b25–35, 8.1 156a27–b3.

5. See also *Met.* Θ 7 1049a30–4.

6. 'Just' is predicated as an accident of substance terms such as 'man'. So if 'virtue' were a genus of 'just', 'virtue' would be predicated of these substance

paronymous term 'virtuous' is a genus of the paronymous term 'just'. However, Aristotle seems to deny that paronymous terms can be the genus of anything:

> Being a color does not happen by accident to white, but color is its genus.... The predication of a genus is never said of the species paronymously, but all genera are predicated of their species synonymously.... A man therefore who says that the white is colored has not rendered it as a genus, since he has said it paronymously. (*Top.* 2.1 109a37–b9)

This passage implies that no paronymous term can be the genus of anything, and hence also not of another paronymous term.[7] In sum, then, the supposed genus of a nonsubstance paronymous term cannot be a substance term, nor can it be a nonsubstance noun or paronymous term. This suggests that nonsubstance paronymous terms do not have a genus and hence are not subjects of essential predications. If so, they can be predicated of substance terms and be predicates of cross-categorial predications without violating S18.

'WHITE' VS. 'WHITENESS'. I have argued that no nonsubstance paronymous term is the subject of an essential predication. Now, Aristotle states that 'white' is a paronymous term, corresponding to the noun 'whiteness' (*Cat.* 8 10a27–32). So 'color' should be predicated essentially as a genus of 'whiteness', but not of 'white'. On the other hand, Aristotle repeatedly says that 'color' is a genus of 'white', using the paronymous term 'white' instead of the noun 'whiteness'.[8] But when he says this, I submit, he is not attentive to the distinction between nonsubstance nouns and the corresponding paronymous terms.[9] If he did, he should have said that 'color' is a genus of 'whiteness' instead of 'white'. Thus, it is not the same term that is the predicate of a cross-categorial predication and of which 'color' is predicated as a genus: the former term is 'white', the latter is 'whiteness'. Both terms belong to the

---

terms as well (S4). But as we have seen, nonsubstance nouns such as 'virtue' cannot be predicated of substance terms.

7. See Alexander *in Top.* 136.26–8.

8. See the passages mentioned on p. 132n28 above.

9. In one of these passages, Aristotle switches from 'white' (*Top.* 2.2 109a37–8) to 'whiteness' (109b3).

nonsubstance category of quality, but they differ as to the predicative relations in which they can stand to other terms.

This distinction also helps solve the problem discussed above in connection with the claim in S9, that no subject of an essential predication can be the predicate of a nonessential predication (p. 132). For paronymous terms like 'white' can serve as predicates of nonessential predications, but not as subjects of essential predications; on the other hand, nouns like 'whiteness' can be taken to be predicated essentially of everything of which they are predicated.

ESSENTIAL SELF-PREDICATION. The claim that paronymous terms are not subjects of essential predications is in tension with a passage from *Topics* 5, in which Aristotle states that everything is predicated essentially of itself:

> See if he has stated a thing as a proprium of itself; for then what has been stated to be a proprium will not be a proprium. For a thing itself always indicates its own essence, and what indicates the essence is not a property but a definition. (*Top.* 5.5 135a9–12)

In this passage, Aristotle argues that nothing is a proprium of itself, on the grounds that everything is predicated essentially of itself, and a proprium is not predicated essentially of its subject.[10] Aristotle describes this essential self-predication as that of a definition of its subject. He holds that a definition and its subject are identical in that they signify numerically the same item (*Top.* 1.7 103a23–7, 7.2 152b36–153a1). Thus, Aristotle regards predications in which a definition is predicated of its subject as self-predications. For example, if the term 'biped terrestrial animal' is predicated as a definition of the term 'man', this is regarded as an essential self-predication; for the two terms signify the same item. Likewise, if 'animal' is predicated of 'animal', Aristotle seems to regard this as an essential self-predication as well, treating it as a predication of a definition.[11]

The passage just quoted implies that every term is predicated essentially of itself and hence is the subject of an essential predication.

---

10. See Barnes (1970: 151).

11. Alexander *in Top.* 67.5–11; see also p. 144 below.

This is problematic because of S18, according to which predicates of cross-categorial predications are not subjects of essential predications. Given Aristotle's commitment to S18, his claim that everything is predicated essentially of itself should be restricted. Thus, we may assume that all terms are predicated essentially of themselves except those that are predicates of cross-categorial predications:

S19: B is predicated essentially of B if and only if there is no C such that B is predicated of C and B is not in the same category as C

For example, the paronymous term 'just', being the predicate of cross-categorial predications, is not predicated essentially of itself. On the other hand, the corresponding noun 'justice' is predicated essentially of itself. When Aristotle claims that everything is predicated essentially of itself, I submit, he is not attentive to the distinction between paronymous terms and nouns, and therefore does not make it clear that this claim is true for the latter but not for the former.

SELF-A_N-PREDICATION. S18 and S19 entail that every term that is the subject of an essential predication is predicated essentially of itself. Thus, they entail

S20: B is predicated essentially of B if and only if there is an A that is predicated essentially of B

Now, essential predication implies a_N-predication (S10). Moreover, every subject of a_N-predication is also the subject of an essential predication (S11). Thus, S10, S11, and S20 entail

S21: Ba_NB if and only if there is an A such that Aa_NB

A term is a_N-predicated of itself if and only if it is the subject of some a_N-predication. Or equivalently, a term is a_N-predicated of itself if and only if it is the subject of an essential predication. Thus, not every term is a_N-predicated of itself. Otherwise, Aristotle would be committed to the implausible view that every a_X-proposition implies the corresponding

a_N-proposition, as shown by the following instance of Barbara NXN:[12]

Aa_NA (major premise)
Aa_XB (minor premise)
Aa_NB (conclusion)

On the other hand, there are no such problems with the view expressed by S21, that a term is a_N-predicated of itself just in case it is the subject of some a_N-predication.

ESSENCE TERMS VS. NONESSENCE TERMS. Let us say that a term is an *essence term* just in case it is the subject of an essential predication, that is, just in case it has an essence. More precisely, B is an essence term just in case there is an A that is predicated essentially of B. All other terms are *nonessence terms.* The class of essence terms includes exactly those terms that are predicated essentially of themselves (S20). The class of nonessence terms includes exactly those terms that are predicates of cross-categorial predications (S19). Also, every term that is the predicate of a nonessential predication is a nonessence term (S9).

Nonsubstance terms typically come in pairs, one of them being an essence term and the other a nonessence term: nouns such as 'whiteness' and 'justice' are essence terms, whereas paronyms such as 'white' and 'just' are nonessence terms. However, the class of nonessence terms does not consist exclusively of paronyms. It also contains nonparonymous nonsubstance terms:

> But in some cases, because there are no names for the qualities, it is impossible for things to be called paronymously from them. For example, the runner or the boxer, so called in virtue of a natural capacity, is not called paronymously from any quality; for there are no names for the capacities in virtue of which these men are said to be qualified. (*Cat.* 8 10a32–b2)

12. See Łukasiewicz (1957: 189–90), McCall (1963: 50), and Brenner (2000: 342); similarly, Rescher (1974: 7–8).

The terms 'runner' (δρομικός) and 'boxer' (πυκτικός) are nonsubstance terms. They belong to the category of quality. Nevertheless, they are predicated of substance terms, for example, of 'Kallias' and 'Mikkalos'. So 'runner' and 'boxer' are predicates of cross-categorial predications and therefore are nonessence terms. However, they are not paronymous terms, since there is no corresponding noun from which they are called paronymously. Hence they are nonparonymous nonessence terms.

So far we have not discussed how substance terms fit into this picture. Are they essence terms or nonessence terms? Or do they, like nonsubstance terms, come in paronymous pairs consisting of an essence term and a nonessence term? In what follows, I argue that substance terms do not come in pairs, but that they are all essence terms. To this end, I consider Aristotle's treatment of categories in *Topics* 1.9. This will help us determine the status of substance terms and will also confirm the above findings concerning nonsubstance terms.

TOPICS' CATEGORIES VERSUS *CATEGORIES'* CATEGORIES. At the beginning of *Topics* 1.9, Aristotle gives a list of ten categories: essence (τί ἐστι), quantity, quality, relation, and so on (103b22–3). This list differs from that given in chapter 4 of the *Categories* in that the first category is labeled 'essence' (τί ἐστι) instead of 'substance' (οὐσία). Alexander and others neglect this terminological difference, taking 'essence' here simply to mean 'substance' and thus identifying the *Topics'* first category with the *Categories'* category of substance.[13] However, as we will see shortly, there is reason to think that the *Topics'* category of essence does not coincide with the category of substance.

In order to compare the ten categories introduced in *Topics* 1.9 with those introduced in the *Categories,* it will be helpful to refer to the former as T-categories and to the latter as C-categories. Our discussion so far has been concerned only with the C-categories, and I will sometimes continue referring to the C-categories simply as 'categories.' Likewise, the term 'cross-categorial' will always be understood to mean 'cross-C-categorial'.

---

13. Alexander *in Top.* 65.14–19 and 66.26, Waitz (1846: 447), Arpe (1938: 11–14), Bocheński (1956: 63), Ackrill (1963: 79), Kapp (1968: 225), Mansion (1968: 198–9), Kahn (1978: 237), Oehler (1986: 245–6), Irwin (1988: 502), and Smith (1997: 75).

That the first T-category does not coincide with the first C-category is suggested by a passage in which Aristotle states that someone who signifies essence may thereby signify substance, or quantity, or one of the other categories:

> He who signifies essence signifies sometimes substance, sometimes quantity, sometimes quality, sometimes one of the other categories. (*Top.* 1.9 103b27–9)

The second instance of the verb 'signify' in this passage is combined with a list of categories whose first three members are substance, quantity, and quality. This seems to be a list of the C-categories.[14] If so, then the phrase 'essence' (τί ἐστι) cannot refer to the C-category of substance in the present passage, given that the C-categories are mutually disjoint and that one therefore cannot at the same time signify the C-category of substance and another C-category.[15] Consequently, the traditional view is that the phrase 'essence' (τί ἐστι) is used in two different ways in *Topics* 1.9: when Aristotle gives the list of categories at the beginning of the chapter, the phrase is taken to mean 'substance' and to refer to the C-category of substance (103b22 and 26); in the present passage and in the remainder of the chapter, the phrase is taken to be used in a different way (103b27, 30, 32, 34, 37, 38).

Michael Frede and others avoid this double use of 'essence' by denying that 'essence' means 'substance' in *Topics* 1.9.[16] On their view, the T-category of essence is different from the C-category of substance, inasmuch as an instance of signifying the T-category of essence can be an instance of signifying not only the C-category of substance but also any other C-category. Aristotle illustrates this difference between the first T-category and the first C-category by the following series of examples:

---

14. Malcolm (1981: 665), Ebert (1985: 132). Thus, the word 'categories' (κατηγοριῶν) refers to the C-categories in the present passage at 103b29, whereas it refers to the T-categories at 103b25.

15. I take it that the C-categories provide an exhaustive and disjoint classification of the class of items under consideration, so that every item belongs to exactly one C-category (*pace* Morrison 1992: 20), and likewise for the T-categories.

16. Frede (1981: 9–12), Malcolm (1981: 664–8), Ebert (1985: 125–32).

> For when man is set out and he states that what is set out is man or animal, he says what it is and signifies substance. And when white color is set out and he states that what is set out is white or color, he says what it is and signifies quality. Likewise, also, when foot-long length is set out and he states that what is set out is foot-long length, he says what it is and signifies quantity. And likewise with the other categories. (*Top.* 1.9 103b29–35)

In this passage, Aristotle considers predications such as 'Man is man', 'Man is animal', 'White color is color', and so on. These are presumably essential predications, namely, essential self-predications and genus-predications.[17] As is clear from 103b27–9, each of these predications involves signifying the T-category of essence. At the same time, these predications involve signifying the C-categories of substance, quality, and quantity, respectively. Thus, the examples suggest that the first T- and C-categories are not coextensive, and that the T-category of essence includes not only items from the C-category of substance but also items from the nine nonsubstance C-categories.

THE NINE NONESSENCE T-CATEGORIES. Having presented the above examples, Aristotle goes on to explain when something signifies the T-category of essence and when it signifies the other nine T-categories:

> Each of such items, when it is said of itself or when the genus is said of it, signifies essence. But when it is said of another, it does not signify essence, but quantity, or quality, or one of the other categories. (*Top.* 1.9 103b35–9)

---

17. Alexander *in Top.* 67.3–11. It is worth emphasizing that in this passage, nonsubstance terms like 'white color' and 'foot-long length' are accepted as subjects of essential predications, and hence are taken to have an essence. In *Metaphysics* Z, by contrast, Aristotle tends to think that only items that belong to the C-category of substance have a definition and essence in a proper and unqualified way (Z 4 1030a17–27, Z 5 1031a1–2). If nonsubstance terms do not have an essence, the T-category of essence presumably coincides with the C-category of substance (Frede 1981: 19–21). But this is not true for the *Topics,* where nonsubstance terms are accepted as subjects of essential predications in the same way as substance terms (see p. 132n28 above).

The first sentence of this passage raises several problems of translation and interpretation.[18] Nevertheless, it is clear that the sentence picks up the foregoing examples, each of which consisted of an essential self-predication and a genus-predication. The sentence confirms that each of the examples involves an instance of signifying essence, that is, the T-category of essence.

The second sentence concerns the nine nonessence T-categories.[19] It states that if a term is predicated 'of another', then it does not belong to the T-category of essence, but to one of the other T-categories.[20] The condition of being predicated 'of another' should not be taken to mean that a term is predicated of a numerically distinct term, for this is presumably also true for many terms that do not belong to a nonessence T-category. Since Alexander, the condition has been understood to mean that a term be predicated of a term that belongs to another C-category.[21] On this interpretation, every term that is the predicate of a cross-categorial predication belongs to a nonessence T-category (as mentioned above, the term 'cross-categorial' means 'cross-C-categorial').[22] Thus, nonsubstance paronymous terms such as 'white' and 'just', being the predicates of cross-categorial predications, belong to one of the nonessence T-categories.

---

18. For a discussion of this sentence, see Malink (2007: 276–83).

19. Although the sentence might, in principle, also be taken to concern the nine nonsubstance C-categories, it is preferable to take it to concern the nine nonessence T-categories; for these constitute the natural complement to the first T-category, which is referred to by the phrase 'not signify essence' in the sentence at 103b38.

20. The T-categories can be viewed as a classification of terms (Ebert 1985: 114–23) and also as a classification of predications (Frede 1981: 5–9). The two views are not incompatible; for example, one may assume that a term belongs to a given T-category just in case every predication in which this term occurs as the predicate belongs to that T-category (see Alexander *in Top.* 66.7–10). For present purposes, it is useful to regard the T-categories as a classification of terms.

21. Alexander *in Top.* 67.11–16, Granger (1984: 6), and de Rijk (2002: 486); similarly, Smith (1997: 8 and 76).

22. In other words, if there is a B such that A is predicated of B and B is not in the same C-category as A, then A is in a nonessence T-category.

EVERY SUBSTANCE TERM IS AN ESSENCE TERM. It seems clear that substance terms do not belong to any of the nine nonessence T-categories of quantity, quality, relation, and so on. Aristotle would presumably not want to say that terms such as 'man' or 'animal' signify quantity, quality, or another nonessence T-category. Thus, substance terms cannot be predicates of cross-categorial predications, given that every predicate of a cross-categorial predication belongs to a nonessence T-category.[23] In other words, no substance term is predicated of a nonsubstance term.

As Alexander points out (*in Top.* 67.20–2), one may object that there are what he calls unnatural predications such as 'The white is a man', in which substance terms like 'man' seem to be predicated of nonsubstance terms like 'the white'. However, it is questionable whether unnatural predications count as predications in the proper sense. In the *Topics,* Aristotle deals with predications whose predicate is a definition, genus, differentia, proprium, or accident of the subject (S1). Aristotle gives no indication that unnatural predications are predications in this sense. Hence we may accept that substance terms are predicated—in the *Topics*' sense of the predicables—only of substance terms:

S22: If A is in the category of substance and A is predicated of B, then B is in the category of substance

This, in conjunction with S19, entails that every substance term is predicated essentially of itself:

S23: If A is in the category of substance, then A is predicated essentially of A

Consequently, every substance term is an essence term (that is, a term that is the subject of an essential predication).[24] Unlike nonsubstance

23. This means that substance terms are not 'said of another' (*Top.* 1.9 103b37; see Ebert 1985: 134). It is in accordance with some passages in which Aristotle suggests that substance terms are not 'said of another subject' (*Phys.* 1.7 190a36, *APost.* 1.4 73b5–8, 1.22 83a24–32).

24. According to S8 (p. 123), every subject of an essential predication has a genus. Hence every substance term has a genus. This genus is again a substance term (S17, p. 135), and hence has a genus. Now, it seems reasonable

terms, substance terms do not come in pairs consisting of an essence term and a nonessence term.

THE T-CATEGORY OF ESSENCE. Which terms belong to the T-category of essence? First, all substance terms seem to be in the T-category of essence.[25] For, as we have seen, no substance term is in a nonessence T-category. So given that the T-categories are exhaustive, every substance term is in the T-category of essence.

Second, there is reason to think that nonsubstance nouns such as 'justice' and 'whiteness' are in the T-category of essence. As we saw above, Aristotle holds that every term that is the predicate of a cross-categorial predication is in one of the nonessence T-categories. The fact that he does not mention other conditions for being in these categories suggests that the nine nonessence T-categories include no terms other than those that are predicates of cross-categorial predications. Thus, since nonsubstance nouns are not predicates of cross-categorial predications, they do not belong to any of the nonessence T-categories. Given that the T-categories are exhaustive, all nonsubstance nouns are in the T-category of essence.

---

to assume that every genus is distinct from any item of which it is a genus. If so, it follows that there is a sequence of infinitely many distinct items A, B, C, . . . , such that B is a genus of A, C is a genus of B, and so on. But in the *Posterior Analytics* Aristotle denies the existence of such infinite chains of predication (*APost.* 1.22 82b37–83a1, 83b24–31, 84a25–8; cf. also *APr.* 1.27 43a36–7; see Barnes 1994: 174–5). There are several ways to solve this problem. One is to deny that a genus must be distinct from that of which it is a genus and to assume that in some cases a highest genus is a genus of itself. Another solution would be to reject S8 and to replace it by S8*: if there is an A that is predicated essentially of B, then there is a C such that either C is a genus of B or B is a genus of C. This option allows for highest genera that do not themselves have any genus although they are subjects of essential predications. In the derivation of S1–23 above, S8 is only used to establish S9 and S18. It is not difficult to verify that these last two statements can also be established by using S8* instead of S8 (the modified proofs will rely on S3). Thus, the line of reasoning from S1 to S23 would remain intact if S8 is replaced by S8*.

25. Cf. Ebert (1985: 134).

essence	quantity	quality	relation	...	action	passion
'man'	'equal'	'white'	'small'		'destroy'	'frightened'
'equality'						
'whiteness'						
'smallness'						
⋮						
'destruction'						
'fright'						

Table 9.1. The T-categories (*Topics* 1.9)

This is in accordance with the interpretation of T-categories given by John Malcolm and Theodor Ebert, who hold that the T-category of essence includes substance terms such as 'man' and nonsubstance nouns such as 'color' and 'blindness', whereas the nonessence T-categories include nonsubstance paronyms such as 'colored' and 'blind'.[26] Thus, the T-categories—the ten categories introduced in chapter 1.9 of the *Topics*—can be represented as shown in Table 9.1. By contrast, the classification given by the C-categories—the ten categories introduced in chapter 4 of the *Categories*—is shown in Table 9.2.

Every substance term is an essence term (that is, a term that is the subject of an essential predication). More generally, it can be shown that all terms in the T-category of essence are essence terms.[27] At the same time, it can be shown that every essence term is in the T-category of essence.[28] Thus, the T-category of essence includes all and only essence

26. Malcolm (1981: 666) and Ebert (1985: 125–6 and 137–8). As mentioned above (pp. 141–142), the nine nonessence T-categories include not only paronyms but also nonparonymous terms such as 'runner' and 'boxer'.

27. Consider a term that belongs to the T-category of essence. Given that the T-categories are disjoint, this term does not belong to any of the nonessence T-categories. Consequently, this term is not the predicate of a cross-categorial predication (103b37–9) and hence is the subject of an essential predication (S19). Thus, both substance terms and nonsubstance nouns such as 'justice' are essence terms.

28. Consider a term that is the subject of an essential predication. This term is not the predicate of a cross-categorial predication (S19). Given that

terms. In other words, the first T-category includes all and only terms that are the subject of an essential predication.

substance	quantity	quality	relation	. . .	action	passion
'man'	'equal'	'white'	'small'		'destroy'	'frightened'
	'equality'	'whiteness'	'smallness'		'destruction'	'fright'

Table 9.2. The C-categories (*Categories* 4)

REMARK ON DIFFERENTIAE. Every subject of an essential predication belongs to the T-category of essence. In many cases, the predicate of an essential predication, too, will belong to this T-category. For example, 'animal' is predicated essentially of 'man' and belongs to the T-category of essence. However, there is reason to think that not all predicates of essential predications belong to it. This can be seen by considering differentiae of substance terms. A differentia is predicated essentially of that of which it is a differentia.[29] For example, 'biped' is predicated essentially as a differentia of 'man'.[30] At the same time, Aristotle holds that differentiae are nonsubstance terms, belonging to the C-category of quality:

> The differentia always signifies a quality of the genus. (*Top.* 4.6 128a26–7)

> The differentia is generally held to signify a quality. (*Top.* 6.6 144a18–19)[31]

Thus, 'biped' is a nonsubstance term, whereas 'man' is a substance term. So 'biped', being a differentia of 'man', is the predicate of a

---

the nonessence T-categories include only terms that are predicates of cross-categorial predications (103b37–9), the term under consideration does not belong to these categories. Since the T-categories are exhaustive, the term belongs to the T-category of essence.

29. See pp. 116–117 above; cf. also *Top.* 7.3 153a17–18 and 7.5 154a27–8.

30. *Top.* 4.2 122b16–17 and 6.6 144b16–24.

31. Further passages in which differentiae are treated as qualities include *Top.* 4.2 122b16–17, 6.6 144a21–2, *Met.* Δ 14 1020a33–b2, Δ 28 1024b5–6, K 12 1068b19, and *Phys.* 5.2 226a28; see also Code (2010: 94).

cross-categorial predication. More generally, when a differentia is predicated of a substance term, this is a cross-categorial essential predication. Such a differentia is in one of the nonessence T-categories (103b37–9) and hence is not in the T-category of essence. It is the predicate of an essential predication but does not belong to the T-category of essence. Moreover, such a differentia is not the subject of an essential predication; that is, it does not have an essence.[32] Such a differentia also need not be predicated essentially of everything of which it is predicated (see S9). Thus, a term may be predicated essentially as a differentia of a subject while being predicated nonessentially as an accident of another subject.[33] For example, 'four-sided' may be predicated as a differentia of 'square' but as an accident of 'house'.

Aristotle's treatment of differentiae is connected with various problems, partly because he treats them differently in different works. For present purposes, we may set these issues aside. It suffices to note that in the *Topics,* differentiae of substance terms are regarded as nonsubstance terms and as predicates of cross-categorial essential predications.[34]

BACK TO THE MODAL SYLLOGISTIC. We have now completed our discussion of categories in the *Topics* and are going to apply the results to the modal syllogistic. One of the main results is that no essence term is the predicate of a cross-categorial predication (S18). Since the relation of a_X-predication is restricted by the predicables (S14), the same is true for a_X-predication: no essence term is the predicate of a cross-categorial a_X-predication. As we have seen, there are two kinds of essence terms, namely, substance terms (such as 'man') and nonsubstance essence terms

32. Nevertheless, for any differentia (like 'biped') there may be a corresponding term (like 'bipedness') that has an essence. This may help explain the fact that Aristotle sometimes speaks of a definition of differentiae (*Top.* 6.11–12 149a25–33 and *Cat.* 5 3a25–8).

33. See Kung (1977: 367 and 372–4; 1978: 155). By contrast, if something is a genus of something, then it is predicated essentially of everything of which it is predicated (S3, p. 119).

34. For further discussion of this account of differentiae, see Malink (2007: 283–7).

(such as 'whiteness' and 'justice'). None of these terms can be the predicate of a cross-categorial a_X-predication.

Since every subject of an a_N-predication is an essence term (S11), no subject of an a_N-predication is the predicate of a cross-categorial a_X-predication. In other words, S11, S14, and S18 entail

S24: If there is an A such that Aa_NB, and Ba_XC, then B and C are in the same category

As we will see in Chapter 10, S24 can help explain the validity of Barbara NXN. In particular, it can be used to reject some counterexamples to Barbara NXN put forward by Theophrastus and Eudemus; for in each of these counterexamples, the middle term is taken to be both the subject of an a_N-predication and the predicate of a cross-categorial a_X-predication.

10

Essence Terms and Substance Terms

PLAN OF THIS CHAPTER. As we have seen, Aristotle's account of categories in the *Topics* relies on the distinction between essence terms like 'justice' and nonessence terms like 'just'. The purpose of the present chapter is to show that his modal syllogistic, too, relies on it. Aristotle makes use of this distinction in *Prior Analytics* 1.34 to reject a purported counterexample to Celarent NXN. I argue that the distinction can be used in a similar way to reject two of Theophrastus's and Eudemus's counterexamples to Barbara NXN.

I then discuss in detail the role of essence terms in the modal syllogistic. Every subject of an a_N-predication is an essence term (S11). Moreover, it can be shown that essence terms are a_N-predicated of everything of which they are a_X-predicated. This will allow us to reject a further argument of Theophrastus and Eudemus against the validity of Barbara NXN.

Finally, since substance terms are essence terms, they are a_N-predicated of everything of which they are a_X-predicated, and they are not a_X-predicated of nonsubstance terms. With this at hand, we will be able to reject another of Theophrastus's and Eudemus's counterexamples to Barbara NXN.

TWO PUTATIVE COUNTEREXAMPLES TO BARBARA NXN. To justify their claim that Barbara NXN is invalid, Theophrastus and Eudemus gave three counterexamples to this mood. Two of them are as follows (we will consider the other one later in this chapter):[1]

1. Alexander *in APr.* 124.26–30; similarly, Philoponus *in APr.* 124.24–8.

'having knowledge' is a_N-predicated of 'literate'
'literate' is a_X-predicated of 'man'
but 'having knowledge' is not a_N-predicated of 'man'

'moving by means of legs' is a_N-predicated of 'walking'
'walking' is a_X-predicated of 'man'
but 'moving by means of legs' is not a_N-predicated of 'man'

In both counterexamples, the major and middle terms are nonsubstance terms, whereas the minor term is a substance term. The minor premises are presumably not actually true, but it suffices for Theophrastus and Eudemus to assume counterfactually that they are true.

If Aristotle is to defend his endorsement of Barbara NXN, he should be able to explain what is wrong with the two counterexamples. Now, Aristotle does not discuss these two counterexamples in the *Prior Analytics*. But he discusses a similar apparent counterexample to Celarent NXN in *Prior Analytics* 1.34, in one of the few places where modal syllogisms are addressed outside of chapters 1.1–22. Since his treatment of this counterexample can help us see how he would respond to the above two counterexamples to Barbara NXN, let us consider it in some detail.

AN APPARENT COUNTEREXAMPLE TO CELARENT NXN. Chapter 1.34 of the *Prior Analytics* begins as follows:

> Mistakes frequently will happen because the terms in the premise have not been well set out, as, for example, if A is health, B stands for illness, and C for man. For it is true to say that it is not possible for A to belong to any B, for health belongs to no illness, and again that B belongs to every C, for every man is susceptible of illness. It might seem to result, then, that it is not possible for health to belong to any man. (*APr.* 1.34 47b40–48a8)

In this passage, Aristotle introduces a counterexample to Celarent NXN, which is similar to the above two counterexamples to Barbara NXN in that the major and middle terms are nonsubstance terms:[2]

2. Aristotle justifies the minor premise of this counterexample by pointing out that "every man is susceptible of illness" (48a5–6). The modal expression 'susceptible' (δεκτικός) is sometimes taken to indicate that the minor

'health' is e_N-predicated of 'illness'
'illness' is a_X-predicated of 'man'
but 'health' is not e_N-predicated of 'man'

Aristotle rejects this counterexample as follows:

> The cause of this is that the terms are not set out well as a matter of language, since there will not be a deduction if terms according to the condition are substituted, i.e., if healthy is put in place of health and ill in place of illness. For it is not true to say that being healthy cannot belong to the ill; but if this is not assumed no deduction comes about. ... Thus, it is clear that in the case of such premises the term according to the condition must always be substituted for the condition and put as a term. (*APr.* 1.34 48a8–13 and 48a26–8)

In this passage, Aristotle appeals to a distinction between 'condition' (ἕξις) and 'according to the condition' (κατὰ τὴν ἕξιν). The nouns 'health' and 'illness' come under the heading of 'condition', whereas the paronymous terms 'healthy' and 'ill' come under the heading of 'according to the condition'.[3]

In chapter 8 of the *Categories*, Aristotle takes conditions (ἕξεις) to be a special kind of qualities, and he distinguishes qualities from that

premise is an a_Q-proposition instead of an a_X-proposition (Smith 1989: 163, Nortmann 1996: 245). If this were correct, Aristotle would discuss a purported counterexample to Celarent NQN and would thus be ignoring his earlier claim that Celarent NQN is invalid (1.16 35b34–6; for this claim, see Alexander *in APr.* 207.28–34, Moraux & Wiesner 2001: 109, and Ebert & Nortmann 2007: 591). It is therefore preferable to follow those who take the minor premise of Aristotle's counterexample to be an a_X-proposition (Alexander *in APr.* 353.26–8, Maier 1900a: 311, Ross 1949: 403, Mignucci 1969: 476, de Rijk 2002: 583, and Striker 2009: 219). The expression 'susceptible' only indicates that nothing prevents us from assuming counterfactually the truth of an assertoric premise that is not actually true; for a similar use of 'it is possible' (ἐνδέχεται), see *APr.* 1.10 30b35, 1.11 31b6, 31b30, 32a3. At 1.34 48a15–17, Aristotle discusses a counterexample to Cesare NXN that is very similar to the present counterexample to Celarent NXN (cf. Alexander *in APr.* 354.35–355.7, Maier 1900a: 311n6, and Malink 2009b: 392–3).

3. Similarly *Top.* 5.4 133b24–31, 5.6 135b33–136a4, and *EN* 1.13 1103a8–10.

which is "said paronymously according to these" qualities (8b25–7 and 10a27–32). He states that 'whiteness' and 'justice' come under the heading of qualities, whereas 'white' and 'just' come under the heading of paronymous terms. As we saw above, this corresponds to the distinction between nonsubstance essence terms and nonessence terms (p. 141). Thus, Aristotle's rejection of the counterexample to Celarent NXN can be understood in terms of the latter distinction.

Aristotle rejects the counterexample by stating that the terms 'health' and 'illness' must be replaced by 'healthy' and 'ill', respectively. In other words, nonsubstance essence terms should be replaced by nonessence terms. Aristotle does not explain why they should be so replaced, but, as pointed out by Alexander, the reason seems to be that nonsubstance essence terms like 'illness' cannot be truly a_X-predicated of substance terms like 'man'.[4] As we saw above, essence terms cannot be predicates of cross-categorial a_X-predications (pp. 150–151). So if the minor premise of the counterexample is to be true, the middle term cannot be an essence term like 'illness', but should be a nonessence term like 'ill'.

One might try to construe the counterexample in such a way that the middle term is 'ill' and the major term is the essence term 'health'. However, as shown by Alexander, this does not yield a successful counterexample to Celarent NXN.[5] Alternatively, the counterexample might be construed in such a way that the major and middle terms are the nonessence terms 'healthy' and 'ill'. But Aristotle makes it clear that, in this case, the major premise would be false. Aristotle accepts that 'health' is e_N-predicated of 'illness', on the grounds that no illness can be health.[6] But he does not accept that 'healthy' is e_N-predicated of 'ill'. The reason for this seems to be that something that is ill can become

---

4. Alexander *in APr.* 353.16–24 and 354.2–9.

5. One might argue that 'health' is e_N-predicated of 'ill', on the grounds that nonsubstance essence terms like 'health' can be predicated only of nonsubstance essence terms and hence are e_N-predicated of all other terms. But in this case, 'health' would also be e_N-predicated of 'man', so that the counterexample would fail (Alexander *in APr.* 354.15–16).

6. *APr.* 1.34 48a4–5; see Alexander *in APr.* 353.24–6 and 354.9–10.

healthy and vice versa.[7] By contrast, nothing that is an illness can become health.

If the counterexample to Celarent NXN is to be successful, then in order for the major premise to be true, the middle term must be the essence term 'illness'. But in order for the minor premise to be true, the middle term must be the nonessence term 'ill'. This latter term is distinct from the essence term 'illness'. So if both premises are to be true, then there is not a single middle term, and the counterexample must be rejected because of a *quaternio terminorum* (Alexander *in APr.* 354.10–12).

REJECTING THEOPHRASTUS'S AND EUDEMUS'S COUNTEREXAMPLES TO BARBARA NXN. Let us now return to the two counterexamples to Barbara NXN put forward by Theophrastus and Eudemus. Given the above account of a_N- and a_X-predication, these two counterexamples can be rejected in the same way Aristotle rejects the counterexample to Celarent NXN. If their major premise is true, the middle term is the subject of an a_N-predication and hence is the subject of an essential predication (S11). Thus, the middle term is a nonsubstance essence term. On the other hand, if the minor premise is true, the middle term is a_X-predicated of the substance term 'man'. Since nonsubstance essence terms are not a_X-predicated of substance terms (pp. 150–151), the minor premise requires the middle term to be a nonessence term. Thus, the two counterexamples must be rejected because of a *quaternio terminorum,* even though Theophrastus and Eudemus represent the middle term by the same word in both premises (namely, 'literate' and 'walking').

This rejection of the two counterexamples is based on the principle that no subject of an a_N-predication is the predicate of a cross-categorial a_X-predication (S24). This principle can also be used to exclude the following apparent counterexample to Barbara NXN:

'athlete' is a_N-predicated of 'boxer'
'boxer' is a_X-predicated of 'man'
but 'athlete' is not a_N-predicated of 'man'

7. See *APr.* 1.34 48a11–12. See also: "it is possible that the healthy should become ill, and the white should become black" (*Cat.* 10 13a20–1).

One of Theophrastus's and Eudemus's counterexamples assumed, counterfactually, that 'walking' is a_X-predicated of 'man'. The present counterexample assumes that 'boxer' is a_X-predicated of 'man'. As we saw above, Aristotle takes 'boxer' (πυκτικός) to be a nonsubstance term belonging to the category of quality (p. 142). Hence 'boxer' is the predicate of a cross-categorial a_X-predication. At the same time, however, the counterexample assumes that 'boxer' is the subject of an a_N-predication. This contradicts S24.

Given that 'boxer' is a_X-predicated of 'man' or another substance term, it is a nonessence term. In this case, it cannot be the subject of an a_N-predication. Thus, 'athlete' is not a_N-predicated of 'boxer', just as 'healthy' is not e_N-predicated of 'ill', and just as 'moving by means of legs' is not a_N-predicated of 'walking' (if 'walking' is a nonessence term).

PARONYMOUS PAIRS IN *PRIOR ANALYTICS* 1.1–22. Aristotle's discussion in *Prior Analytics* 1.34 shows that the modal syllogistic relies on the distinction between nouns like 'illness' and paronymous terms like 'ill'. However, Aristotle does not seem to pay attention to this distinction in chapters 1.1–22 of the *Prior Analytics*.[8] There, he sometimes switches from a nonsubstance noun to a corresponding paronym within one and the same counterexample.[9] Also, he often uses nonsubstance nouns like 'motion' and 'health' as predicates of a_X-predications whose subject is a substance term.[10] Thus, Aristotle uses nouns where he should have used paronyms. This looseness in his usage of nonsubstance terms may partly be explained by the fact that in the *Prior Analytics* categorical propositions are typically formulated by means of the verb 'belong' (ὑπάρχειν). When categorical propositions are expressed in terms of the verb 'be', it is obviously inappropriate to say "Every man is health."

---

8. There seem to be no systematic rules governing Aristotle's use of 'wakefulness' (31b28, 31b41–32a4) and 'awake' (31b9, 31b31, 38a41–b2); or 'motion' (30a29–30, 30b5–6, 38a42) and 'moving' (32a5, 34b11–12, 34b38); or 'sleep' (the noun, 40a37) and 'sleeping' (31b9).

9. For instance, he switches from 'motion' (30a29–30) to 'moving' (30a31), and from 'wakefulness' (31b28) to 'awake' (31b31).

10. For example, Aristotle takes 'motion' to be a_X-predicated of 'man' (30a29–30), 'wakefulness' of 'animal' (31b28–30), and 'health' of 'horse' (37b37–8).

By contrast, it sounds more appropriate to say "Health belongs to every man." Nevertheless, I take it that if such categorical propositions are true, the predicate really is a paronymous term, even if Aristotle uses a noun.[11]

Aristotle's looseness in *Prior Analytics* 1.1–22 does not mean that the distinction between nonsubstance nouns and paronymous terms plays no role in the modal syllogistic. On the contrary, if I am correct, it plays a crucial role. As argued above, this distinction corresponds to that between nonsubstance essence terms and nonessence terms. When a nonsubstance term is the predicate of a cross-categorial a_X-predication in the *Prior Analytics,* I regard it as a nonessence term. When a nonsubstance term is the subject of an a_N-predication, I regard it as an essence term. In all other cases, I remain neutral as to whether a nonsubstance term is an essence term or a nonessence term.

ANOTHER OBJECTION OF THEOPHRASTUS AND EUDEMUS TO THE VALIDITY OF BARBARA NXN. Alexander informs us that in addition to their counterexamples, Theophrastus and Eudemus put forward the following objection to the validity of Barbara NXN:[12]

> If B belongs to all C but not by necessity, it is possible that B sometime be disjoined from C. But when B has been disjoined from C, A will also be disjoined from it. And if this is so, A will not belong to C by necessity. (Alexander *in APr.* 124.18–21)

Theophrastus's and Eudemus's argument relies on the notion of being at some time disjoined from something. It is not clear whether, and how, this notion can be properly applied to Aristotle's modal syllogistic. It is clear, however, that Theophrastus and Eudemus are assuming that both premises of Barbara NXN can be true while the middle term is not a_N-predicated of the minor term.[13] Although this assumption may be true for their own version of the modal syllogistic, it is not true, I submit, for Aristotle's modal syllogistic. In any case, it is not true within the

11. See Alexander *in APr.* 354.30–1 and 355.14–16; cf. also Thom (1981: 270n3).

12. See also Philoponus *in APr.* 124.11–24.

13. Mueller (1999a: 119n32).

interpretation of the modal syllogistic we have developed so far. This can be seen as follows.

Suppose that A is a_N-predicated of B, and B is a_X-predicated of C. Since B is the subject of an a_N-predication, B is a_N-predicated of itself (S21). So B is a_N-predicated of B and B is a_X-predicated of C. This is an instance of Barbara NXN in which the major term is identical with the middle term. It follows via Barbara NXN that B is a_N-predicated of C (S15).

Thus, S15 and S21 entail

S25: If Aa_NB and Ba_XC, then Ba_NC

In other words, every subject of an a_N-predication is a_N-predicated of everything of which it is a_X-predicated. So if the two premises of Barbara NXN are true, the middle term is a_N-predicated of the minor term. This suffices to reject Theophrastus's and Eudemus's argument against the validity of Barbara NXN.

The minor premise of Barbara NXN is an a_X-proposition of the form 'B belongs to all C'. In conjunction with the major premise, the minor premise entails the corresponding a_N-proposition. This does not mean that the minor premise *is* an a_N-proposition, for the a_X-proposition 'B belongs to all C' and the a_N-proposition 'B necessarily belongs to all C' are two distinct linguistic items. Not every a_X-proposition entails the corresponding a_N-proposition. An a_X-proposition may be true while the corresponding a_N-proposition is false if its predicate is not the subject of an a_N-predication. But within the context given by the a_N-premise of Barbara NXN, the a_X-premise implies the corresponding a_N-proposition. This is in line with the interpretation of Barbara NXN given by Jeroen van Rijen, according to which

> the assertoric premise can be proved to entail its apodeictic variant within the specific context in which it occurred. (van Rijen 1989: 211)

Although van Rijen's account as to why the a_X-premise of Barbara NXN implies its a_N-variant is somewhat different from my own, I agree with him that there is this implication.

ANOTHER PUTATIVE COUNTEREXAMPLE TO BARBARA NXN. As mentioned above, Theophrastus and Eudemus gave three counterexamples to

Barbara NXN. We have already considered two of them; the other one is as follows:[14]

'animal' is a_N-predicated of 'man'
'man' is a_X-predicated of 'moving'
but 'animal' is not a_N-predicated of 'moving'

This counterexample differs from the other two in that the middle term is a substance term. We rejected those two counterexamples by means of the thesis that no subject of an a_N-predication is the predicate of a cross-categorial a_X-predication (S24). The same thesis can also be used to reject the present counterexample. For the middle term 'man' is a substance term, and the minor term 'moving' is a nonsubstance term; so the middle term is the predicate of a cross-categorial a_X-predication while also being the subject of an a_N-predication.

More generally, every substance term is an essence term, that is, a term that is the subject of an essential predication (S23) and hence also the subject of an a_N-predication (S10). Consequently, no substance term is a_X-predicated of a nonsubstance term. Another consequence is that every substance term is a_N-predicated of everything of which it is a_X-predicated (S25). These are two remarkable consequences concerning substance terms. In the remainder of this chapter, I discuss these consequences in more detail and give further justification for them.

SUBSTANCE TERMS ARE NOT A_X-PREDICATED OF NONSUBSTANCE TERMS. In the course of his syllogistic, Aristotle gives numerous examples of terms used in categorical propositions. Among the examples he gives in *Prior Analytics* 1.1–22, I take exactly the following to be substance terms: 'animal', 'man', 'horse', 'raven', 'swan', 'cloak', 'snow', 'pitch', 'stone', 'substance', 'line', 'unit', and 'number'.[15] I take all other terms that appear in chapters 1.1–22 to be nonsubstance terms, for example, 'moving',

14. Alexander *in APr.* 124.24–5.

15. One may doubt whether Aristotle really thought that 'line', 'unit', and 'number' are substance terms. Even if he did not, he may have regarded them as substance terms in the context of the syllogistic. He seems to do this when he assumes that 'substance' is a_X-predicated of 'number' (*APr.* 1.5 27a20).

'health', 'awake', and 'white'. Aristotle often assumes that a nonsubstance term is a_X-predicated of a substance term, for example, that 'white' is a_X-predicated of 'man', and 'moving' of 'animal'.[16]

On the other hand, Aristotle virtually never assumes that a substance term is a_X-predicated of a nonsubstance term. There is only a single passage in *Prior Analytics* 1.1–22 where he seems to accept such an a_X-predication: in chapter 1.19, he writes that "all the awake is animal" (38b1–2). This example is part of a complex argument that constitutes one of the major difficulties of the modal syllogistic. As we will see in Chapter 14, the argument is problematic because it implies that a_N-propositions are compatible with e_Q-propositions. I will argue that it is preferable to reject the a_X-predication between 'animal' and 'awake' and to modify Aristotle's argument in such a way that it does not rely on this a_X-predication (pp. 218–220).

Apart from this passage, Aristotle never assumes that a substance term is a_X-predicated of a nonsubstance term in *Prior Analytics* 1.1–22.[17] Thus, there is a striking contrast between Aristotle's frequent acceptance of a_X-predications such as 'Moving belongs to all animal' and his avoidance of a_X-predications such as 'Animal belongs to all moving'. This contrast cannot be explained by the orthodox *dictum* semantics, on which A is a_X-predicated of B if and only if every individual that falls under B also falls under A. For on this interpretation, there would seem to be no relevant difference between assuming the truth of 'Moving belongs to all animal' and that of 'Animal belongs to all moving'. Rather, the contrast can be explained by the interpretation of the modal syllogistic pursued here, on which Aristotle's notion of a_X-predication is governed by the *Topics'* theory of predication (as described by S14). For on this interpretation, nonsubstance terms can be a_X-predicated of substance terms, but not vice versa.

'ANIMAL BELONGS TO ALL FOOTED'. The thesis that substance terms are not a_X-predicated of nonsubstance terms is largely in accordance with

---

16. For instance, 1.4 26b13–14, 1.5 27b23–7, 27b32–4, 1.6 28b22–4, 1.9 30a29–30, 30b5–6, 1.11 31b28–31, 32a5, 1.18 37b37–8. See also pp. 327 and 329 below.

17. See pp. 327 and 329. Also, Aristotle never assumes that a substance term is a_N-predicated of a nonsubstance term (p. 330).

Aristotle's practice in *Prior Analytics* 1.1–22. However, in *Prior Analytics* 2.2–3, Aristotle assumes that the substance term 'animal' is a_X-predicated of the non-substance term 'footed'.[18] He refers to the former as a genus and to the latter as a differentia. Thus, he seems to accept that a genus of a species can be a_X-predicated of a differentia of that species. In this context, a_X-predication does not meet the requirements imposed by S14, and cannot be identified with the reflexive and transitive closure of a subclass of the *Topics*' notion of predication; for according to the *Topics*' theory of predicables, no genus of a species can be predicated of a differentia of that species (see p. 120).

Now, chapters 2.2–3 deal exclusively with the assertoric syllogistic. Unlike in the modal syllogistic, in the assertoric syllogistic, a_X-predication need not be restricted by the predicables (see p. 130). According to the preorder semantics, any preorder can serve as the primitive relation of a_X-predication in the assertoric syllogistic. In particular, the relation of set-theoretic inclusion in a domain of sets of individuals can serve as a_X-predication. Thus, the nonempty set semantics introduced in Chapter 5 can be viewed as a special instance of the preorder semantics (p. 81). In this semantics, A is a_X-predicated of B if and only if every individual that falls under B falls under A. In this sense, 'animal' is a_X-predicated of 'footed'. Such models of the preorder semantics may be used in the assertoric syllogistic, and they may also be used to account for Aristotle's a_X-predications in chapters 2.2–3. But they cannot be used, I suggest, in the modal syllogistic, since in the modal syllogistic substance terms cannot be a_X-predicated of nonsubstance terms. Given this, a_X-predications like 'Animal belongs to all footed' are ruled out in the modal syllogistic as well as those like 'Man belongs to all moving'.[19]

---

18. *APr.* 2.2 54b4–7 and 2.3 56a26–9.

19. At *APr.* 1.15 34b7–14, Aristotle states that 'man' is not a_X-predicated of 'moving', on the grounds that this would be an a_X-predication not without qualification but 'determined with respect to time' (see pp. 233–238 below). This passage might be used to reject Theophrastus's and Eudemus's counterexample to Barbara NXN, whose minor premise states that 'man' is a_X-predicated of 'moving'. Thus, Huby (2002: 96–7; 2007: 87) suggests that the passage may be a reply to Theophrastus's and Eudemus's counterexample. One might then still assume that 'animal' is a_X-predicated of 'footed', because this a_X-predication would not be 'determined with respect to time'.

UNNATURAL PREDICATION IN THE *POSTERIOR ANALYTICS.* In the modal syllogistic, substance terms cannot be a_X-predicated of nonsubstance terms. This thesis is grounded in the *Topics*' thesis that substance terms cannot be predicated, in the sense of the predicables, of nonsubstance terms (S22).[20] Evidence for this latter thesis is found not only in the *Topics* but also in the *Posterior Analytics.* There, Aristotle distinguishes between what later came to be called natural and unnatural predication.[21] Examples of unnatural predication are 'The white is a man', 'That large is a log', and 'The white is walking'. Examples of natural predications are 'The man is white', 'The man is walking', and 'The log is large'. Aristotle emphasizes that only natural predication is predication in the proper sense; unnatural predication is

> either not predicating at all or else predicating not *simpliciter* but predicating accidentally. (*APost.* 1.22 83a15–17)

In order to better understand the difference between natural and unnatural predication, it is useful to consider Philoponus's account of these two kinds of predication (*in APost.* 235.10–236.22). Philoponus distinguishes between two mutually exclusive and jointly exhaustive classes of

---

However, 34b7–14 is concerned with a specific mood of the problematic syllogistic, namely, Barbara XQM. If Aristotle intended the passage as a defense of Barbara NXN, we would expect it to occur within the apodeictic syllogistic in chapter 1.9. Thus, I will not use the passage to reject Theophrastus's and Eudemus's counterexample.

20. Aristotle holds that substance terms can occur in the definition of nonsubstance terms. For example, 'number' occurs in the definition of 'odd', 'nose' in the definition of 'snub', and 'animal' in the definition of 'female' (*APost.* 1.4 73a37–b3, 22 84a12–17, *Met.* Z 5 1030b16–28; cf. also *APost.* 1.6 74b8–10, and *Phys.* 1.3 186b22–3). He typically expresses this by saying that the substance term is present in (ἐνυπάρχειν) the definition of the nonsubstance term (*APost.* 1.4 73a37–8, 73b1–2, 22 84a15–16, *Met.* Z 1 1028a35–6). However, this does not mean that the substance term is predicated, in the sense of the predicables, of the nonsubstance term. If 'evil' is defined as 'the contrary of good', then 'good' is present in (ἐνυπάρχει) the definition of 'evil' (*Top.* 6.9 147b17–22). But this does not mean that 'good' is predicated of 'evil'. Similarly, the fact that 'animal' is present in the definition of 'female' does not imply that the former is predicated of the latter.

21. *APost.* 1.19 81b25–9 and 1.22 83a1–18; cf. also *APr.* 1.27 43a33–6.

beings: substances and accidents (235.13–15). Thus, the accidents are exactly the nonsubstances. When a substance is said of a substance, this is a natural predication; for example, 'Man is an animal'. When a nonsubstance is said of a substance, this is also a natural predication; for example, 'Man is white'. On the other hand, when a substance is said of a nonsubstance, this is an unnatural predication; for example, 'The white is a log'. The fourth case, when a nonsubstance is said of a nonsubstance, is more complicated. According to Philoponus, some such predications are unnatural while others are natural; for example, 'That baldheaded is white' is an unnatural predication, while 'White is a color' is a natural predication (235.21–236.8). There are various ways to describe the difference between these two subcases. Within the framework introduced above, we may say that the predication is unnatural if the subject is a nonessence term, and natural if the subject is an essence term.

Now, Aristotle denies that unnatural predications can occur in scientific demonstrations (ἀποδείξεις):

> Let us suppose that what is predicated is always predicated without qualification, and not per accidens, of what it is predicated of; for this is the way in which demonstrations demonstrate. (*APost.* 1.22 83a18–21)

Although this passage does not explicitly mention a_X-predication, Jonathan Barnes and David Bostock take it to apply to a_X-predication.[22] If they are right, Aristotle denies that unnatural a_X-predications can occur in demonstrations.

A demonstration is a certain kind of deduction (συλλογισμός): it is a deduction by possessing which we know something.[23] How the modal syllogistic is related to the *Posterior Analytics*' theory of demonstration is a difficult question, and it is not my intention here to enter into a discussion of it.[24] At any rate, the present interpretation of the modal

22. Barnes (1994: 176), Bostock (2004: 150).

23. See *APr.* 1.4 25b28–31 and *APost.* 1.2 71b17–19.

24. Several commentators have argued that there is a close relation between them (Rescher 1964: 169–72; 1974: 8–10; van Rijen 1989: 201–11; Striker 1994: 41). However, unlike the modal syllogistic, the *Posterior Analytics* is not primarily concerned with modally qualified propositions such as

syllogistic is similar to the theory of demonstration in that it excludes unnatural a_X-predications whose predicate is a substance term and whose subject is a nonsubstance term. There are, however, also dissimilarities. Aristotle considers predications such as 'The white is walking' as unnatural (*APost.* 1.22 83a1–17). In this example, 'the white' and 'walking' are presumably nonessence terms rather than essence terms. Thus, Aristotle seems to hold that a predication is unnatural when both terms are nonessence terms. On the other hand, I am assuming that any term is a_X-predicated of itself, including every nonessence term. It follows from this that there are a_X-predications in which both terms are nonessence terms.

SUBSTANCE TERMS CAN BE E_X-, I_X-, O_X-, E_N-, I_N-, AND O_N-PREDICATED OF NONSUBSTANCE TERMS. I have argued that, in the modal syllogistic, substance terms cannot be a_X-predicated of nonsubstance terms. Hence there are no true a_X-propositions whose predicate is a substance term and whose subject is a nonsubstance term. But this does not mean that there are no other true categorical propositions with such predicates and subjects. In the course of the modal syllogistic, Aristotle often assumes that nonsubstance terms are a_X- and a_N-predicated of substance terms (see pp. 329–330). Consequently, his conversion rules commit him to accepting that substance terms are i_X- and i_N-predicated of nonsubstance terms. Moreover, Aristotle explicitly assumes that 'animal' is e_X-, i_X-, o_X-, i_N-, o_N-, and i_Q-predicated of 'white'.[25] These are not unnatural predications. It is an unnatural predication when a substance term is a_X-predicated of a nonsubstance term. But the notion of unnatural predication does not apply to any of the various kinds

---

'A necessarily belongs to all B'. Rather, it is concerned with assertoric propositions such as 'A belongs to all B' which are true by necessity (see Barnes 1994: xxi–ii; 2007: 484–7).

25. See pp. 329–332 below. Also, whenever Aristotle assumes that a nonsubstance term is e_X-, i_X-, i_N-, or i_Q-predicated of a substance term, his conversion rules commit him to accepting also the converse e_X-, i_X-, i_N-, or i_Q-predication. Likewise, Aristotle assumes that 'white' is e_N-predicated of 'raven' (p. 331) and hence given his rule of e_N-conversion, that 'raven' is e_N-predicated of 'white'.

of particular and negative predication Aristotle considers in the modal syllogistic.[26]

SUBSTANCE TERMS ARE A_N-PREDICATED OF WHATEVER THEY ARE A_X-PREDICATED OF. As mentioned above, substance terms are a_N-predicated of everything of which they are a_X-predicated. This is confirmed by a passage from *Posterior Analytics* 1.22, which follows shortly after Aristotle's discussion of unnatural predication:

> That which signifies substance signifies just what or just a subspecies of what is that of which it is predicated.[27] (*APost.* 1.22 83a24–5)

The passage is not as clear as one would like. But the phrase 'that which signifies substance' may be taken to pick out the class of substance terms. Also, the phrase 'exactly what or exactly a subspecies of what' (ὅπερ ἐκεῖνο ἢ ὅπερ ἐκεῖνό τι) seems to refer to essential predications.[28] Consequently, the passage seems to state that substance terms are predicated essentially of everything of which they are predicated. Aristotle does not specify exactly what kinds of predications the phrase 'is predicated' (κατηγορεῖται) refers to in the present passage, but the predications in question may be taken to include a_X-predications. Thus, Jonathan Barnes (1994: 177) takes the passage to state that substance terms are predicated essentially of everything of which they are a_X-predicated. Since essential predication implies a_N-predication (S10), it follows that substance terms are a_N-predicated of everything of which they are a_X-predicated. This is another case where the interpretation of the modal syllogistic pursued here is in accordance with the *Posterior Analytics*' theory of predication and demonstration.

SUMMARY. Our discussion of Barbara NXN has now come to an end. I have argued that the validity of this mood cannot properly be explained by the heterodox apodeictic *dictum de omni* (Chapter 7). Rather, it can

26. *Pace* Striker (1994: 47) and Rini (1998: 560), who hold that if 'animal' is i_X- or i_N-predicated of 'white', this is an unnatural predication.

27. τὰ μὲν οὐσίαν σημαίνοντα ὅπερ ἐκεῖνο ἢ ὅπερ ἐκεῖνό τι σημαίνει καθ' οὗ κατηγορεῖται.

28. For this use of ὅπερ phrases, see Brunschwig (1967: 154–5), Kung (1977: 362), Furth (1988: 45), Barnes (1994: 176), and Bostock (2004: 144).

be explained by means of the *Topics*' theory of predicables and categories (Chapters 8–10). I have suggested that in the modal syllogistic, a_N- and a_X-predication should be treated as primitive relations governed by the *Topics*' theory of predication. This theory is encapsulated in S1–4, S6, S8, S17, S19, and S22. The relations of a_N- and a_X-predication and their connection to the *Topics*' theory of predication are characterized by S10–12, S14, and S16. All other statements of the form S*n* considered above follow logically from the statements just mentioned. In particular, these statements imply that Barbara NXN is valid (S15) and thereby explain why Aristotle took this mood to be valid.

In Chapters 11 and 12, we will have a look at the remaining three perfect NXN-moods, namely, Celarent, Darii, and Ferio NXN.

11

Universal Negative Necessity Propositions

Having considered a_N-predication, we now turn to the relations of e_N-, i_N-, and o_N-predication. Aristotle does not discuss the nature of these relations in any detail. However, his claims of validity and invalidity in the apodeictic syllogistic impose several requirements on them. In particular, they require these relations to validate both the perfect NXN-moods stated in *Prior Analytics* 1.9 and the conversion rules stated in chapter 1.3. It is not obvious whether and how these requirements can be satisfied simultaneously. Some commentators have thought that this is not possible. By contrast, I argue that it is possible. I want to show how e_N-, i_N-, and o_N-predication can be defined in terms of the primitive relations of a_N- and a_X-predication in such a way that they satisfy those requirements. The present chapter will do this for e_N-predication, and Chapter 12 will deal with i_N- and o_N-predication.

THE HETERODOX APODEICTIC *DICTUM DE NULLO*. One of Aristotle's requirements on e_N-predication is that it validate Celarent NXN. Aristotle justifies the validity of this mood in the same passage in which he also justifies the validity of Barbara NXN, as follows:

> Since A belongs or does not belong by necessity to all B and C is one of the Bs, it is evident that one or the other of these will also apply to C by necessity. (*APr.* 1.9 30a21–3)

As argued above, the justification of Barbara NXN in this passage relies on the heterodox apodeictic *dictum de omni:* A is a_N-predicated of B just in case A is a_N-predicated of everything of which B is a_X-predicated

(p. 110). Given this, Aristotle's justification of Celarent NXN should be taken to rely on the heterodox apodeictic *dictum de nullo:* A is e_N-predicated of B just in case A is e_N-predicated of everything of which B is a_X-predicated.

Similarly to what we saw in connection with the heterodox apodeictic *dictum de omni* and Barbara NXN, the heterodox apodeictic *dictum de nullo* entails that Celarent NXN is valid but cannot explain why Aristotle took it to be valid. Now, Aristotle's endorsement of Barbara NXN can be explained by the *Topics'* theory of predication. In what follows, I want to show that this theory can also explain his endorsement of Celarent NXN.

DEFINING E_N-PREDICATION. One of Aristotle's examples of e_N-predication is that 'animal' is e_N-predicated of 'snow' (*Prior Analytics* 1.16 36a30–1). Both 'animal' and 'snow' are substance terms, and they are e_X-predicated of one another. More generally, it seems clear that any two substance terms that are e_X-predicated of one another are also e_N-predicated of one another. For example, e_N-predication holds between any two of the following terms: 'swan', 'cloak', 'snow', 'pitch', 'stone', 'line'.

Aristotle also accepts e_N-predications that hold between nonsubstance terms. Recall his discussion of the apparent counterexample to Celarent NXN in *Prior Analytics* 1.34. As we have seen, this discussion turns on the distinction between essence terms like 'health' and nonessence terms like 'healthy'. Aristotle denies that e_N-predication holds between the nonessence terms 'healthy' and 'ill', but he accepts that it holds between the essence terms 'health' and 'illness' (see pp. 155–156):

> It is not true to say that being healthy cannot belong to the ill. (*APr.* 1.34 48a11–12)

> It is true to say that it is not possible for A [health] to belong to any B [illness], for health belongs to no illness. (*APr.* 1.34 48a3–5)

This suggests that, for Aristotle, any two nonsubstance essence terms that are e_X-predicated of one another are also e_N-predicated of one another. Hence since every substance term is an essence term (p. 146), e_N-predication holds between any two essence terms that are

e_X-predicated of one another. This, I suggest, can be regarded as an exhaustive description of all e_N-predications:[1]

$$Ae_NB \quad \text{if and only if} \quad Ae_XB \text{ and A and B are essence terms}$$

According to this account, e_N-predication can only hold between essence terms. This is in accordance with Aristotle's view that e_N-predication does not hold between the nonessence terms 'healthy' and 'ill'.

Now, essence terms are exactly those terms that are subjects of a_N-predications: A is an essence term just in case there is a C that is a_N-predicated of A.[2] Moreover, given the heterodox assertoric *dictum de nullo,* e_X-predication amounts to the condition that there is nothing of which both terms are a_X-predicated. Thus, the above equivalence can be rewritten as follows:

$$Ae_NB \quad \text{if and only if} \quad \begin{array}{l}\text{for every Z, if } Ba_XZ \text{ then not } Aa_XZ,\\ \text{and there are C and D such that}\\ Ca_NA \text{ and } Da_NB\end{array}$$

This equivalence can be taken as a definition of e_N-predication in terms of a_N- and a_X-predication.[3]

PAIRS OF CONTRARY TERMS. I have suggested that e_N-predication obtains between essence terms such as 'justice' and 'injustice' but not between nonessence terms such as 'just' and 'unjust'. This is in line with some of Aristotle's views about contrariety. In his lost work *On Opposites,* Aristotle took 'justice' and 'injustice' to be contraries but

1. It is somewhat similar to Patterson's (1995: 50) characterization of what he calls strong e_N-predication.

2. By definition, A is an essence term just in case there is a D that is predicated essentially of A (p. 141). This latter condition holds just in case there is a C that is a_N-predicated of A (see S10 and S11).

3. In the context of the problematic syllogistic, we will consider a slightly stronger definition of e_N-predication, which in addition requires that at least one of the two argument-terms be a substance term (see p. 266). In the context of the apodeictic syllogistic, however, we may ignore this complication (see p. 312).

denied that 'just' and 'unjust' are contraries in the proper sense.[4] An explanation of why the last two terms are not contraries may be found in the following passage from the *Categories:*

> All contraries must either be in the same genus or in contrary genera, or be themselves genera. For white and black are in the same genus (since color is their genus), but justice and injustice are in contrary genera (since the genus of one is virtue, of the other vice), while good and bad are not in a genus but are themselves actually genera of certain things. (*Cat.* 11 14a19–25)

This passage implies that in every pair of contrary terms, each term either has a genus or is a genus of something. As we have seen, the nonessence terms 'healthy' and 'ill' are not subjects of essential predications and hence do not have a genus. Moreover, we have seen that 'healthy' and 'ill', being paronymous terms, are not genera of anything (p. 138). Hence these two terms are not contraries. For basically the same reasons, I suggest, they are also not e_N-predicated of one another. In this respect, the above account of e_N-predication fits with Aristotle's account of contrariety.[5]

THE VALIDITY OF CELARENT NXN. The above definition of e_N-predication entails that Celarent NXN is valid. This can be seen as follows:

1. Ae_NB	(major premise)
2. Ba_XC	(minor premise)
3. For every Z, if Ba_XZ then not Aa_XZ	(from 1; by definition of e_N-predication)

4. As reported by Simplicius *in Cat.* 389.5–18.

5. This does not mean that e_N-predication implies contrariety; for example, 'redness' and 'greenness' are e_N-predicated of one another but are not contraries (instead, they are intermediates between the contraries 'white' and 'black'). Conversely, it can be shown that, under certain assumptions, contrariety implies e_N-predication. According to Aristotle, some contraries are highest genera that "are not in a genus," which can be taken to mean that they do not have a genus distinct from themselves. If it is assumed that highest genera are subjects of essential predications even though they have no genus distinct from themselves (see pp. 146–147n24), then contrariety implies e_N-predication as defined above.

4. For every Z, if Ca_XZ then not Aa_XZ (from 2, 3; by Barbara XXX)
5. There are D and E such that Da_NA and Ea_NB (from 1; by definition of e_N-predication)
6. There are D and E such that Da_NA and Ea_NC (from 2, 5; by Barbara NXN)
7. Ae_NC (from 4, 6; by definition of e_N-predication)

The steps in lines 4 and 6 of this proof are justified by assertoric Barbara and by Barbara NXN, along with some rules of classical propositional and quantifier logic. All other steps are justified by the definition of e_N-predication. Thus, this definition allows us to reduce the validity of Celarent NXN to that of assertoric Barbara and of Barbara NXN. Given that the *Topics*' theory of predication explains the validity of Barbara NXN, it thereby also explains the validity of Celarent NXN. This is how I think the *Topics*' theory of predication helps explain why Aristotle took Celarent NXN to be valid.

THE VALIDITY OF E_N-CONVERSION. Aristotle also requires e_N-predication to validate the rule of e_N-conversion: if Ae_NB then Be_NA. He justifies this conversion rule as follows:

> If it is necessary for A to belong to no B, then it is necessary for B to belong to no A; for if it is possible for it to belong to some, then it would be possible for A to belong to some B. (*APr.* 1.3 25a29–32)

Aristotle's argument in this passage seems to rely on the conversion of propositions of the form 'it is possible for A to belong to some B'. Later on in chapter 1.3, however, Aristotle seems to establish this latter conversion by means of e_N-conversion (25a40–b3). Thus, some commentators hold that Aristotle's argument for the validity of e_N-conversion in the present passage is circular.[6] Others hold that Aristotle's arguments for modal conversion rules in chapter 1.3, including the present argument, do not constitute proofs in the proper sense, but are only meant to make

6. Becker (1933: 90), Ross (1949: 294–5).

those conversion rules plausible in a loose way.[7] In any case, it is not clear what Aristotle's argument in the present passage is supposed to be and what assumptions it relies on.

In the assertoric syllogistic, the rule of e_X-conversion is validated by the heterodox assertoric *dictum de nullo*. But the heterodox apodeictic *dictum de nullo* does not validate e_N-conversion. This is not a specific problem of the heterodox interpretation: the orthodox apodeictic *dictum de nullo* does not validate it either. By contrast, the above definition of e_N-predication obviously validates e_N-conversion. For this definition is, on logical grounds, symmetric. Thus, the definition accounts for the validity of both Celarent NXN and the rule of e_N-conversion.

AN ALLEGED AMBIGUITY OF E_N-PROPOSITIONS. It is often thought that the validity of Celarent NXN is in tension with the validity of e_N-conversion. Commentators have argued that Aristotle's asserting the validity of both leads to an ambiguity of e_N-propositions. The alleged ambiguity is usually described in terms of what is known as the *de re* and the *de dicto* readings: the validity of Celarent NXN is taken to require a *de re* reading of e_N-propositions, whereas the validity of e_N-conversion is taken to require a *de dicto* reading.[8]

The distinction between *de re* and *de dicto* readings of Aristotle's categorical propositions is based on two assumptions. The first assumption is that Aristotle endorsed the orthodox *dictum* semantics of assertoric propositions, according to which the quantifiers of the *dictum* semantics range over individuals. For example, an e_X-proposition is taken to be true just in case no individual falls under both the subject and the predicate. The second assumption is that the modality of Aristotle's modalized propositions can be adequately represented by modal sentential operators such as 'it is necessary that'.

Given these two assumptions, there are two natural ways to specify the semantics of e_N-propositions. On one of them, an e_N-proposition is taken to be true just in case for every individual that falls under the subject, it is necessary that this individual does not fall under the predicate. This is the *de re* reading. On this reading, Celarent NXN is valid but

7. Wieland (1980: 111), Ebert & Nortmann (2007: 266).

8. This view is held by the authors mentioned on p. 10n8 above.

e_N-conversion is not valid. To see that e_N-conversion is not valid, suppose that 'moving' is e_X-predicated of 'animal' (an example used by Aristotle in *Prior Analytics* 1.9[9]). In this case, for every individual that falls under 'moving', it is necessary that this individual does not fall under 'animal'. Hence 'animal' is e_N-predicated of 'moving' on the *de re* reading. But it is not the case that for every individual that falls under 'animal', it is necessary that this individual does not fall under 'moving'. Hence 'moving' is not e_N-predicated of 'animal' on the *de re* reading, and the rule of e_N-conversion is invalid on this reading.

The other way of specifying the semantics of e_N-propositions is the *de dicto* reading. According to it, an e_N-proposition is true just in case it is necessary that no individual falls under both the subject and the predicate term. On this reading, e_N-conversion is valid but Celarent NXN is not valid. To see that Celarent NXN is not valid, consider the terms 'healthy' and 'ill'. It is necessary that no individual falls under both of them. Hence 'ill' is e_N-predicated of 'healthy' on the *de dicto* reading. Moreover, we may assume that 'healthy' is a_X-predicated of 'man' (an example used by Aristotle at 1.18 37b35–8). But it is not necessary that no individual falls under both 'man' and 'ill'. Hence 'ill' is not e_N-predicated of 'man' on the *de dicto* reading, and Celarent NXN is invalid on this reading.

In sum, the *de dicto* reading validates e_N-conversion, and the *de re* reading validates Celarent NXN, but neither reading validates both. This leads to the view that e_N-propositions are ambiguous in the modal syllogistic: when Aristotle appeals to the validity of Celarent NXN, he tacitly relies on the *de re* reading, and when he appeals to the validity of e_N-conversion, he relies on the *de dicto* reading.

AVOIDING THE AMBIGUITY. If the *de re* and *de dicto* readings were the only possible interpretations of e_N-propositions, we would have reason to think that e_N-propositions are ambiguous in Aristotle's modal syllogistic. There are, however, numerous alternative interpretations. Several authors have suggested interpretations of e_N-propositions on which both Celarent NXN and e_N-conversion are valid.[10] Another such

9. *APr.* 1.9 30a28–33, 30b5–6; see Alexander *in APr.* 131.4–7.

10. An example is Brenner (2000: 336), whose interpretation can be described as follows: an e_N-proposition is true just in case each individual

interpretation is given by the definition of e_N-predication I suggested above. This interpretation rejects the two assumptions on which the distinction between *de re* and *de dicto* readings of categorical propositions is based: it rejects the orthodox *dictum* semantics, and it does not make use of modal sentential operators. Nevertheless, the above definition of e_N-predication is also similar in some respects to each of the two readings. For example, it is similar to the *de re* reading in that 'healthy' is not e_N-predicated of 'ill'; and it is similar to the *de dicto* reading in that, even if only animals are moving, 'stone' is not e_N-predicated of 'moving'.[11] More generally, the similarities between them can be described as follows.

The *de re* reading of e_N-propositions can be viewed as stating that the predicate term is incompatible with the individuals that fall under the subject term. However, it does not state that the predicate term is incompatible with the subject term itself. The *de dicto* reading, on the other hand, can be viewed as stating that the predicate term is incompatible with the subject term, without stating that it is incompatible with the individuals that fall under the subject term.

Now, the definition of e_N-predication I suggested above implies that everything of which the predicate term is a_X-predicated is e_N-predicated

---

concept that falls under the predicate term is incompatible with each individual concept that falls under the subject term. It is obvious that e_N-conversion is valid on this interpretation. Celarent NXN is also valid, given that an a_X-proposition is true just in case every individual concept that falls under the subject term falls under the predicate term. A somewhat similar interpretation of e_N-propositions is given by Thom (1991: 149; 1993: 202). Others accept the *de re* reading of e_N-propositions and stipulate the validity e_N-conversion by means of certain restrictions on the set of admissible models (Johnson 1989: 274–5; 2004: 272; Thomason 1993: 123). Another strategy is to assume a complex set-theoretic interpretation of e_N-propositions that makes both Celarent NXN and e_N-conversion valid (Johnson 1993: 172; 1995: 3; Thom 1996: 146). Nortmann (1996: 122–6) proposes a solution in terms of modal propositional and quantifier logic (though he does not give a uniform interpretation of e_N-propositions; see p. 16n12 above).

11. Likewise, the above account of a_N-predication is similar to the *de re* reading of a_N-propositions in that 'healthy' is not a_N-predicated of itself; and it is similar to the *de dicto* reading in that, even if only animals are moving, 'animal' is not a_N-predicated of 'moving'.

of—and hence incompatible with—everything of which the subject term is a_X-predicated. In particular, since every term is a_X-predicated of itself, the predicate term itself is incompatible with the subject term itself. In this respect, the above definition of e_N-predication is similar to the *de dicto* reading. At the same time, terms such as 'man' can be taken to be a_X-predicated of categorical singular terms such as 'Socrates' or 'Kallias', which stand for individuals.[12] According to the above definition of e_N-predication, the predicate term is incompatible with every categorical singular term of which the subject term is a_X-predicated. Thus the predicate term may be taken to be incompatible with every individual that falls under the subject term. In this respect, the above definition of e_N-predication is similar to the *de re* reading.

Hence the present interpretation of e_N-propositions contains both *de dicto*–like aspects and *de re*–like aspects. Accordingly, the interpretation validates both Celarent NXN and e_N-conversion and thereby avoids the alleged ambiguity of e_N-propositions.

---

12. Cf. pp. 48 and 83 above.

12

Particular Necessity Propositions

Finally, let us consider the relations of i_N- and o_N-predication in Aristotle's apodeictic syllogistic. I begin with i_N-predication, adopting what is known as the disjunctive strategy for defining i_N-predication. With regard to o_N-predication, a major problem is Aristotle's claim that Baroco XNN and Bocardo NXN are invalid. As we will see, the invalidity of these two moods is in some tension with Aristotle's use of the method of ecthesis in the apodeictic syllogistic.

THREE REQUIREMENTS ON I_N-PREDICATION. Aristotle does not mention an apodeictic *dictum de aliquo* that would characterize the relation of i_N-predication. Nor is it obvious what such an apodeictic *dictum de aliquo* would look like. However, Aristotle implicitly characterizes i_N-predication by imposing three requirements on it in the apodeictic syllogistic:

(i) The rule of i_N-conversion should be valid (1.3 25a32–4)
(ii) Aa_NB should imply Ai_NB (1.3 25a32–4)
(iii) Darii NXN should be valid (1.9 30a37–30b1)

The question is: how can i_N-predication be defined in such a way that all three requirements are satisfied?

A TENTATIVE DEFINITION OF I_N-PREDICATION. The heterodox apodeictic *dictum de omni* states that A is a_N-predicated of B just in case A is a_N-predicated of everything of which B is a_X-predicated. Correspondingly, the heterodox apodeictic *dictum de aliquo* might be taken to state that

A is i_N-predicated of B just in case A is a_N-predicated of *something* of which B is a_X-predicated:

$$\text{A}i_N\text{B} \quad \text{if and only if} \quad \text{for some Z, B}a_X\text{Z and A}a_N\text{Z}$$

On this tentative definition, i_N-predication would meet requirement (ii); for if A is a_N-predicated of B, then due to the reflexivity of a_X-predication, there is something (namely, B itself) of which B is a_X-predicated and A is a_N-predicated. Moreover, i_N-predication would on this tentative definition meet requirement (iii), the validity of Darii NXN. This can be seen as follows:

1. Aa_NB	(major premise)
2. Bi_XC	(minor premise)
3. For some Z, Ca_XZ and Ba_XZ	(from 2; by heterodox assertoric *dictum de aliquo*)
4. For some Z, Ca_XZ and Aa_NZ	(from 1, 3; by Barbara NXN)
5. Ai_NC	(from 4)

The step in line 4 is justified by Barbara NXN, along with rules of classical propositional and quantifier logic. The step in line 5 relies on the tentative definition of i_N-predication. On this tentative definition, however, i_N-predication would fail to meet requirement (i), the validity of i_N-conversion; for if A is a_N-predicated of something of which B is a_X-predicated, it does not follow that B is a_N-predicated of something of which B is a_X-predicated.

THE DISJUNCTIVE DEFINITION OF I_N-PREDICATION. Aristotle justifies the rule of i_N-conversion as follows:

> And if A belongs to all or to some B by necessity, then it is necessary for B to belong to some A; for if it is not necessary, then neither would A belong to some B by necessity. (*APr.* 1.3 25a32–4)

Similarly to what we saw above in connection with Aristotle's justification of e_N-conversion, it is not clear exactly what his argument in the present passage is supposed to be. The passage offers little explanation as to why i_N-conversion is valid. Because of this, commentators have suggested alternative ways to verify its validity. A common strategy is to build conversion as it were into the definition of i_N-predication, either by disjunction or by conjunction. Thus, A is taken to be i_N-predicated of B just in case B is predicated assertorically of something of which A is predicated necessarily *or* / *and* vice versa. The conjunctive strategy (that is, the *and* option) has difficulties in accounting for the validity of Darii NXN.[1] Many authors therefore prefer the disjunctive strategy.[2] In accordance with this strategy, I adopt the following disjunctive definition of i_N-predication:

Ai_NB	if and only if	for some Z (Ba_XZ and Aa_NZ) or
		for some Z (Aa_XZ and Ba_NZ)

The definiens of this disjunctive definition is weaker than that of the earlier tentative definition. Consequently, this definition continues to validate Darii NXN and the implication from Aa_NB to Ai_NB. Moreover, it also validates i_N-conversion, for the disjunctive definition is symmetric. On the disjunctive definition, then, i_N-predication meets all three requirements mentioned above.

The disjunctive definition implies that i_N-propositions may be true even if the predicate term is not a_N-predicated of anything. It suffices that the subject term is a_N-predicated of something of which the predicate term is a_X-predicated. For example, if 'moving' is a_X-predicated of something of which 'animal' is a_N-predicated, then 'moving' is i_N-predicated of 'animal'. This is confirmed by Aristotle's claim that the mood Darapti XNN in the third figure is valid; for if 'moving' is a_X-predicated of something of which 'animal' is a_N-predicated, it follows by Darapti XNN that 'moving' is i_N-predicated of 'animal'.

In addition to Darapti XNN, Aristotle also endorses Darapti NXN. His endorsement of these two moods implies that he took the above disjunctive definiens to be at least a *sufficient* condition for i_N-predication.

1. Authors who adopt the conjunctive strategy have to make additional stipulations to guarantee the validity of Darii NXN (Johnson 1989: 274; 2004: 272; Thomason 1993: 116).

2. Thom (1991: 146; 1993: 202; 1996: 146), Johnson (1993: 172; 1995: 3), Brenner (2000: 336), and Schmidt (2000: 43).

My suggestion is that he took it also to be a *necessary* condition for i_N-predication. Put another way, the suggestion is that he endorsed the above disjunctive definition of i_N-predication and relied on it as a criterion to determine whether a given mood or conversion rule is valid. This accounts for all of his claims about i_N-predication in the apodeictic syllogistic.

I_N-PREDICATION AND SUBSTANCE TERMS. As I argued above, substance terms are a_N-predicated of everything of which they are a_X-predicated (p. 160). Consequently, given the disjunctive definition of i_N-predication, i_X-predication implies i_N-predication whenever at least one of the two argument-terms is a substance term. In other words, if an i_X-proposition is to be true while the corresponding i_N-proposition is false, both argument-terms must be nonsubstance terms. Now, there are only three passages in the modal syllogistic where Aristotle needs an i_X-proposition to be true while the corresponding i_N-proposition is false. In all of them, both terms are nonsubstance terms: 'Motion belongs to some white', 'Wakefulness belongs to some biped', and 'Biped belongs to some wakefulness'.[3] This is in accordance with our disjunctive definition of i_N-predication.

3. As we saw above (pp. 149–150), Aristotle regards 'biped' as a nonsubstance term, even though it is predicated essentially of substance terms. The first proposition mentioned above is from a counterexample to Darii XNN (*APr.* 1.9 30b5–6; cf. Alexander *in APr.* 134.28–31); the other two propositions are from counterexamples to Datisi XNN and Disamis NXN (1.11 31b27–33; cf. Alexander *in APr.* 149.1–4 and 149.13–20). In all three propositions, the predicate is the major term of the counterexample, and the subject is the minor term. In all three counterexamples, the premise pair presumably implies an i_X-conclusion. But the corresponding i_N-proposition must be false in order to establish the invalidity of the moods in question. However, there seems to be a problem with the last two i_X-propositions. In the counterexamples in which they occur, Aristotle needs to assume that 'biped' is not a_N-predicated of anything of which 'wakefulness' is a_X-predicated (although 'biped' is presumably i_X-predicated of 'wakefulness'). Otherwise, it would follow by Darapti NXN that 'biped' is i_N-predicated of 'wakefulness'. In any case, Aristotle could replace his counterexamples to Datisi XNN and Disamis NXN by the unproblematic counterexample he gave to Darii XNN, using the terms 'motion', 'animal', and 'white'.

FOUR REQUIREMENTS ON o_N-PREDICATION. Let us now turn to the relation of o_N-predication. Aristotle imposes four requirements on this relation in the apodeictic syllogistic, two of them concerning validity and two concerning invalidity:

(i) o_N-conversion should be invalid (1.3 25a34–6)
(ii) Ferio NXN should be valid (1.9 30b1–2)
(iii) Baroco NNN and Bocardo NNN should be valid (1.8 30a6–14)
(iv) Baroco XNN and Bocardo NXN should be invalid (1.10 31a15–17 and 1.11 32a4–5)

The question is again: how can o_N-predication be defined in such a way that all these requirements are satisfied?

A TENTATIVE DEFINITION OF o_N-PREDICATION. The heterodox apodeictic *dictum de nullo* states that A is e_N-predicated of B just in case A is e_N-predicated of everything of which B is a_X-predicated. Correspondingly, the heterodox apodeictic *dictum de aliquo non* might be taken to state that A is o_N-predicated of B just in case A is e_N-predicated of *something* of which B is a_X-predicated:

$$Ao_NB \quad \text{if and only if} \quad \text{for some Z, } Ba_XZ \text{ and } Ae_NZ$$

On this definition, o_N-predication would meet requirements (i)–(iii); but, as we will see shortly, it would fail to meet requirement (iv), the invalidity of Baroco XNN and Bocardo NXN. To see why it would fail to meet this requirement, it is helpful first to have a look at Aristotle's proof of the validity of Baroco NNN and Bocardo NNN. We will then see that, given the above tentative definition, this proof can be modified so as to establish also the validity of the former two moods.

PROOF BY ECTHESIS OF BAROCO NNN AND BOCARDO NNN. Aristotle holds that Baroco NNN and Bocardo NNN are valid. Unlike the other valid NNN-moods in the second and third figures, these two moods cannot be proved to be valid solely by means of conversion rules and the four perfect NNN-moods. Instead, Aristotle establishes their validity by means of the method of ecthesis. When applied to an o_N-proposition Ao_NB, the step of ecthesis consists, roughly speaking, in setting out an item to which B 'belongs' and to which A 'necessarily does not belong'. Aristotle describes the ecthetic proofs of Baroco NNN and Bocardo NNN as follows:

> It is necessary for us to set out something to which each of the two terms [that is, the predicate term of the o_N-premise of Baroco NNN and Bocardo NNN] does not belong, and produce the deduction about this. For it will be necessary in application to these; and if it is necessary of what is set out, then it will be necessary of something of that former term; for what is set out is just a certain 'that'. Each of these deductions occurs in its own figure. (*APr.* 1.8 30a9–14)

According to this passage, each of the two ecthetic proofs involves an auxiliary deduction. In the last sentence of the passage, Aristotle states that this auxiliary deduction is in the same figure as the mood whose validity is to be proved. Thus, the auxiliary deduction used in the proof of Baroco NNN is in the second figure, and that used in the proof of Bocardo NNN is in the third figure. It is widely assumed that the former auxiliary deduction is Camestres NNN, and the latter is Felapton NNN.[4] Accordingly, the step of ecthesis is taken to yield an e_N-proposition that can serve as a premise in these two deductions. The ecthetic proof of Baroco NNN can then be reconstructed as follows:

1. Ba_NA	(major premise)
2. Bo_NC	(minor premise)
3. For some Z, Ca_XZ and Be_NZ	(from 2)
4. Ca_XD and Be_ND	(from 3; by existential instantiation)
5. Ca_XD and Ae_ND	(from 1, 4; by Camestres NNN)
6. Ao_NC	(from 5; by Felapton NXN)

4. Alexander *in APr.* 121.26–122.16 and 123.10–18 (see Mueller 1999a: 118n23), Pacius (1597: 143–4), Maier (1900a: 106n1), Jenkinson (1928: ad 30a14), Ross (1949: 317), Wolff (1998: 150–3), Mueller (1999b: 13 and 62), and Drechsler (2005: 217); similarly, Patzig (1968: 166–7).

The corresponding proof for Bocardo NNN is:

1. Ao_NB	(major premise)
2. Ca_NB	(minor premise)
3. For some Z, Ba_XZ and Ae_NZ	(from 1)
4. Ba_XD and Ae_ND	(from 3; by existential instantiation)
5. Ca_ND and Ae_ND	(from 2, 4; by Barbara NXN)
6. Ao_NC	(from 5; by Felapton NNN)

Several details in these proofs could be reconstructed in a slightly different way than I have done here. Setting aside these details, the important step in both proofs is in line 3. Accepting this step is, for Aristotle, tantamount to accepting the above tentative definition of o_N-predication.[5] Thus, the tentative definition justifies the step in line 3 and hence validates Baroco NNN and Bocardo NNN.

PROOF BY ECTHESIS FOR BAROCO XNN AND BOCARDO NXN? Unfortunately, the above two ecthetic proofs can easily be modified so as to yield proofs of Baroco XNN and Bocardo NXN. The auxiliary deductions employed in the modified proofs are Camestres XNN and Felapton NXN instead of the corresponding NNN-moods. The modified proof for Baroco XNN is as follows:[6]

1. Ba_XA	(major premise)
2. Bo_NC	(minor premise)
3. For some Z, Ca_XZ and Be_NZ	(from 2)
4. Ca_XD and Be_ND	(from 3; by existential instantiation)

5. Strictly speaking, the step in line 3 is only tantamount to accepting that Ao_NB implies the existence of a Z such that Ba_XZ and Ae_NZ. But the converse implication is guaranteed by Felapton NXN, which Aristotle holds to be valid.

6. Similar ecthetic proofs of Baroco XNN are given by Alexander *in APr.* 144.23–145.4 (see Mueller 1999a: 130n153), Henle (1949: 99n27), van Rijen (1989: 195), and Thom (1993: 198; 1996: 133).

5. $\text{Ca}_\text{X}\text{D}$ and $\text{Ae}_\text{N}\text{D}$ (from 1, 4; by Camestres XNN)

6. $\text{Ao}_\text{N}\text{C}$ (from 5; by Felapton NXN)

Similarly, the modified proof of Bocardo NXN is[7]

1. $\text{Ao}_\text{N}\text{B}$	(major premise)
2. $\text{Ca}_\text{X}\text{B}$	(minor premise)
3. For some Z, $\text{Ba}_\text{X}\text{Z}$ and $\text{Ae}_\text{N}\text{Z}$	(from 1)
4. $\text{Ba}_\text{X}\text{D}$ and $\text{Ae}_\text{N}\text{D}$	(from 3; by existential instantiation)
5. $\text{Ca}_\text{X}\text{D}$ and $\text{Ae}_\text{N}\text{D}$	(from 2, 4; by Barbara XXX)
6. $\text{Ao}_\text{N}\text{C}$	(from 5; by Felapton NXN)

In both proofs, the steps in lines 5 and 6 are justified by moods that Aristotle explicitly states to be valid. So if Aristotle accepted the step in line 3, that is, if he accepted our tentative definition of o_N-predication, he would be committed to the validity of Baroco XNN and Bocardo NXN. Since Aristotle denies the validity of these two moods, he is committed to rejecting the tentative definition of o_N-predication.

Given this, however, the above reconstructions of his ecthetic proofs of Baroco NNN and Bocardo NNN are incorrect, since the step in line 3 is not justified. It is unclear whether these two ecthetic proofs can be reconstructed in such a way that they cannot be transformed into unwanted proofs of Baroco XNN and Bocardo NXN. Thus, it is unclear whether Aristotle's ecthetic proofs of the two NNN-moods are consistent with his claim that the two XNN- and NXN-moods are invalid. I am not in a position to offer a positive answer to this question. On the other hand, it must be emphasized that Aristotle's *assertion* of the validity of the two NNN-moods is consistent with the claim that the two XNN- and NXN-moods are invalid. This is proved in Appendix B. It is only his ecthetic *proofs* of the two NNN-moods of which it is doubtful whether

7. Similar ecthetic proofs of Bocardo NXN are given by Alexander *in APr.* 151.22–8 (see Mueller 1999a: 134n205) and Thom (1993: 198; 1996: 133).

they are consistent with the latter claim. For present purposes, however, we may set this issue aside and focus instead on Aristotle's claim that the two XNN- and NXN-moods are invalid.

ARISTOTLE'S COUNTEREXAMPLES TO BAROCO XNN AND BOCARDO NXN. Aristotle claims that the invalidity of Baroco XNN can be established by means of a counterexample consisting of the terms 'animal', 'man', and 'white'.[8] However, it is not clear how these terms can be arranged so as to yield a suitable counterexample to Baroco XNN. Alexander and others argue that the terms are not suitable to establish the invalidity of that mood.[9]

Aristotle also gives a counterexample to establish the invalidity of Bocardo NXN (1.11 32a4–5):

'biped' is o_N-predicated of 'animal'
'moving' is a_X-predicated of 'animal'
but 'biped' is not o_N-predicated of 'moving'

In this counterexample, 'biped' is o_N-predicated of 'animal' but not of 'moving', although 'moving' is a_X-predicated of 'animal'. Aristotle does not explain why 'biped' is not o_N-predicated of 'moving' under these circumstances. In the absence of such an explanation, his counterexample does not really help us see why he regarded Bocardo NXN as invalid. Thus, Aristotle does not give a satisfactory justification for his claim that Baroco XNN and Bocardo NXN are invalid. Moreover, this claim is in tension with his treatment of the two corresponding NNN-moods.

---

8. *APr.* 1.10 31a17 (referring back to the terms used to establish the invalidity of Camestres NXN and Baroco NXN at 30b33–8 and 31a14–15).

9. Alexander *in APr.* 144.4–15, Thom (1991: 147–8; 1996: 133–4), Johnson (2004: 277). By contrast, van Rijen (1989: 195–6) and Schmidt (2000: 75–6) try to construct a suitable counterexample out of Aristotle's terms. Schmidt takes 'white' to be the major term, 'man' the middle term, and 'animal' the minor term. Thus, he assumes that 'man' is a_X-predicated of 'white'. This is incompatible with the account of the modal syllogistic pursued in this study, according to which substance terms cannot be a_X-predicated of nonsubstance terms (see p. 160).

Alexander expressed puzzlement over the claim.[10] Others think that it is simply a mistake on Aristotle's part.[11]

By contrast, I attempt not to treat the claim as a mistake, but to integrate it into a coherent interpretation of Aristotle's modal syllogistic. Later on (p. 267), I propose a definition of o_N-predication that meets the four requirements (i)–(iv), including the invalidity of Baroco XNN and Bocardo NXN. The existence of such a definition shows that Aristotle's treatment of the Barocos and Bocardos in the apodeictic syllogistic is not logically inconsistent. However, my definition of o_N-predication will be somewhat complex and technical; it provides more a technical ad hoc solution than an independently motivated notion of o_N-predication. Thus, even though Aristotle's treatment of the apodeictic Barocos and Bocardos is logically consistent, it may still seem to be erratic and confused. In the remainder of this chapter, I argue that this is not so but that his treatment of these moods is well motivated by a criterion based on certain monotonicity properties of N-propositions.

MONOTONICITY OF X-PROPOSITIONS WITH RESPECT TO A_X-PREDICATION. Let me first explain the monotonicity properties of assertoric propositions. According to assertoric Barbara, a true a_X-proposition remains true when the subject term is replaced by a term of which it is a_X-predicated. In other words, a_X-propositions are downward monotonic in the subject term with respect to a_X-predication.[12] Also, a true a_X-proposition remains true when the predicate term is replaced by a term that is a_X-predicated of it. In other words, a_X-propositions are upward monotonic in the predicate term with respect to a_X-predication.

Similarly, i_X-predications are upward monotonic in both arguments, as indicated by Darii and Disamis. On the other hand, e_X-propositions are downward monotonic in both arguments, as indicated by Camestres

10. Alexander *in APr.* 144.23–145.4 and 151.22–30.

11. Henle (1949: 99n 27), Sainati (1988: 45), van Rijen (1989: 195–8), Thom (1991: 147–8; 1993: 195; 1996: 132–5), and Patterson (1995: 85; 162 and 263).

12. Monotonicity is related to the traditional notion of distribution: an argument-term of an assertoric proposition is called 'distributed' just in case it is downward monotonic with respect to a_X-predication (see van Eijck 1985: 16–17, van Benthem 1986: 111, and Hodges 1998: 226–7).

and Celarent. The monotonicity properties of o_X-propositions are indicated by Baroco and Bocardo: o_X-propositions are downward monotonic in the predicate term and upward monotonic in the subject term:

Aa_XB	Ai_XB	Ae_XB	Ao_XB
↑↓	↑↑	↓↓	↓↑
Barbara XXX	Darii XXX	Camestres XXX	Baroco XXX
Barbara XXX	Disamis XXX	Celarent XXX	Bocardo XXX

Table 12.1. Monotonicity of X-propositions with respect to a_X-predication

MONOTONICITY OF N-PROPOSITIONS WITH RESPECT TO A_N-PREDICATION. In *Prior Analytics* 1.8, Aristotle states that an NNN-mood is valid just in case the corresponding purely assertoric mood is valid. This means that N-propositions have the same monotonicity properties with respect to a_N-predication as X-propositions with respect to a_X-predication:

Aa_NB	Ai_NB	Ae_NB	Ao_NB
↑↓	↑↑	↓↓	↓↑
Barbara NNN	Darii NNN	Camestres NNN	Baroco NNN
Barbara NNN	Disamis NNN	Celarent NNN	Bocardo NNN

Table 12.2. Monotonicity of N-propositions with respect to a_N-predication

As we saw above, Aristotle establishes the validity of Baroco NNN and Bocardo NNN by means of the method of ecthesis. It is not clear whether these ecthetic proofs can be constructed in such a way that they cannot be transformed into proofs of Baroco XNN and Bocardo NXN. Aristotle may or may not have had an answer to this question. If not, he may be taken to stipulate rather than demonstrate the validity of Baroco NNN and Bocardo NNN in *Prior Analytics* 1.8. Wanting NNN-moods to be perfectly parallel to XXX-moods, he would simply stipulate the existence of suitable ecthetic proofs for these two moods rather than really construct such proofs within a well-defined system of ecthesis.

MONOTONICITY OF N-PROPOSITIONS WITH RESPECT TO A_X-PREDICATION. Let us now consider the monotonicity properties of N-propositions with respect to a_X-predication. These monotonicity properties are determined by Aristotle's claims about NXN- and XNN-moods in *Prior Analytics* 1.9–11. Some of these moods are held to be invalid by Aristotle (indicated by italics). Hence N-propositions fail to be monotonic with respect to a_X-predication in some positions (indicated by an asterisk):

Aa_NB	Ai_NB	Ae_NB	Ao_NB
⋆ ↓	⋆⋆	↓↓	⋆⋆
Barbara XNN	*Darii XNN*	Camestres XNN	*Baroco XNN*
Barbara NXN	*Disamis NXN*	Celarent NXN	*Bocardo NXN*

Table 12.3. Monotonicity of N-propositions with respect to a_X-predication

As is clear from Table 12.3, monotonicity is lost in most positions, but it is retained in three: in the subject term of a_N-propositions, and in the subject and predicate terms of e_N-propositions. If Baroco XNN and Bocardo NXN were valid, o_N-propositions would be monotonic with respect to a_X-predication in both the subject and the predicate terms. Thus, if we can explain why Aristotle did not take o_N-propositions to be monotonic with respect to a_X-predication, we can also explain why he took Baroco XNN and Bocardo NXN to be invalid.

EXPLAINING THE MONOTONICITY PROPERTIES OF N-PROPOSITIONS. Consider the three positions in which monotonicity is preserved with respect to a_X-predication. According to the interpretation of a_N- and e_N-propositions I suggested above, each of these positions requires the terms that occupy it to be essence terms: the subject of any true a_N-proposition is required to be an essence term, and so are the subject and predicate terms of any true e_N-proposition. By contrast, none of the two terms in a true i_N-proposition is required to be an essence term according to the above disjunctive definition of i_N-predication. Similarly, we may take it that none of the two terms in a true o_N-proposition is required to be an essence term.[13] If so, then the three positions in which monotonicity of N-propositions is preserved with respect to a_X-predication are exactly those positions in which terms are required to be essence terms. Given the rule of e_N-conversion, these are exactly those positions in which terms are required to be the subject of a universal (as opposed to particular) N-predication.

Now, essence terms are a_N-predicated of everything of which they are a_X-predicated.[14] As a result, downward monotonicity of an essence term with respect to a_X-predication is equivalent to downward monotonicity with respect to a_N-predication. Since each of the three positions under consideration is downward monotonic, the monotonicity properties of N-propositions can thus be explained as follows. In principle, N-propositions are monotonic only with respect to a_N-predication, not with respect to a_X-predication. But in some positions, monotonicity with respect to a_N-predication is equivalent to monotonicity with respect to a_X-predication. These are downward monotonic positions in which terms are required to be essence terms. So it is only in these positions that N-propositions are monotonic with respect to a_X-predication. It follows from this that Baroco NNN and Bocardo NNN are valid and that Baroco XNN and Bocardo NXN are invalid.

Thus Aristotle's treatment of the apodeictic Barocos and Bocardos can be explained by appealing to the monotonicity properties of N-propositions. The explanation is based on the assumption that the subject terms of true universal N-propositions are required to be essence terms. This assumption, in turn, might be grounded in the view that the subject terms of true universal (but not particular) N-propositions are required to be subjects of modal predications in an especially strict sense and therefore need to be essence terms.

On this explanation, then, Aristotle's treatment of the apodeictic Barocos and Bocardos appears more reasonable and coherent. If I am correct, his treatment is not determined by investigating whether or

13. See the definition of o_N-predication given on p. 267 below.

14. See S25, p. 159 (in conjunction with S10, p. 125).

not there are ecthetic proofs for these moods within a well-defined system of ecthesis for the apodeictic syllogistic, nor is it determined by investigating whether or not there are counterexamples to them within a well-defined framework for constructing counterexamples. Rather, it is determined by considerations as to what monotonicity properties N-propositions should have.

CONCLUSION. We have now completed our discussion of Aristotle's apodeictic syllogistic. I have suggested an explanation of why Aristotle took Barbara NXN to be valid, based on the *Topics*' theory of predicables and categories (Chapters 8–10). Both a_N- and a_X-predication, I have argued, should be viewed as primitive relations governed by this theory. These two primitive relations suffice to define e_N- and i_N-predication in such a way as to verify Aristotle's claims in the apodeictic syllogistic. I have argued that Aristotle endorsed these definitions of e_N- and i_N-predication and used them as a criterion to determine whether a given mood or conversion rule is valid or not. Thus, the resulting interpretation can explain most of Aristotle's claims of validity and invalidity in the apodeictic syllogistic. For all moods and conversion rules that do not involve an o_N-proposition, the interpretation explains why Aristotle took them to be valid or invalid, respectively.

I have also offered a tentative definition of o_N-predication. The resulting interpretation of Aristotle's apodeictic syllogistic matches all of his claims about the validity and invalidity of moods and conversion rules except his claim that Baroco XNN and Bocardo NXN are invalid. This last problem is addressed in Chapter 18, where I offer a revised definition of o_N-predication on which the two moods in question are invalid (and everything else remains as it is). The result will be an interpretation that matches all of Aristotle's claims of validity and invalidity in the apodeictic syllogistic.

III

The Problematic Syllogistic

AN OVERVIEW OF THE PROBLEMATIC SYLLOGISTIC. Aristotle's apodeictic syllogistic is followed by what is known as the problematic syllogistic (*Prior Analytics* 1.3 and 1.13–22). The problematic syllogistic is concerned with possibility propositions, that is, with propositions that contain a modal qualifier such as 'possibly'. Aristotle distinguishes two kinds of possibility propositions, traditionally referred to as one-sided and two-sided possibility propositions. Being two-sided possible means being neither impossible nor necessary, while being one-sided possible simply means being not impossible. Thus two-sided possibility precludes necessity, whereas one-sided possibility does not. For example, the statement 'Possibly no man is a horse' is true if understood as a one-sided possibility proposition, but not if understood as a two-sided possibility proposition. In modern notation, two- and one-sided possibility propositions are usually indicated by the letters 'Q' and 'M', respectively. Thus the four kinds of two- and one-sided possibility propositions are represented as follows:

$\text{Aa}_{\text{Q}}\text{B}$	A two-sided-possibly belongs to all B
$\text{Ae}_{\text{Q}}\text{B}$	A two-sided-possibly belongs to no B
$\text{Ai}_{\text{Q}}\text{B}$	A two-sided-possibly belongs to some B
$\text{Ao}_{\text{Q}}\text{B}$	A two-sided-possibly does not belong to some B

$\text{Aa}_{\text{M}}\text{B}$	A one-sided-possibly belongs to all B
$\text{Ae}_{\text{M}}\text{B}$	A one-sided-possibly belongs to no B
$\text{Ai}_{\text{M}}\text{B}$	A one-sided-possibly belongs to some B
$\text{Ao}_{\text{M}}\text{B}$	A one-sided-possibly does not belong to some B

Aristotle usually does not explicitly indicate whether a given possibility proposition is of the two-sided or one-sided kind, but in most cases it is clear from the context which of the two he means. Q-propositions prevail in his modal syllogistic. M-propositions occur only in specific contexts, mainly as conclusions of moods whose validity is established by indirect proofs (*reductio ad absurdum*).

In *Prior Analytics* 1.3, Aristotle states the conversion rules for possibility propositions. For M-propositions and affirmative Q-propositions, the conversion rules are entirely parallel to those for assertoric propositions.[1] For negative Q-propositions, they are different: Aristotle holds that $\mathrm{Ae_QB}$ cannot be converted to $\mathrm{Be_QA}$, but that $\mathrm{Ao_QB}$ can be converted to $\mathrm{Bo_QA}$. This is because Aristotle takes negative Q-propositions to be equivalent to affirmative ones, so that $\mathrm{Ae_QB}$ is equivalent to $\mathrm{Aa_QB}$, and $\mathrm{Ao_QB}$ to $\mathrm{Ai_QB}$ (*Prior Analytics* 1.13).

Like the assertoric and apodeictic syllogistic, the problematic syllogistic is based on a number of perfect moods, namely, on the four standard first-figure moods of the form QQQ, QXQ, and QNQ (*Prior Analytics* 1.14–16). Aristotle also accepts as valid some imperfect first-figure moods. Examples are the four standard first-figure moods of the form XQM, and Celarent and Ferio NQX (1.15–16). He establishes the validity of these moods by means of indirect proofs, among which the proofs for the first-figure XQM-moods are especially complex. The moods Aristotle accepts as valid in the second and third figures are mostly proved in a straightforward way by means of the first-figure moods and conversion rules (1.17–22).

The problematic syllogistic also includes numerous claims about the invalidity of moods and the inconcludence of premise pairs (where inconcludence means that a given premise pair does not yield any conclusion in the figure to which it belongs). As we will see, some of these claims are troublesome and not easy to account for.

PLAN OF THE PART. Aristotle's problematic syllogistic raises many questions, partly logical and partly interpretive in character. It is beyond

1. This is to say, $\mathrm{Aa_MB}$ and $\mathrm{Ai_MB}$ can each be converted to $\mathrm{Bi_MA}$, and $\mathrm{Ae_MB}$ can be converted to $\mathrm{Be_MA}$; $\mathrm{o_M}$-propositions are not convertible. Also, $\mathrm{Aa_QB}$ and $\mathrm{Ai_QB}$ can each be converted to $\mathrm{Bi_QA}$.

the scope of this study to address them all. Instead, I focus on one central question, namely, whether the problematic syllogistic is consistent. More precisely, I consider whether the body of Aristotle's claims of validity, invalidity, and inconcludence in the problematic syllogistic is consistent. This is a substantive question, and there is prima facie evidence that the answer might be negative. Nevertheless, my aim is to explore the conditions for an interpretation on which the body of these claims is consistent. I specify some requirements that such an interpretation must meet, and based on this construct an example of such an interpretation. Unlike in the discussion of the apodeictic syllogistic, I do not attempt to give a full explanation of why Aristotle made the claims of validity, invalidity, and inconcludence he did make (though I sometimes offer partial explanations). Thus, I do not attempt to explain how these claims might be motivated by the *Topics*' theory of predication introduced above. Although it would, of course, be desirable to have such an explanation, I am not in a position to give one. Instead, I take on the more modest task of gathering Aristotle's claims, examining their various consequences, and organizing them into a coherent model.

I begin by identifying some places where the problematic syllogistic might seem to be inconsistent. These concern principles of modal opposition, that is, principles of incompatibility and contradictoriness between possibility propositions and necessity propositions. It is clear from some passages in the problematic syllogistic that Aristotle accepts some such principles, for example, the following:

Aa_QB	is incompatible with	Ao_NB
Ai_QB	is incompatible with	Ae_NB

At the same time, however, Aristotle's claims of invalidity and inconcludence commit him to denying other principles of modal opposition. For example, as we will see, he is committed to denying the following:

Ae_MB	is contradictory to	Ai_NB
Ae_QB	is incompatible with	Aa_NB

Although these two principles are very plausible, they are violated by some of Aristotle's claims of invalidity and inconcludence in the problematic syllogistic. Thus, his claims conflict with principles of modal opposition whose validity one would naturally expect him to endorse. As a result, the modal syllogistic seems to be inconsistent. By contrast, I argue that we should not attribute to Aristotle these and the other principles of modal opposition that he is committed to denying, but should regard them as invalid in his modal syllogistic. On this interpretation, his treatment of modal opposition is unusual and asymmetric, but not inconsistent (Chapter 13).

We will also consider how Aristotle justifies the claims of inconcludence that commit him to denying those principles of modal opposition. Although his justification fails fully to explain his asymmetric treatment of modal opposition, it does provide some clues as to what led him to it (Chapter 14). Recognition of this asymmetry will enable us to specify an adequate deductive system for the whole of Aristotle's modal syllogistic (Chapter 15).

Next, we will consider Aristotle's treatment of XQM-moods in *Prior Analytics* 1.15. I will argue that his claim of the validity of Ferio XQM commits him to a principle about the realization of Q-predications. This principle states that the possibility indicated by certain i_Q-predications is at least partially realized. More precisely, it states that if two terms are i_Q-predicated of each other and one of them is a substance term, then they are also i_X-predicated of each other (Chapter 16).

Based on this principle and the asymmetry in modal opposition, we will finally be in a position to specify an interpretation of Aristotle's possibility propositions. This will require some additional technical resources. In addition to the two primitive relations of a_X- and a_N-predication used so far, we will employ a third primitive relation, namely, a strengthened version of a_N-predication that is incompatible with o_M-predication and whose subject term is required to be a substance term. These three primitive relations will suffice to formulate interpretations of Aristotle's Q-propositions (Chapter 17), and of his M- and negative N-propositions (Chapter 18). I will call the resulting interpretation the predicable semantics of the modal syllogistic. It will be shown in Appendix B that the predicable semantics verifies all of Aristotle's claims of validity, invalidity, and inconcludence in the modal syllogistic. Thus the body of these claims is proved to be consistent.

13

Modal Opposition

I begin by considering evidence that suggests that Aristotle endorses a number of principles of modal opposition. I then show that, although these principles are intrinsically plausible, Aristotle's claims of invalidity and inconcludence in the problematic syllogistic commit him to denying some of them. This leads to various interpretive problems, for which I suggest solutions.

TWO-SIDED AND ONE-SIDED POSSIBILITY. At the beginning of the problematic syllogistic, in *Prior Analytics* 1.13, Aristotle introduces the distinction between two- and one-sided possibility. He regards the former as the primary notion of possibility, and the latter as a secondary notion resting on an equivocation of the term 'possible':

> I say 'to be possible' and 'possible' of that which is not necessary but through which nothing impossible will result if it is put as being the case. For it is only equivocally that we say that what is necessary is possible. (*APr.* 1.13 32a18–21)

In the first sentence of this passage, Aristotle characterizes two-sided possibility. Given that something is impossible just in case something "impossible will result if it is put as being the case," Aristotle states that something is two-sided possible just in case it is neither necessary nor impossible. In the second sentence of the passage, Aristotle briefly characterizes the secondary notion of one-sided possibility, which includes what is necessary: something is one-sided possible just in case it is not impossible.

Two-sided possibility is so called because it is, as it were, bounded at two sides by impossibility and necessity, whereas one-sided possibility is only bounded at one side by impossibility:

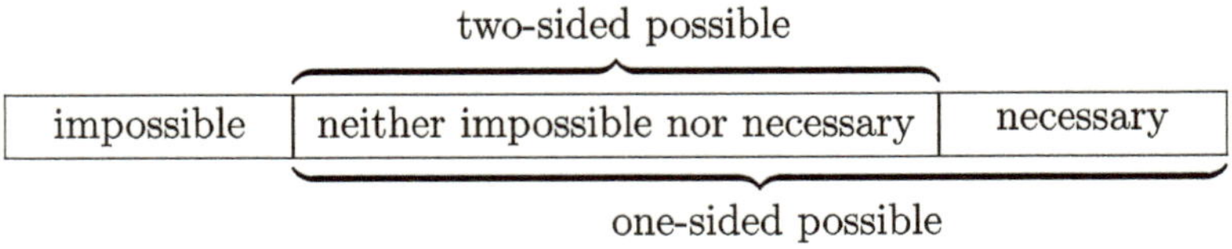

The passage just quoted is followed by a further explanation of one-sided possibility:

> That this [that is, one-sided possibility] is what is possible is evident from opposed pairs of denials and affirmations. For 'it is not possible to belong' and 'it is impossible to belong' and 'it is necessary not to belong' are either the same or follow from one another, and thus their opposites 'it is possible to belong' and 'it is not impossible to belong' and 'it is not necessary not to belong' will also either be the same or follow from one another; for either the affirmation or the denial is true of everything. (*APr.* 1.13 32a21–8)

It is sometimes thought that this passage is not about one-sided, but about two-sided possibility.[1] However, it is difficult to make sense of the passage on that assumption. For the passage evidently states that the three expressions on the left are mutually equivalent, and that so are the three expressions on the right:

'it is not possible to belong'	'it is possible to belong'
'it is impossible to belong'	'it is not impossible to belong'
'it is necessary not to belong'	'it is not necessary not to belong'

These equivalences are valid for one-sided but not for two-sided possibility.[2] The passage should therefore be taken to be about one-sided

1. Waitz (1844: 403), Becker (1933: 11), Ross (1949: 327), Hintikka (1973: 32), Smith (1989: 125), Nortmann (1996: 162–3), Huby (2002: 92), and Ebert & Nortmann (2007: 474–5).

2. Consequently, some authors who take the passage to be about two-sided possibility regard it as an inept interpolation (Becker 1933: 11–14, Ross

possibility.[3] From Aristotle's point of view, the secondary and somewhat artificial notion of one-sided possibility is more in need of explanation than the primary and more natural two-sided notion. Thus Aristotle is arguing that the one-sided notion is a reasonable notion of possibility as well as the two-sided one.

He does so by inferring the equivalence of the three expressions on the right in the above table from the equivalence of the three expressions on the left. The latter equivalence is apparently more obvious to Aristotle than the former. Aristotle takes the two expressions in each row of the table to constitute a contradictory pair of affirmation and denial. For example, 'it is not possible to belong' and 'it is possible to belong' are contradictory. Thus, of the two sentences 'It is not possible for A to belong to all B' and 'It is possible for A to belong to all B' one must be true and the other false. Given these relations of contradictoriness, the equivalence of the three expressions on the left implies the equivalence of their contradictories on the right. In particular, it implies that 'it is possible to belong' is equivalent to 'it is not impossible to belong'.

This last equivalence is a defining feature of the one-sided use of 'possible' as opposed to the two-sided use. Thus, Aristotle's argument shows that there are good reasons to adopt the uncommon one-sided use of 'possible'. It thereby introduces and establishes the secondary, one-sided notion of possibility.

THE PRINCIPLES OF M-N-CONTRADICTORINESS. Aristotle's argument just described is based on the assumption that 'it is possible to belong' is contradictory to 'it is necessary not to belong'. The assumption presumably

---

1949: 327–8, Seel 1982: 163, Smith 1989: 125–6, Huby 2002: 92). Others deny, implausibly, that the passage states equivalences between those modal expressions (Nortmann 1996: 162–6, Ebert & Nortmann 2007: 474–5, Striker 2009: 18 and 128–9; similarly, Hintikka 1973: 33).

3. Authors who hold this view include Gohlke (1936: 66), Hintikka (1977: 81), Buddensiek (1994: 21–3), and Thom (1996: 13). By contrast, the sentence "Therefore, what is possible will not be necessary and what is not necessary will be possible" at 32a28–9 is again about two-sided possibility (see Maier 1900a: 140n1 and Buddensiek 1994: 21–23). The phrase 'not necessary' in the second half of this sentence can be understood as shorthand for 'neither necessarily being nor necessarily not being', which is tantamount to 'neither necessary nor impossible' (Hintikka 1973: 32–4).

implies that 'it is possible to belong to all' is contradictory to 'it is necessary not to belong to all'. This suggests that a_M-propositions are contradictory to o_N-propositions, that is, that of any two propositions of the form Aa_MB and Ao_NB, exactly one must be true and the other false.

Similar results can be obtained when 'belong' is substituted by 'belong to some', 'belong to none', and 'not belong to some'. Thus, Aristotle's argument in *Prior Analytics* 1.13 suggests the following four principles of what we may call M-N-contradictoriness:[4]

Aa_MB	is contradictory to	Ao_NB
Ai_MB	is contradictory to	Ae_NB
Ae_MB	is contradictory to	Ai_NB
Ao_MB	is contradictory to	Aa_NB

Although Aristotle does not explicitly assert these principles in the modal syllogistic, there is further evidence for them in his indirect proofs. These proofs involve a step of assuming, for *reductio ad absurdum,* the contradictory of the desired conclusion. So if the desired conclusion is an M-proposition, the proof involves the contradictory of an M-proposition. Aristotle typically expresses the contradictories of M-propositions by means of the same phrases he uses to express N-propositions. For example, the contradictory of an e_M-proposition is expressed by 'necessarily belong to some', and the contradictory of an o_M-proposition is expressed by 'necessarily belong to all' (see 1.15 34b27–31 and 1.21 39b34–6).

Conversely, the contradictory of an e_N-proposition is expressed by 'possibly belong to some' (1.3 25a30–1). Similarly, M-propositions are expressed by phrases that are typically used to refer to the contradictories of N-propositions. For example, o_M-propositions are expressed by 'not to all by necessity', and e_M-propositions by 'to none by necessity' (1.15 33b29–32, 34b27–31). In sum, then, there seems to be good evidence that Aristotle accepts the above four principles of M-N-contradictoriness.[5]

4. See Buddensiek (1994: 29–30) and Thom (1996: 13).

5. Authors who hold this view include McCall (1963: 35–7), Buddensiek (1994: 29–31), Thom (1996: 13–15 and 157–60), and Schmidt (2000: 102–4).

THE PRINCIPLES OF Q-N-INCOMPATIBILITY. Two-sided possible is that which is neither necessary nor impossible, and one-sided possible is that which is not impossible. Thus everything two-sided possible is also one-sided possible. This suggests that Q-M-subordination is valid, that is, that every Q-proposition implies the corresponding M-proposition. If so, then the above principles of M-N-contradictoriness imply several principles of what we may call Q-N-incompatibility. For example, they imply that a_Q-propositions are incompatible with o_N-propositions, that is, that no two propositions of the form Aa_QB and Ao_NB can be simultaneously true.

As mentioned above, Aristotle holds that affirmative Q-propositions are equivalent to negative Q-propositions.[6] Given this equivalence, the principles of Q-N-incompatibility can be written as follows:

Aa/e_QB	is incompatible with	Ao_NB
Aa/e_QB	is incompatible with	Ai_NB
Ai/o_QB	is incompatible with	Ae_NB
Ai/o_QB	is incompatible with	Aa_NB

These principles are supported by a passage from *Prior Analytics* 1.17, in which Aristotle states that e_Q-propositions are incompatible with 'necessarily belong to some' and with 'necessarily not belong to some'.[7]

APPLYING Q-N-INCOMPATIBILITY IN INDIRECT PROOFS. Some of the above principles of Q-N-incompatibility are used by Aristotle in indirect proofs

---

The principles are also accepted by Angelelli (1979: 182) and Johnson (2004: 265).

6. Aristotle justifies this equivalence at *APr.* 1.13 32a29–b1, arguing that what is two-sided possible is not necessary and may therefore not be the case. He infers from this that for two-sided possibility, 'possible to belong' is equivalent to 'possible not to belong' and hence that 'possible to belong to all' is equivalent to 'possible to belong to none', and 'possible to belong to some' is equivalent to 'possible not to belong to some'.

7. *APr.* 1.17 37a15–30; cf. Alexander *in APr.* 225.22–7, Thom (1996: 15), and Mueller (1999b: 33). This passage may be taken to state that e_Q-propositions are contradictory to the disjunction of 'necessarily belong to some' and 'necessarily not belong to some'; cf. Becker (1933: 25–7), Mignucci (1969: 362), Smith (1989: 135–6), and Buddensiek (1994: 33).

in the problematic syllogistic. Specifically, they are used to establish the validity of NQX-moods that have an e_N-premise and a negative assertoric conclusion, such as Celarent and Ferio NQX. Consider, for example, Aristotle's proof of Ferio NQX:

> If it is not possible for A to belong to any of the Bs and possible for B to belong to some C, then it is necessary for A not to belong to some of the Cs. For if it belongs to all C but it is not possible for A to belong to any B, then neither is it possible for B to belong to any A. Consequently, if A belongs to all C, then it is not possible for B to belong to any of the Cs. But it was assumed to be possible for it to belong to some. (*APr.* 1.16 36a34–9)

In the first sentence of this passage, Aristotle states the e_N-premise, the i_Q-premise, and the o_X-conclusion of Ferio NQX. Next he assumes, for *reductio,* the contradictory of the o_X-conclusion. Given Aristotle's claims about contradictory pairs of assertoric propositions, this contradictory is an a_X-proposition (see p. 32 above). This proposition is combined with the converted version of the e_N-premise so as to obtain the premise pair of Celarent NXN, as follows:

1. Ae_NB	(major premise)
2. Bi_QC	(minor premise)
3. Aa_XC	(assumption for *reductio*)
4. Be_NA	(from 1; by e_N-conversion)
5. Be_NC	(from 3, 4; by Celarent NXN)

Aristotle takes the proposition in line 5 to be incompatible with the premise in line 2, thereby assuming that e_N-propositions are incompatible with i_Q-propositions. Based on this principle of Q-N-incompatibility, the indirect proof is completed, and the validity of Ferio NQX is established.

Aristotle's proofs of other NQX-moods are based on similar principles of Q-N-incompatibility. In particular, Aristotle employs the following principles:[8]

8. See McCall (1963: 83), Wieland (1975: 83–4), Angelelli (1979: 193–4), and Thom (1996: 17). The incompatibility of Aa_QB and Ao_NB is used in the proof of Celarent NQX (1.16 36a10–15) and of Cesare NQX (1.19 38a21–5).

Aa/e_QB	is incompatible with	Ae_NB
Aa/e_QB	is incompatible with	Ao_NB
Ai/o_QB	is incompatible with	Ae_NB

It is worth noting that all these principles involve negative N-propositions. Aristotle does not employ any principles of Q-N-incompatibility involving affirmative N-propositions in the modal syllogistic.[9]

ARISTOTLE IS COMMITTED TO DENYING SOME PRINCIPLES OF M-N-CONTRADICTORINESS AND Q-N-INCOMPATIBILITY. As we have seen, Aristotle accepts as valid several NQX-moods with a negative N-premise. On the other hand, he denies the validity of any NQX-moods with an affirmative N-premise, such as Barbara and Darii NQX. This denial leads to problems. Many of the moods in question could be proved to be valid by means of principles of modal opposition (that is, of M-N-contradictoriness and Q-N-incompatibility). The proofs would proceed by *reductio ad absurdum,* much like Aristotle's indirect proof of Ferio NQX. For example, Darii NQX could be proved to be valid as follows:

1. Aa_NB	(major premise)
2. Bi_QC	(minor premise)
3. Ae_XC	(assumption for *reductio*)
4. Ci_QB	(from 2; by i_Q-conversion)
5. Ao_MB	(from 3, 4; by Ferio XQM)

In line 3, the contradictory of the i_X-conclusion of Darii NQX is assumed for *reductio.* Given Aristotle's claims about contradictory pairs of assertoric propositions, this contradictory is an e_X-proposition (p. 32). The

Curiously, this passage contains a proof not only of Celarent NQX but also of the subaltern mood Celaront NQX (cf. Alexander *in APr.* 208.22–5). The proof of this latter mood is based on the incompatibility of Aa_QB and Ae_NB.

9. *Pace* Thom (1996: 70), who thinks that the incompatibility of Ae_QB and Ai_NB is used in an indirect proof of Ferio QNM at 1.16 36a39–b2. However, this passage is not about Ferio QNM but about Ferio QNQ, a perfect mood whose justification does not involve an indirect proof (see Alexander *in APr.* 212.29–213.11, Ross 1949: 348, Smith 1989: 234, and Mueller 1999b: 67).

conversion rule applied in line 4 is explicitly endorsed by Aristotle, and so is the mood Ferio XQM applied in line 5. Thus, if the o_M-proposition in line 5 is contradictory to or incompatible with the a_N-premise in line 1, the indirect proof would be successful, and Darii NQX would be valid.

If Aristotle accepted that o_M-propositions are incompatible with a_N-propositions, he would also have to accept the validity of Darii NQX. His claim that this mood is invalid therefore commits him to the view that o_M-propositions are not incompatible with, and hence not contradictory to, a_N-propositions. This means that for some A and B, both Ao_MB and Aa_NB are true.

In the same way, Aristotle's claims about the invalidity of other NQX- and QNX-moods commit him to denying further principles of M-N-contradictoriness. In particular, Aristotle is committed to the following:[10]

Ae_MB	is not incompatible with	Aa_NB
Ae_MB	is not incompatible with	Ai_NB
Ao_MB	is not incompatible with	Aa_NB

Moreover, Aristotle is committed to denying some of the principles of Q-N-incompatibility mentioned above. This follows from his treatment of the premise pairs Be_QA, Ba_NC and Ba_NA, Be_QC. We may label these premise pairs ea-2-QN and ae-2-NQ, respectively, where the number 2 indicates that they are in the second figure. Aristotle claims that these two premise pairs are inconcludent, that is, that they do not yield any conclusion in the figure to which they belong. This means that neither premise pair gives rise to a valid mood in the second figure.

The inconcludence of ea-2-QN implies that the mood eao-2-QNX is invalid (that is, the mood in which an o_X-conclusion is inferred from

10. Aristotle holds that the following moods are invalid: Darii, Darapti, Disamis, and Datisi of the form NQX and QNX. But if Ae_MB was incompatible with Aa_NB, then Darapti QNX and Darapti NQX would be provable via Celarent XQM (see Striker 2009: 168). If Ae_MB was incompatible with Ai_NB, then Darii QNX, Datisi QNX, and Disamis NQX would be provable via Celarent XQM. Finally, if Ao_MB was incompatible with Aa_NB, then Darii NQX, Disamis QNX, and Datisi NQX would be provable via Ferio XQM.

ea-2-QN in the second figure). However, if e_Q-propositions were incompatible with a_N-propositions, this mood could be proved to be valid by *reductio*, as follows:

1. Be_QA	(major premise)
2. Ba_NC	(minor premise)
3. Aa_XC	(assumption for *reductio*)
4. Be_QC	(from 1, 3; by Celarent QXQ)

Similarly, aeo-2-NQX could be proved to be valid via Barbara NXN. Thus, Aristotle's claim that ea-2-QN and ae-2-NQ are inconcludent commits him to denying that e_Q-propositions are incompatible with a_N-propositions. This means that for some B and C, both Be_QC and Ba_NC are true. In sum, Aristotle is committed to the following:[11]

Aa/e_QB	is not incompatible with	Aa_NB
Aa/e_QB	is not incompatible with	Ai_NB
Ai/o_QB	is not incompatible with	Aa_NB

It is worth noting that all principles of modal opposition that Aristotle is committed to denying concern affirmative N-propositions. None of them concerns negative N-propositions.

IS THE MODAL SYLLOGISTIC INCONSISTENT? The fact that Aristotle's claims in the modal syllogistic commit him to denying some principles of modal opposition is puzzling. The principles in question are very plausible in themselves, and the default assumption should be that Aristotle accepted them. But if he accepted them, his modal syllogistic would involve serious errors and would be inconsistent. Many commentators have argued that this is in fact the natural conclusion to draw from

11. If Ae_QB was incompatible with Ai_NB, then eao-2-QNX would be provable via Darapti NXN, and aee-2-NQX would be provable via Darii NXN. Similarly, eio-2-QNX would be provable via Celarent QXQ. Moreover, if Ao_QB was incompatible with Aa_NB, then eae-2-QNX would be provable via Ferio QXQ, and aeo-2-NQX would be provable via Felapton QXQ. Similarly, aoo-2-NQX would be provable via Barbara NXN.

Aristotle's claim that ea-2-QN and ae-2-NQ are inconcludent.[12] This conclusion can be avoided only if Aristotle is not taken to accept all principles of modal opposition. This is the only way to try to understand the body of his claims about validity and invalidity in the modal syllogistic as consistent and correct. It is therefore appropriate, I think, to explore in more detail what an interpretation of the modal syllogistic would look like on which those principles of modal opposition are not valid.

There is clear evidence that in his proofs of moods like Celarent NQX, Aristotle employs principles of modal opposition for *negative* N-propositions. These principles are not violated by any of his claims in the modal syllogistic. Otherwise we would have strong reason to think that the modal syllogistic is inconsistent. The principles of modal opposition that are violated by Aristotle's claims all concern *affirmative* N-propositions. Thus Aristotle may be taken to accept principles of modal opposition for negative, but not for affirmative, N-propositions. On this view, Aristotle's treatment of modal opposition is unusual and asymmetric but not necessarily inconsistent.

It is true that there are a number of passages in which Aristotle seems to rely on principles of modal opposition for affirmative N-propositions. We have already discussed some such passages, and we will encounter more of them shortly. My purpose in the rest of this chapter is to explore whether and how these passages can be interpreted in such a way that they do not rely on principles of modal opposition for affirmative N-propositions. As we will see, such interpretations are not always straightforward. But they help us determine what the costs are of an interpretation on which the modal syllogistic is consistent. Whether these costs are higher than those of taking the modal syllogistic to be inconsistent remains for the reader to judge.

WHAT ARE THE CONTRADICTORIES OF NEGATIVE M-PROPOSITIONS? As mentioned above, many of Aristotle's indirect proofs in the problematic syllogistic involve contradictories of M-propositions. Now, on the

12. McCall (1963: 93), Patterson (1995: 194–8), Thom (1996: 128–9), Nortmann (1996: 279), Ebert & Nortmann (2007: 655 and 667–8), and Striker (2009: 161); similarly, Johnson (2004: 303).

interpretation I wish to explore here, principles of M-N-contradictoriness are not valid for affirmative N-propositions. Consequently, the contradictories of negative M-propositions cannot be identified with affirmative N-propositions. For example, the contradictory of an o_M-proposition cannot be identified with an a_N-proposition. It is true that Aristotle expresses the contradictory of o_M-propositions by means of the phrase 'necessarily belong to all' (see p. 198). But when he does so, I submit, this phrase does not express an a_N-proposition, but merely the contradictory of an o_M-proposition. While this contradictory may be similar in meaning to an a_N-proposition, it is not equivalent to it. Likewise, when Aristotle expresses the contradictory of an e_M-proposition by 'necessarily belong to some', this phrase should not be taken to indicate an i_N-proposition. One might say that within the language used to talk about the modal syllogistic, Aristotle helps himself to principles of modal opposition that are not valid in the modal syllogistic itself.[13]

In the assertoric syllogistic, the contradictory of every assertoric proposition can be identified with another X-proposition. For example, Aristotle identifies the contradictory of an o_X-proposition with the corresponding a_X-proposition, and so on (see p. 32 above). In the modal syllogistic, however, the contradictories of negative M-propositions cannot be identified with N-propositions. At least, they cannot be so identified if the modal syllogistic is to be consistent. More generally, they cannot be identified with any of the sixteen kinds of categorical propositions discussed in the modal syllogistic (four kinds for each of the four modalities X, N, M, and Q). Thus, when Aristotle appeals to contradictories of negative M-propositions in indirect proofs, these contradictories are not among his sixteen kinds of categorical propositions. They are merely contradictories of negative M-propositions. They occur in indirect proofs of moods that have an M-conclusion, but not elsewhere in the modal syllogistic.

As a result, Aristotle's indirect proofs of these moods cannot be represented by means of the usual sixteen kinds of categorical propositions. To represent them, the language must be extended to include contradictories of M-propositions. There is, however, a question about the syntactic

13. Similarly, Wieland (1980: 115).

status of these contradictories. One option would be to assume that they are categorical propositions just as the other sixteen kinds of propositions. As such, they would be simple declarative sentences consisting of a subject term, a predicate term, and a copula. For example, the contradictory of an o_M-proposition would be obtained by applying a copula '$\overline{o_M}$', say, to two terms A and B, yielding $A\overline{o_M}B$. Likewise, the contradictory of an e_M-proposition would be the categorical proposition $A\overline{e_M}B$. On this view, Aristotle recognizes altogether twenty kinds of categorical propositions (sixteen plus the contradictories of the four kinds of M-propositions). An alternative option would be to deny that the contradictories of M-propositions are categorical propositions. Instead, they might be taken to be complex propositions that result from prefixing an M-proposition with a phrase such as 'it is not the case that'.

For our purposes, it is not necessary to give a precise account of the syntax of the contradictories of M-propositions. In what follows, I indicate them by formulae such as 'not Ao_MB' and 'not Ae_MB', which are meant to be neutral about their syntactic status. For example, when Aristotle gives an indirect proof of a mood with an o_M-conclusion, I indicate the assumption for *reductio* by 'not Ao_MB'.

AN AMBIGUITY OF 'NECESSARY'. The term 'necessary' and its cognates are used in different ways in Aristotle's modal syllogistic. First of all, there is a distinction between a relative sense indicating the necessity of deductive inferences (*necessitas consequentiae*) and an absolute sense specifying the modality of a single proposition. But on the interpretation pursued here, we need to make a further distinction within the absolute sense: sometimes 'necessary' indicates an N-proposition, and sometimes the contradictory of an M-proposition. For example, 'necessarily belong to all' sometimes indicates an a_N-proposition, and sometimes the contradictory of an o_M-proposition.

Once we are aware of this distinction, the context often makes it clear where 'necessary' is used in which way. Many syllogistic moods have N-propositions as their premises or as their conclusion. When 'necessary' is used to express such a proposition, it is clear that it indicates an N-proposition. On the other hand, 'necessary' is also used to refer to the assumption for *reductio* in indirect proofs of moods that have an M-conclusion. In this case, it is clear that 'necessary' indicates the contradictory of an M-proposition. In other cases, we should be careful and sometimes refrain from deciding exactly what is meant by 'necessary' in a given passage. In particular, this applies to Aristotle's remarks that what is necessary is not two-sided possible and that e_Q-propositions are incompatible with 'necessarily belong to some'. These remarks should not be understood to mean that Q-propositions are incompatible with N-propositions. Rather, they may be taken to mean that Q-propositions are incompatible with the contradictory of the corresponding M-proposition. On this reading, the remarks amount to an assertion of the principles of Q-M-subordination, according to which any Q-proposition implies the corresponding M-proposition. So understood, Aristotle's remarks are not in tension with any claims he makes elsewhere in the modal syllogistic.

INDIRECT PROOF OF BOCARDO QXM. I now want to discuss two passages in which Aristotle seems to accept principles of modal opposition for affirmative N-propositions. One of them concerns the mood Bocardo QXM:

Ao_QC	(major premise)
Ba_XC	(minor premise)
Ao_MB	(conclusion)

Aristotle establishes the validity of this mood by *reductio ad absurdum.* The assumption for *reductio* is the contradictory of the o_M-conclusion and is expressed by 'A necessarily belongs to all B':

> Let B belong to all C, and let it be possible for A not to belong to some C. Then it must be possible for A not to belong to some B. For if A necessarily belongs to all B, and B is put as belonging to all C, then A will necessarily belong to all C; for this has been proved earlier. But it was assumed that A may possibly not belong to some C. (*APr.* 1.21 39b33–9)

The remark "for this has been proved earlier" clearly refers to the justification of Barbara NXN in *Prior Analytics* 1.9. Thus, the auxiliary mood used in the proof seems to be Barbara NXN, whose major premise and conclusion are a_N-propositions. If this is correct, the assumption for *reductio* is an a_N-proposition, and the proof can be represented as follows:[14]

1. Ao_QC	(major premise)
2. Ba_XC	(minor premise)
3. Aa_NB	(assumption for *reductio*)
4. Aa_NC	(from 2, 3; by Barbara NXN)

In line 3 of this proof, the assumption for *reductio* is identified with an a_N-proposition. This identification relies on the principle that o_M-propositions are contradictory to a_N-propositions. Moreover, line 4 is taken to be incompatible with line 1, based on the principle that o_Q-propositions are incompatible with a_N-propositions.[15] Thus, the above reconstruction attributes to Aristotle principles of modal opposition for affirmative N-propositions.

As we have seen, however, these principles lead to serious problems in Aristotle's modal syllogistic. In order to avoid an appeal to these principles, the assumption for *reductio* should be taken to be not an a_N-proposition, but merely the contradictory of an o_M-proposition, indicated by the formula 'not Ao_MB'. Accordingly, the auxiliary mood employed in the proof should be taken to be not Barbara NXN, but rather a contraposed version of Bocardo MXM, yielding the conclusion 'not Ao_MC':

1. Ao_QC	(major premise)
2. Ba_XC	(minor premise)
3. not Ao_MB	(assumption for *reductio*)
4. not Ao_MC	(from 2, 3; by contraposed version of Bocardo MXM)

As before, line 4 of this reconstruction is taken to be incompatible with line 1. This incompatibility is justified by Q-M-subordination, according to which any Q-proposition implies the corresponding

14. See Ross (1949: 366), Mueller (1999b: 43), and Ebert & Nortmann (2007: 706).

15. Buddensiek (1994: 31), and Thom (1996: 17).

M-proposition. Thus, the proof relies on Q-M-subordination instead of the incompatibility of a_N- and o_Q-propositions.

Aristotle's remark "for this has been proved earlier" may then be taken to indicate that the justification of Barbara NXN carries over to the contraposed version of Bocardo MXM so that the latter is valid as well as the former.[16] It must be admitted that this is not a straightforward interpretation of Aristotle's proof of Bocardo QXM. But it is, I think, required if the modal syllogistic is to be viewed as consistent. This is one of the costs of an interpretation on which the modal syllogistic is consistent.

ESTABLISHING E_M-CONVERSION. Now for the second passage in which Aristotle seems to accept a principle of modal opposition for affirmative N-propositions. The passage is from chapter 1.3 and contains Aristotle's justification of the rule of e_M-conversion:

> If it is possible for white to belong to no coat, then it is also possible for coat to belong to no white. For if it is necessary for it to belong to some, then white will also belong to some coat by necessity; for this has been proved earlier. (*APr.* 1.3 25b10–13)

The remark "for this has been proved earlier" clearly refers to the justification of i_N-conversion earlier in chapter 1.3 (25a32–4). Thus Aristotle seems to take the contradictories of e_M-propositions to be i_N-propositions, endorsing a principle of M-N-contradictoriness for affirmative N-propositions.

This problem can be addressed in the same way as the problem with his proof of Bocardo QXM: Aristotle's argument should not be taken to involve i_N-propositions, but merely the contradictories of e_M-propositions. The remark "for this has been proved earlier" may then be taken to indicate that the justification of i_N-conversion (whatever it is) carries over to the contradictories of e_M-propositions so that these contradictories are convertible as well as i_N-propositions. As before, this is

16. Aristotle does not mention Bocardo MXM elsewhere in the modal syllogistic. Nevertheless, this mood will be valid in the semantics presented in Appendix B; see Fact 144, p. 324.

not a straightforward interpretation. It is one of the costs of interpreting the modal syllogistic as consistent.

WHAT LED ARISTOTLE TO HIS ASYMMETRIC TREATMENT OF MODAL OPPOSITION? I have argued that Aristotle is committed to an asymmetric treatment of modal opposition and have pointed out some costs of accepting this asymmetry. It is unclear whether Aristotle was aware of these costs. It is even unclear whether he was aware of the fact that he was committing himself to that asymmetry by his claims of invalidity and inconcludence. Nevertheless, the question arises what led Aristotle to that asymmetry in modal opposition.

To answer this question, a natural place to start is by considering how Aristotle justifies the claims that commit him to the asymmetry. As we have seen, there are two groups of claims that commit him to denying principles of modal opposition for affirmative N-propositions. The first group, which implies a denial of principles of M-N-contradictoriness, concerns the invalidity of moods such as Darii NQX and Darapti NQX (p. 202n10). Curiously, however, Aristotle does not explain why these moods are invalid. In particular, he does not give a counterexample to justify their invalidity.[17] He simply states their invalidity without proof.

The second group of claims, which implies a denial of principles of Q-N-incompatibility, concerns the inconcludence of ea-2-QN and ae-2-NQ (p. 203n11). Unlike with the moods of the first group, Aristotle does provide arguments to justify the inconcludence of these two premise pairs. I discuss these arguments in Chapter 14.

17. The invalidity of Darii NQX and QNX is asserted at 1.16 35b26–8 and 36a39–b2 but is not justified by a counterexample. Likewise for the invalidity of Darapti QNX and NQX (1.22 40a11–18), and for the invalidity of Datisi QNX and NQX, and Disamis QNX and NQX (1.22 40a39–b2).

14

Establishing Inconcludence

Aristotle's justification of the inconcludence of ea-2-QN and ae-2-NQ may be approached with high expectations; for in order to be convincing, it should explain how e_Q-propositions can be compatible with a_N-propositions. It should thereby also help us see what led Aristotle to his asymmetric treatment of modal opposition. Now, Aristotle's justification of the inconcludence of these two premise pairs is more complex than his usual proofs of inconcludence in the modal syllogistic. We will therefore first have a look at his usual method of establishing inconcludence and then consider the more complex case of ea-2-QN and ae-2-NQ.

INCONCLUDENCE IN PRIOR ANALYTICS 1.1–22. A premise pair is inconcludent if, as Aristotle puts it, "there will be no deduction" from it. This means that a premise pair in a given figure is inconcludent just in case it does not yield a valid mood in this figure: there is no categorical proposition that could serve as the conclusion of such a valid mood.[1]

In the assertoric syllogistic, Aristotle focuses on four kinds of categorical propositions, namely, the four kinds of X-propositions. His standard method to establish inconcludence consists in giving two counterexamples (see *Prior Analytics* 1.4 26a2–8). In both counterexamples, the

1. A premise pair may be inconcludent although it yields a valid mood outside the figure to which it belongs. For example, Aristotle states that the third-figure premise pair consisting of the major premise Aa_QB and the minor premise Ce_NB is inconcludent (1.22 40a35–8). The conclusion Co_MA follows from this premise pair via Felapton NQM (which Aristotle accepts as valid). But the resulting mood is not in the third figure.

premise pair is true, but in one of them the major term is a_X-predicated of the minor term, and in the other the major term is e_X-predicated of the minor term. The first counterexample shows that no negative X-proposition follows from the premise pair, and the latter shows that no affirmative X-proposition follows from it. Thus, it is proved that no X-proposition follows from the premise pair.

In the apodeictic syllogistic, Aristotle does not make any statements of inconcludence. He only states that moods are invalid, which means that a certain conclusion does not follow from a given premise pair, while another conclusion may follow from it in the figure to which it belongs.

In the problematic syllogistic, Aristotle makes both statements of invalidity and inconcludence. In addition to the four kinds of X-propositions, he now takes into account the twelve kinds of N-, Q-, and M-propositions. Thus, a premise pair is inconcludent just in case no categorical proposition of any of these sixteen kinds follows from it in the figure to which the premise pair belongs.

INCONCLUDENCE IN THE PROBLEMATIC SYLLOGISTIC. In *Prior Analytics* 1.14, Aristotle introduces his standard method of establishing inconcludence in the problematic syllogistic (33b3–17). As in the assertoric syllogistic, this method consists in giving two counterexamples. In one of them, the premise pair is true while the major term 'belongs necessarily to all' of the minor term. In the other one, the premise pair is true while the major term 'belongs necessarily to none' of the minor term (or 'does not belong possibly to any' of it).

Aristotle states that the first counterexample rules out negative X- and N-conclusions and that the second rules out affirmative X- and N-conclusions (33b10–13). He also states that the two counterexamples rule out possibility conclusions (33b13–17). However, he does not distinguish here between one-sided and two-sided possibility and does not explain which counterexample is meant to rule out which kind of possibility conclusion. Instead, he merely writes: "for what is necessary was not possible" (33b17). This remark is true for two-sided possibility but not for one-sided possibility.[2] So Aristotle seems to be concerned here

2. This kind of remark typically occurs in contexts dealing exclusively with two-sided possibility; see *APr.* 1.13 32a18–19, 32a36, 1.17 37a8–9, 37b9–10, 1.19 38a35–6.

only with Q-conclusions but not with M-conclusions.[3] As we will see shortly, this problem also occurs elsewhere: Aristotle gives no explicit indication that he intends to rule out M-conclusions in his proofs of inconcludence in the problematic syllogistic.

Of course, this does not mean that he does not intend to rule them out. For it is obvious enough how the two counterexamples might rule out M-conclusions: the counterexample in which the major term 'belongs necessarily to all' of the minor term can be meant to rule out any negative M-conclusions, and the counterexample in which the major term 'belongs necessarily to none' of the minor term can be meant to rule out any affirmative M-conclusions. On this account, 'belong necessarily to all' rules out any negative conclusions of all four modalities, and 'belong necessarily to none' rules out any affirmative conclusions.

SUBALTERNATION AND MODAL SUBORDINATION. If this account of Aristotle's proofs of inconcludence is correct, the phrase 'belong necessarily to all' in these proofs should not be taken to indicate an a_N-proposition. For given Aristotle's asymmetric treatment of modal opposition, a_N-propositions are not incompatible with negative M-propositions, and therefore cannot rule them out in proofs of inconcludence. Rather, the phrase 'belong necessarily to all' may be taken to indicate the contradictory of an o_M-proposition in these proofs. So understood, the phrase obviously rules out o_M-conclusions. Moreover, it can naturally be taken to rule out any other negative conclusions, provided that the principles of subalternation and modal subordination shown in Figure 3 are accepted:

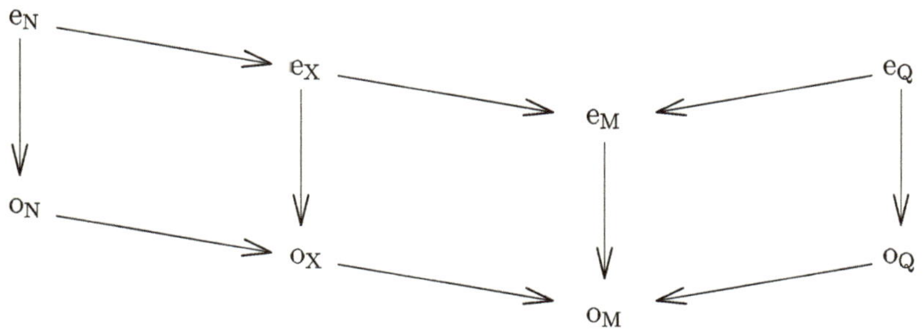

Figure 3

3. See Nortmann (1996: 187).

In this diagram, subalternation is indicated by the four vertical arrows, and modal subordination is indicated by the six diagonal arrows. Given the implications indicated by these arrows, every negative categorical proposition implies the corresponding o_M-proposition. Consequently, the contradictory of an o_M-proposition rules out any negative conclusions.

Analogous assumptions can be made for affirmative propositions, as shown in Figure 4:

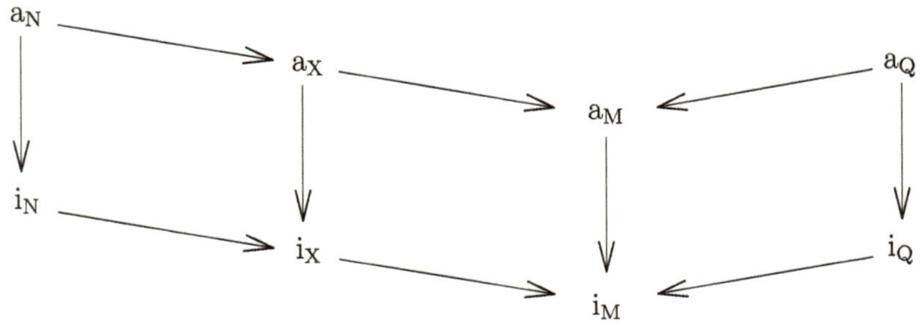

Figure 4

Given this, every affirmative categorical proposition implies the corresponding i_M-proposition. The phrase 'belong necessarily to none' in Aristotle's proofs of inconcludence can then be taken to indicate the contradictory of an i_M-proposition, ruling out any affirmative conclusions.

In sum, given the above principles of subalternation and modal subordination for negative and affirmative propositions, the inconcludence of a premise pair can be established by two counterexamples ruling out an o_M- and i_M-conclusion, respectively.

DOES ARISTOTLE ENDORSE SUBALTERNATION AND MODAL SUBORDINATION?

In interpreting Aristotle's proofs of inconcludence, it is convenient to attribute to him the above principles of subalternation and modal subordination. It must be admitted that Aristotle does not explicitly assert these principles. Nevertheless, each of the principles is intrinsically plausible. Also, at least for some of them, there is independent evidence that he endorsed them. First, his conversion rules commit him to the principles of subalternation for affirmative propositions. For he holds that for all four modalities, a-propositions convert to i-propositions, and i-propositions are convertible.[4] Consequently, Aristotle is also committed to subalternation for negative Q- and X-propositions. For negative Q-propositions are equivalent to affirmative ones, and subalternation for negative X-propositions can be established by *reductio ad absurdum* by means of subalternation for affirmative X-propositions (given the contradictoriness between o_X- and a_X-propositions, and between e_X- and i_X-propositions). There is, as far as I can see, no decisive evidence that Aristotle endorsed subalternation for negative N- and M-propositions. But there is also no reason to think that he would deny it.

As to modal subordination, we have already seen that Aristotle does not assert or use principles of N-X-subordination in the modal syllogistic (p. 131). But he does occasionally use some principles of Q-M- and X-M-subordination. As discussed above, he seems to appeal to Q-M-subordination in his indirect proof of Bocardo QXM.[5] He also seems to appeal to X-M-subordination in a passage in which he infers the validity of Celarent NQM from that of Celarent NQX:[6]

> And it is evident that a deduction of being possible not to belong also comes about since there is one of not belonging. (*APr.* 1.16 36a15–17)

So although Aristotle does not explicitly assert all the above principles of subalternation and modal subordination, he also does not reject them. Nor does the body of his claims in the modal syllogistic commit him to denying any of them. There may be room for interpretations of the modal syllogistic on which not all of these principles are valid. But since these principles are intrinsically plausible and there is no reason not to accept them in the context of Aristotle's modal syllogistic, it seems more natural to adopt an interpretation on which they are valid. The interpretation developed in the present study will validate all of them.[7]

---

4. *APr.* 1.2 25a17–22, 1.3 25a32–4, 25a39–b2.

5. *APr.* 1.21 39b35–9; see pp. 208–209 above.

6. See McCall (1963: 41), Wieland (1975: 83), Nortmann (1990: 78; 1996: 65), Buddensiek (1994: 31), and Thom (1996: 45). In *Prior Analytics* 1.15 (34a40–1, 34b4–5), Aristotle seems to infer an a_M-proposition from the corresponding a_X-proposition by means of X-M-subordination; see Angelelli (1979: 196–7), Striker (2009: 146), and Malink & Rosen (forthcoming).

7. See Facts 18–22, pp. 291–293 below.

THE INCONCLUDENCE OF AA-2-QN. Most of Aristotle's proofs of inconcludence follow the standard method just described. In chapter 1.19, however, there are two deviant proofs aiming to establish the inconcludence of the premise pairs ea-2-QN (38a26–b4) and aa-2-QN (38b13–23). As we have seen, the claim that the first of these premise pairs is inconcludent commits Aristotle to denying the principles of Q-N-incompatibility for affirmative N-propositions. The same is true for his claim that the second premise pair is inconcludent; for given the equivalence of affirmative and negative Q-propositions, the two premise pairs are strictly equivalent. Nevertheless, Aristotle chooses to establish their inconcludence separately by two different arguments. He begins with ea-2-QN. But since his argument for this premise pair is more complicated than that for the other, it will be useful first to consider his treatment of aa-2-QN before turning to the argument for ea-2-QN.

Unlike his usual proofs of inconcludence, Aristotle's argument for the inconcludence of aa-2-QN contains only one counterexample. In this counterexample, the premise pair is true while the major term 'belongs necessarily to none' of the minor term (38b18–20). The counterexample is uncontroversial, and it rules out any affirmative conclusion. But the more important question is whether the premise pair yields a negative conclusion. Aristotle does not give a counterexample to rule this out. Instead, he introduces an additional claim to the effect that premise pairs that do not include a negative X- or N-proposition cannot yield a negative X- or N-conclusion:

> It is evident (φανερόν) that there will not be a deduction of not belonging or of not belonging of necessity, because a negative premise has not been assumed either in the sense of belonging or in the sense of belonging of necessity. (*APr.* 1.19 38b14–17)

Unfortunately, Aristotle does not prove or justify this claim. As a matter of fact, the claim is true in Aristotle's modal syllogistic. But this does not help to establish Aristotle's statements in the modal syllogistic. Had he chosen to accept that aa-2-QN yields a negative conclusion, the claim would be false.[8] Moreover, the claim does not help rule out negative

8. At 1.24 41b27–31, Aristotle makes a similar claim to the effect that a valid mood with a negative conclusion must have at least one negative

M-conclusions. Again, Aristotle does not seem to take into account M-conclusions in his argument (see 1.19 38b17–20).

It is not easy to supply the missing counterexample in Aristotle's argument for the inconcludence of aa-2-QN; for if such a counterexample is to rule out an o_X-conclusion, the premise pair should be true in it while the major term is a_X-predicated of the minor term. In this case, the middle term would be both a_N- and a_Q-predicated of the minor term (see p. 203). To rule out an o_X-conclusion, Aristotle appeals to the claim about negative N- and X-conclusions. He says that the claim is evident (φανερόν), but in fact it is not. He may have thought that this claim is so attractive that in order for it to be true, aa-2-QN should be taken to be inconcludent. But this does not explain how a_N-propositions can be compatible with a_Q-propositions. As we will see, Aristotle's argument for the inconcludence of ea-2-QN is more helpful in this respect.

THE INCONCLUDENCE OF EA-2-QN. Aristotle's argument for the inconcludence of this premise pair begins with an uncontroversial counterexample ruling out all affirmative conclusions.[9] This counterexample also rules out negative Q-conclusions.[10] Next, Aristotle argues that ea-2-QN does not yield an N-conclusion, and in particular no negative N-conclusion (38a36–8). His argument is problematic and may be unconvincing.[11] But the claim that ea-2-QN does not yield a negative N-conclusion is plausible. At least, it is more plausible than the claim that it does not yield a negative X-conclusion. Aristotle undertakes to justify this latter claim by means of the following counterexample:[12]

---

premise. But Aristotle does not offer an independent justification for this claim either.

9. *APr.* 1.19 38a30–4; cf. 38b3–4 and Alexander *in APr.* 237.29–35.

10. *APr.* 1.19 38a34–6; cf. Alexander *in APr.* 236.21–4.

11. See Smith (1989: 137), Thom (1996: 110), Nortmann (1996: 280), Schmidt (2000: 182n335), and Ebert & Nortmann (2007: 656–61).

12. *APr.* 1.19 38a38–b3; cf. Ross (1949: 360), Thom (1996: 110–11), and Mueller (1999b: 216n423). Aristotle's remark "when the terms are like this it is necessary to belong" at 38b3 might be taken to suggest that 'animal' is not only a_X-predicated of 'awake' but also 'belongs necessarily to all' of it (Alexander *in APr.* 236.40–237.12 and 240.30–2; Philoponus *in APr.* 225.20–1; Maier 1900a: 187; Schmidt 2000: 182). In this case, the counterexample might be intended to rule out not only o_X- but also o_M-conclusions. At

'motion' is e_Q-predicated of 'animal'
'motion' is a_N-predicated of 'awake'
and 'animal' is a_X-predicated of 'awake'

This counterexample is intended to establish that a premise pair of the form ea-2-QN can be true while the major term is a_X-predicated of the minor term. This suffices to show that ea-2-QN does not yield an o_X- or e_X-conclusion. However, Aristotle's counterexample is puzzling. The first and third proposition in it imply, via the perfect mood Celarent QXQ, that 'motion' is e_Q-predicated of 'awake'. Hence, 'motion' is both a_N- and e_Q-predicated of 'awake'. Because of this violation of Q-N-incompatibility, Philoponus considers Aristotle's counterexample defective and rejects it.[13] Others reject the counterexample for other reasons, especially because of the nature of its minor a_N-premise.[14]

I suggest that the counterexample should be rejected for another reason. Since 'man' is a substance term and 'awake' is a nonsubstance term, the counterexample assumes that a substance term is a_X-predicated of a nonsubstance term. As mentioned above (p. 161), this is the only passage in *Prior Analytics* 1.1–22 in which a substance term is taken to be a_X-predicated of a nonsubstance term. On the interpretation of the modal syllogistic pursued in this study, such a_X-predications are not admissible and should be rejected. Thus, I suggest that the counterexample should be modified by replacing the nonsubstance term 'awake' with a substance term (see p. 220).

THE INCONCLUDENCE OF AE-2-NQ. Having completed his discussion of ea-2-QN, Aristotle adds a brief comment on ae-2-NQ, claiming that the proof of inconcludence for this premise pair is similar to that for the former (38b4–5). He presumably means that ae-2-NQ can be proved to

---

any rate, however, Aristotle does not explicitly discuss M-conclusions in his argument.

13. Philoponus *in APr.* 225.21–32; similarly, Thom (1996: 111).

14. Some reject it on the grounds that the minor premise is a mere *de dicto* necessity (Nortmann 1996: 281, Ebert & Nortmann 2007: 662–4, Striker 2009: 161). Thom (1996: 330) suggests rejecting it because the term 'moving', which is a_N-predicated of 'awake', does not stand for a proper species or natural kind.

be inconcludent by counterexamples that consist of the same terms as those he gave for the earlier premise pair. As Alexander points out, when adopting these counterexamples to ae-2-NQ, we have to interchange the major and minor premise.[15]

If this is correct, the counterexample ruling out negative X-conclusions for ae-2-NQ is as follows:

'motion' is a_N-predicated of 'awake'
'motion' is e_Q-predicated of 'animal'
and 'awake' is a_X-predicated of 'animal'

In this counterexample, 'awake' is a_X-predicated of 'animal' instead of the other way around. Given the major premise, it follows via Barbara NXN that 'motion' is a_N-predicated of 'animal'. Hence 'motion' is both a_N- and e_Q-predicated of 'animal'.

The nonsubstance term 'awake' is the subject of an a_N-predication and is at the same time a_X-predicated of the substance term 'man'. This conflicts with the interpretation of the modal syllogistic introduced above, on which subjects of a_N-predications cannot be predicates of cross-categorial a_X-predications (p. 151). The present counterexample is the only place in *Prior Analytics* 1.1–22 where a nonsubstance term is the subject of an a_N-predication while also being a_X-predicated of a substance term.[16] I therefore suggest that the counterexample is an oversight on Aristotle's part and that it should be modified in such a way that 'awake' is replaced by a substance term. For instance, 'awake' may be replaced by 'animal' while the minor term is taken to be 'man':

'motion' is a_N-predicated of 'animal'
'motion' is e_Q-predicated of 'man'
and 'animal' is a_X-predicated of 'man'

---

15. Alexander *in APr.* 238.11–21; cf. Mueller (1999b: 217n431). Similarly, the premises need to be interchanged when Aristotle says, at 1.11 31b31–3, that Disamis NXN is proved to be invalid by the same terms as Datisi XNN (see Alexander *in APr.* 149.13–20).

16. In fact, the present counterexample and the corresponding counterexample for ea-2-QN are the only places in chapters 1.1–22 where a nonsubstance term is taken to be the subject of an a_N-predication; see p. 330 below.

Likewise, the corresponding counterexample for ea-2-QN can be modified by replacing 'awake' with 'man', as follows:

'motion' is e_Q-predicated of 'animal'
'motion' is a_N-predicated of 'man'
and 'animal' is a_X-predicated of 'man'

These two counterexamples do not conflict with any of the claims about a_N- and a_X-predication made earlier in this study. They are compatible with the interpretation of the modal syllogistic pursued here. If it is accepted that 'motion' is both a_N- and e_Q-predicated of 'animal', the counterexamples succeed in showing that the two premise pairs in question do not yield a negative X-conclusion. In fact, since 'animal' is not o_M-predicated of 'man', they succeed in showing that the two premise pairs do not yield any negative conclusion at all. I will use analogous counterexamples to demonstrate the inconcludence of these premise pairs within the formal semantics for the modal syllogistic presented in Appendix B.[17]

WHY IS 'MOTION' BOTH A_N- AND E_Q-PREDICATED OF 'ANIMAL'? Aristotle's claims of inconcludence commit him to the view that some terms are simultaneously a_N- and e_Q-predicated of others. As indicated by his treatment of ae-2-NQ, Aristotle seems to accept that 'motion' and 'animal' are such a pair of terms. Of course, he does not explain why the former is both a_N- and e_Q-predicated of the latter, and it is difficult to find a plausible explanation for it. Still, a possible explanation might be the following. On the one hand, 'motion' may be taken to be a_N-predicated of 'animal' inasmuch as every animal necessarily has a capacity to undergo motion in a certain way, and for every animal this capacity is necessarily realized at some times. On the other hand, there is the idea that for every animal, the capacity in question may or may not be realized at any given time. This could be taken to imply that 'motion' is e_Q-predicated of 'animal'. If so, then we would have an explanation as to why a_N-propositions fail to be incompatible with e_Q-propositions. I do not mean to suggest that this is an entirely satisfactory explanation,

17. See Facts 114 and 115, p. 317.

but it seems conceivable that Aristotle thought something along these lines.

By contrast, this kind of explanation does not work for negative N-propositions. It cannot be used to argue that e_N-propositions fail to be incompatible with e_Q-propositions. For example, 'motion' is e_N-predicated of 'number' since numbers necessarily lack the capacity to undergo motion. But there is no question as to whether this capacity may or may not be realized at a given time for a given number.

This may help explain Aristotle's asymmetric treatment of modal opposition, on which Q-N-incompatibility fails for affirmative but not for negative N-propositions. As we saw above, another possible reason for this asymmetry may be Aristotle's motivation to validate the principle that a negative X- or N-conclusion can only follow from premise pairs that contain a negative X- or N-proposition (pp. 216–217).

TWO STRATEGIES TO EXPRESS NEGATIVE N-PROPOSITIONS. Finally, let me suggest one more possible reason for Aristotle's asymmetric treatment of modal opposition. It has to do with certain linguistic considerations. Greek language has two strategies to express negative N-propositions. One of them uses expressions of necessity. For instance, e_N-propositions can be expressed by phrases like 'necessarily to none' (ἐξ ἀνάγκης οὐδενί). The other strategy uses expressions of possibility. For instance, e_N-propositions can be expressed by phrases like 'to none possibly' (οὐδενὶ ἐνδέχεσθαι). The second strategy is perhaps more natural in Greek than it is in English; it is comparable to English phrases like 'cannot to any'.

Aristotle uses both strategies to express negative N-propositions. However, in those passages in which he appeals to the incompatibility of Q-propositions and negative N-propositions, he uses exclusively the second strategy. This allows him to express the incompatibility in a particularly striking way, as follows:[18]

'to all possibly'	is incompatible with	'to none possibly'
'to all possibly'	is incompatible with	'not to all possibly'
'to all possibly'	is incompatible with	'to some not possibly'
'to some possibly'	is incompatible with	'to none possibly'

18. *APr.* 1.16 36a7–15, 36a34–9, 1.19 38a21–5.

The incompatibility of these expressions is especially evident since it can be recognized without taking into account the modal expression 'possibly'. It suffices to consider the quantifying and negative expressions like 'all' and 'none'. By contrast, the incompatibility of 'to all possibly' and 'necessarily to none' is not equally evident; for in order to recognize it, one needs to take into account the modal expressions 'possibly' and 'necessarily'.

Now, Aristotle does not express affirmative N-propositions in terms of 'possibly'.[19] In principle, he could express them by phrases like 'not possibly to none'. However, such phrases are quite unusual because they contain two occurrences of a negative expression. Since Aristotle always expresses Q-propositions in terms of 'possibly', the supposed incompatibility of Q-propositions and affirmative N-propositions is not as evident as that between Q- and negative N-propositions; for unlike the latter incompatibility, the former cannot be recognized without taking into account the modal expressions 'necessarily' and 'possibly'. These linguistic considerations might be part of the reason why Aristotle took Q-propositions to be incompatible with negative but not with affirmative N-propositions.

This completes our discussion of Aristotle's asymmetric treatment of modal opposition. Chapter 15 will provide a more general overview of the whole of Aristotle's modal syllogistic.

---

19. Wieland (1975: 90n20; 1980: 115).

15

A Deductive System for the Modal Syllogistic

Aristotle's syllogistic can be viewed as a deductive system of categorical propositions. Corcoran (1972) and Smiley (1973) have specified suitable deductive systems for the assertoric syllogistic (see pp. 31–33 above). These systems are based on Aristotle's conversion rules and perfect first-figure moods. They include indirect deductions based on the standard principles of contradictoriness between assertoric propositions. These systems are adequate with respect to Aristotle's assertoric syllogistic in that every assertoric mood held to be valid by Aristotle is deducible in them, and no assertoric mood held to be invalid by him is deducible in them.

On the other hand, no adequate deductive system has so far been specified for the modal syllogistic. Attempts to construct such a system have been made by McCall (1963), Thom (1996), and Johnson (2004).[1] But their systems are not adequate with respect to Aristotle's modal syllogistic. This is partly because they contain the full range of principles of M-N-contradictoriness for both affirmative and negative N-propositions;[2] for, as we saw above, Aristotle is committed to denying the principles of M-N-contradictoriness for affirmative N-propositions. Moreover, their systems conflict with Aristotle's claims

1. McCall and Johnson do not employ the framework of deductive systems used by Corcoran and Smiley, but adopt an axiomatic approach in the style of Łukasiewicz (1957). For present purposes we may neglect this difference, since their systems can be readily translated into the framework of deductive systems.

2. See McCall (1963: 35–7), Thom (1996: 13–15), and Johnson (2004: 265).

because they contain principles of Q-N-incompatibility that he is committed to denying.[3]

Once we refrain from attributing to Aristotle those troublesome principles of modal opposition, the way is cleared for deductive systems that are adequate with respect to the whole of his modal syllogistic. In what follows, I specify several versions of such a system.

A PURELY DIRECT DEDUCTIVE SYSTEM FOR THE MODAL SYLLOGISTIC. Let us begin with a simple deductive system that includes only direct deductions, but no indirect deductions (that is, no method of *reductio ad absurdum*). The system contains three kinds of deduction rules. First, the system includes Aristotle's conversion rules for assertoric and modalized propositions (which are summarized on p. 274 below). Second, the system includes the equivalence of affirmative and negative Q-propositions. Thus, a_Q-propositions can be deduced from the corresponding e_Q-propositions and vice versa; likewise for i_Q- and o_Q-propositions. The rules of this second group are often referred to as rules of *qualitative conversion*. Third, the system includes a number of syllogistic moods regarded as deduction rules. As shown by Table 15.1, there are six moods for the assertoric syllogistic, ten for the apodeictic syllogistic, and seventeen for the problematic syllogistic. Some of them are among Aristotle's perfect moods, and others are imperfect. In sum, then, the deduction rules of the system are shown in Table 15.1.

The deductive system does not include any principles of incompatibility, contradictoriness, or modal subordination, nor does it include any principles of subalternation for negative propositions. Nevertheless, the system is strong enough to capture Aristotle's claims of validity. It can be proved that all moods held to be valid by Aristotle in the assertoric and modal syllogistic are deducible in this system.[4] Moreover, the system is weak enough to match Aristotle's claims of invalidity: no mood held to be invalid by Aristotle is deducible in the deductive system. This can be proved by means of the predicable semantics for the modal syllogistic presented in Appendix B; for all deduction rules of the

3. For the problems that arise in these systems because of the troublesome principles of modal opposition, see McCall (1963: 93), Thom (1996: 128–9), and Johnson (2004: 303).

4. This is proved on pp. 295–302 below.

Aristotle's conversion rules
rules of qualitative conversion

Barbara, Darii, Celarent, Ferio XXX (perfect)
Baroco, Bocardo XXX

Barbara, Darii, Celarent, Ferio NXN and NNN (perfect)
Baroco, Bocardo NNN

Barbara, Darii QQQ, QXQ, and QNQ (perfect)
Barbara, Darii, Celarent, Ferio XQM and NQM
Celarent and Ferio NQX
Bocardo QXM

Table 15.1. A deductive system for the modal syllogistic, direct deductions only

above deductive system will be valid in the predicable semantics, but no mood held to be invalid by Aristotle will be valid in it.[5] It follows from this that also all premise pairs held to be inconcludent by Aristotle are inconcludent in the deductive system (since the inconcludence of a premise pair amounts to the invalidity of several moods that share the same premise pair).

Thus, the deductive system given in Table 15.1 is adequate with respect to Aristotle's assertoric and modal syllogistic.

ADDING PRINCIPLES OF SUBALTERNATION AND MODAL SUBORDINATION. The above deductive system can be extended by principles of subalternation for negative X-, N-, Q-, and M-propositions. Moreover, it can be extended by principles of modal subordination, namely, of N-X-, X-M-, and Q-M-subordination. The resulting strengthened deductive system is still adequate with respect to Aristotle's modal syllogistic. No mood held to be invalid by Aristotle becomes deducible in the strengthened deductive system, for all these principles of subalternation and modal subordination are valid in the predicable semantics.[6]

5. This is proved on pp. 304–322.

6. See Facts 18–22, pp. 291–293.

When the principles of N-X-subordination are added, the moods of the form NNN, QNQ, and NQM can be removed from the list of deduction rules, since they can now be deduced by means of N-X-subordination from the corresponding moods of the form NXN, QXQ, and XQM.[7] Adding the principles of subalternation and of X-M- and Q-M-subordination does not allow us to remove any moods from the list of deduction rules.

ADDING INDIRECT DEDUCTIONS. The deductive system can also be extended to include indirect deductions. In indirect deductions, certain premises are laid down, and then the contradictory of the desired conclusion is assumed for *reductio*. From this assumption for *reductio* and the original premises, we construct a direct deduction and try to go on until the deduction contains a pair of contradictory or incompatible propositions. If successful, we have completed the indirect deduction of the desired conclusion from the original premises.

Indirect deductions rely on the notions of contradictoriness and incompatibility. As we saw above, it is important to be careful about exactly which propositions are taken to be contradictory to and incompatible with each other in Aristotle's modal syllogistic. Our deductive system will contain the usual principles of contradictoriness for X-propositions, and those principles of Q-N-incompatibility that Aristotle uses in his indirect deductions for moods such as Celarent and Ferio NQX (see pp. 200–201):[8]

Aa_XB	is contradictory to	Ao_XB
Ae_XB	is contradictory to	Ai_XB
Aa_QB	is incompatible with	Ao_NB
Ai_QB	is incompatible with	Ae_NB

These are the only principles of contradictoriness and incompatibility that are included in the system. Since all these principles will be valid

7. Aristotle himself, however, does not use principles of N-X-subordination in the modal syllogistic; cf. p. 131 above.

8. Aristotle also relies on the incompatibility of Aa_QB and Ae_NB (p. 201n8). But given his conversion rules, this incompatibility can be reduced to that of Ai_QB and Ae_NB.

contradictoriness of a_X and o_X, and of e_X and i_X
incompatibility of a_Q and o_N, and of i_Q and e_N

Aristotle's conversion rules
rules of qualitative conversion
full subalternation for X-, N-, Q-, and M-propositions
all rules of N-X-, X-M-, and Q-M-subordination

Barbara, Darii, Celarent, Ferio XXX (perfect)

Barbara, Darii, Celarent, Ferio NXN (perfect)
Baroco, Bocardo NNN

Barbara, Darii QQQ and QXQ (perfect)
Barbara, Darii, Celarent, Ferio XQM
Bocardo QXM

Table 15.2. A deductive system for the modal syllogistic, direct and indirect deductions

in the predicable semantics, they do not threaten the adequacy of the system.[9]

Given indirect deductions based on the above principles, four moods can be removed from the list of deduction rules specified in Table 15.1, namely, Baroco and Bocardo XXX, and Celarent and Ferio NQX. For example, the last of these moods can be deduced by means of Celarent NXN as follows (p. 200):

1. Ae_NB (major premise)
2. Bi_QC (minor premise)
3. Aa_XC (assumption for *reductio*)
4. Be_NA (from 1; by e_N-conversion)
5. Be_NC (from 3, 4; by Celarent NXN)

Thus, we obtain the deductive system shown in Table 15.2, including indirect deductions, full subalternation, and full modal subordination.

9. For the validity in the predicable semantics of the above principles of Q-N-incompatibility, see Fact 30, p. 295.

As before, this deductive system is adequate with respect to Aristotle's assertoric and modal syllogistic.

The deduction rules listed in Table 15.2 include seven imperfect moods, namely, Bocardo and Baroco NNN, the four XQM-moods, and Bocardo QXM. According to Aristotle, the first two of these moods should be proved valid by ecthesis. As we saw above, however, it is difficult to construct these ecthetic proofs in such a way that they conform with Aristotle's claims of invalidity in the apodeictic syllogistic (pp. 181–185). I therefore do not undertake to integrate these proofs into an adequate deductive system for the modal syllogistic.[10]

The four XQM-moods are proved valid in *Prior Analytics* 1.15 by indirect deductions involving a special method of modal reasoning. I discuss these moods in Chapter 16, but again I do not try to integrate Aristotle's proofs of them into the deductive system. Finally, there is Bocardo QXM, whose validity is proved in *Prior Analytics* 1.21 by an indirect deduction. This indirect deduction can, in principle, be integrated into the above system.

ADDING THE INDIRECT DEDUCTION FOR BOCARDO QXM. Aristotle's indirect proof of Bocardo QXM can be reconstructed as follows (p. 208):

1. Ao_QC	(major premise)
2. Ba_XC	(minor premise)
3. not Ao_MB	(assumption for *reductio*)
4. not Ao_MC	(from 2, 3; by contraposed version of Bocardo MXM)

Integrating this indirect proof into the deductive system is not straightforward and takes some effort. First, the language must be extended to include contradictories of M-propositions, indicated by 'not Ao_MB'. Second the contraposed version of Bocardo MXM used in the proof must be assumed as an additional deduction rule. Third, the o_Q-proposition in line 1 must be taken to be incompatible with 'not Ao_MC' in line 4. If all these ingredients are added, the resulting deductive system is

10. A deductive system including these ecthetic proofs has been suggested by Johnson (1993: 177–82); see also p. 87n4 above.

still adequate, since the second and third assumptions are valid in the predicable semantics.[11]

REDUCING ARISTOTLE'S CLAIMS OF INVALIDITY AND INCONCLUDENCE. In the above deductive systems, the validity of certain moods is reduced to the validity of others, namely, of those that are taken as the deduction rules of the system. For example, the validity of assertoric Cesare in the second figure is reduced by means of e_X-conversion to the validity of assertoric Celarent in the first figure. In the same way, the invalidity of moods can be reduced to the invalidity of others. For example, the invalidity of Cesare XNN can be reduced by means of e_X-conversion to the invalidity of Celarent XNN. Also, the invalidity of Celarent XQQ can be reduced by means of N-X-subordination to the invalidity of Celarent NQQ. Likewise, the inconcludence of premise pairs can be reduced to the inconcludence of others.

Thus the body of Aristotle's claims of invalidity and inconcludence can be reduced to a certain subset of such claims. Aristotle himself does not engage in such a project. Although he systematically reduces the validity of second- and third-figure moods to the validity of first-figure moods, he does not in general attempt to reduce the invalidity of a mood to that of another mood, or the inconcludence of a premise pair to that of another premise pair. Part of the reason for this may be that performing such reductions would make the presentation of his syllogistic considerably more difficult.

Nevertheless, we may specify a list of moods and premise pairs from whose invalidity and inconcludence all of Aristotle's claims of invalidity and inconcludence are deducible by means of certain deduction rules. Which moods and premise pairs are included in this list will depend on the deduction rules that are allowed. As before, I assume Aristotle's conversion rules, rules of qualitative conversion, full subalternation, and

11. For the validity in the predicable semantics of (the contraposed version of) Bocardo MXM, see Fact 144, p. 324 below. The third assumption is justified by Q-M-subordination, which is valid in the predicable semantics. One might argue that the contraposed version of Bocardo MXM can be regarded as a perfect mood in the same way as Barbara NXN. In this case, we would be left with only six imperfect moods used as deduction rules in the system when Bocardo QXM is removed from the list.

ae-1-XX, ee-1-XX, aa-2-XX, oa-1-XX, oa-2-XX

aaa-1-XNN, aii-1-XNN, aeo-2-NXN, eao-3-XNN
aoo-2-XNN, oao-3-NXN

aaa-1-QNX, eae-1-QNX, eao-3-QNX, aaa-1-NQX, aai-3-NQX
eae-1-NQQ, eio-1-NQQ, eae-1-NQN, eao-3-NQN
aai-3-QXX, eao-3-QXX, eao-3-XQX
aa-2-QQ, ae-1-QN, ao-2-QX, oa-2-NQ, oa-1-NQ, ea-2-QN, ae-2-NQ

conversion of o_X-, o_N-, o_M-, and e_Q-propositions

Table 15.3. Primitive claims of invalidity and inconcludence

full modal subordination (see Table 15.2). Given these rules, all of Aristotle's claims of invalidity and inconcludence in the assertoric and modal syllogistic follow from the invalidity of the moods and conversion rules listed in Table 15.3 and from the inconcludence of the premise pairs listed in it.[12]

Table 15.3 may be regarded as a list of primitive claims of invalidity and inconcludence in the assertoric and modal syllogistic. Most of these primitive claims are not difficult to establish in the predicable semantics. There are some difficult cases, namely, the invalidity of Baroco XNN and Bocardo NXN, and those invalidities and inconcludences that give rise to the asymmetry in modal opposition (Darapti NQX, ea-2-QN, ae-2-NQ). But even these moods and premise pairs can be shown to be invalid and inconcludent in the predicable semantics. Hence this semantics matches all of Aristotle's claims of invalidity and inconcludence.

SUMMARY. We are now in a position to formulate a general criterion for when a given semantics is adequate with respect to Aristotle's assertoric and modal syllogistic. Any semantics is adequate if it meets the following two conditions. First, all deduction rules of the deductive system given in Table 15.2 must be valid in the semantics. Second, all moods and conversion rules mentioned in the list of primitive claims of invalidity

12. This is proved on pp. 304–322 below.

and inconcludence in Table 15.3 must be invalid in the semantics, and all premise pairs mentioned in this list must be inconcludent in the semantics. The predicable semantics meets these two conditions and is therefore adequate with respect to the whole of Aristotle's syllogistic. Thus, every mood and conversion rule held to be valid by Aristotle in the assertoric and modal syllogistic is valid in the predicable semantics, and no mood or conversion rule held to be invalid by him is valid in it.

16

The Validity of XQM-Moods

PLAN OF THIS CHAPTER. In *Prior Analytics* 1.15, Aristotle is concerned with first-figure premise pairs of the form QX and XQ. He holds that premise pairs of the former kind yield a two-sided possibility conclusion, whereas those of the latter kind yield only a one-sided possibility conclusion. Thus, Aristotle asserts the validity of Barbara, Celarent, Darii, and Ferio of the form QXQ and XQM. The four QXQ-moods are regarded as perfect by Aristotle. By contrast, the four XQM-moods are not regarded as perfect, but as being in need of proof. Aristotle establishes their validity by means of indirect proofs that are considerably more complex than those he gives elsewhere in the modal syllogistic. The proofs employ a special modal rule based on the principle that nothing impossible follows from something possible (34a34–b6 and 34b19–31). The rule can be formulated as follows: given the premise that A is possible, and given a deduction of B from the assumption that A is the case, it may be inferred that B is possible (see 34a5–33). Aristotle uses this rule in chapter 1.15 to establish the four XQM-moods, and also seems to appeal to it in chapter 1.16 to establish the corresponding NQM-moods (35b37–36a2).[1]

His proofs of the XQM-moods are controversial and give rise to several problems of interpretation. It is beyond the scope of this book to

1. The rule is not applied elsewhere in the *Prior Analytics,* but is applied in various places throughout Aristotle's other works; see Rosen & Malink (2012).

examine these proofs in any detail.[2] My main aim here is to offer an adequate semantics for the modal syllogistic, that is, a semantics that is in accordance with Aristotle's claims of validity and invalidity. To this end, it is not so much necessary to examine Aristotle's proofs of the four XQM-moods as it is important to determine the semantic and logical consequences of his taking these moods to be valid.

In doing so, I focus on a passage from *Prior Analytics* 1.15 in which Aristotle states that Barbara XQM is valid only if its a_X-premise is "taken not with a limitation in time, but without qualification" (34b7–18). I examine what this requirement means, and I argue that the requirement is also applicable to the e_X-premise of Celarent and Ferio XQM. As we will see, the validity of the last two moods commits Aristotle to a principle about the realization of Q-predications: if a term is i_Q-predicated of a substance term, then it is also i_X-predicated of this term. This principle, I argue, can help us understand the requirement that the major premise of the four XQM-moods be taken 'without qualification'. Finally, I discuss some consequences of the principle about the realization of Q-predications.

IS THE PREMISE PAIR OF BARBARA XQM INCONCLUDENT? Having completed his proof of Barbara XQM, Aristotle proceeds to defend the validity of this mood against a potential objection. He concedes that the mood would be invalid if its major premise were understood 'with a limitation in time', and requires that it be understood not with such a limitation, but without qualification:

> One must take 'belonging to all' not with a limitation of time such as 'now' or 'at this time', but without qualification (ἁπλῶς). For it is through premises of this sort that we produce deductions, since there will not be a deduction if the premise is taken as holding only at a moment. For perhaps nothing prevents man from belonging to all moving at some time (ποτέ), i.e. if nothing else should be moving, and it is possible for moving to belong to all horse, but yet it is not possible for man to belong to any horse. (*APr.* 1.15 34b7–14)

2. For a detailed discussion of these proofs, see Malink & Rosen (forthcoming).

In the last sentence of this passage, Aristotle shows that if the major premise were understood 'with a limitation in time', the validity of Barbara XQM could be refuted by the following counterexample:

'man' is a_X-predicated of 'moving'
'moving' is a_Q-predicated of 'horse'
but 'man' is not i_M-predicated of 'horse'

Aristotle goes on to show that if the major premise is understood 'with a limitation in time', the premise pair of Barbara XQM also does not yield an o_M-conclusion:

> Next, let the first term be animal, the middle term moving, the last term man. The premises will be in the same relationship, then, but the conclusion will be necessary, not possible; for man is by necessity an animal. It is evident, then, that the universal premise should be taken without qualification (ἁπλῶς), and not with a limitation of time. (*APr.* 1.15 34b14–18)

'animal' is a_X-predicated of 'moving'
'moving' is a_Q-predicated of 'man'
but 'animal' is not o_M-predicated of 'man'

If both counterexamples were admissible, they would show that the premise pair of Barbara XQM yields neither an i_M-conclusion nor an o_M-conclusion, and hence that it is inconcludent.[3] Thus, the premise pair would be inconcludent if the major premise were understood 'with a limitation of time'.

THE OMNITEMPORAL INTERPRETATION OF 'WITHOUT QUALIFICATION'. Propositions 'with a limitation of time' seem to be propositions that are

3. See Alexander *in APr.* 190.9–19, Nortmann (1996: 232–4), Fait (1999: 147), Mueller (1999b: 39), and Ebert & Nortmann (2007: 575). I do not agree with those who hold that the second counterexample is somewhat inappropriate on the grounds that it shows the invalidity only of Barbara XQQ but not of Barbara XQM (Becker 1933: 58–9, Ross 1949: 340, Patterson 1995: 175–6; similarly, Rini 2011: 152).

true at some time (ποτέ) but false at other times. For example, the a_X-proposition 'Man belongs to all moving' might be true at a time at which only men are moving, but false at other times. Aristotle rejects such a_X-propositions by requiring that a_X-propositions be taken 'without qualification' (ἁπλῶς). He does, however, not explain what this requirement means. Some commentators suggest an omnitemporal interpretation, according to which an a_X-proposition is without qualification just in case it is true at all times.[4] More specifically, they take an a_X-proposition to be without qualification just in case, at all times, every individual that falls under the subject term falls under the predicate term.

However, this omnitemporal interpretation is in conflict with Aristotle's use of a_X-propositions elsewhere in the modal syllogistic. Both in the apodeictic and in the problematic syllogistic, Aristotle frequently assumes the truth of a_X-propositions such as 'Moving belongs to all animal', 'Wakefulness belongs to all animal', and 'Health belongs to all horse' (see p. 329). These do not seem to be omnitemporally true any more than a_X-propositions such as 'Man belongs to all moving'.[5] Nor is it plausible that Aristotle is making a counterfactual assumption that they are omnitemporally true. Aristotle's remarks in chapter 1.15 suggest that a_X-propositions 'taken with a limitation of time' are excluded throughout the whole modal syllogistic. If these remarks were intended to exclude a_X-propositions that fail to be true omnitemporally, they would also exclude a_X-propositions such as 'Moving belongs to all animal'. Thus, the remarks from 1.15 would conflict with Aristotle's practice elsewhere in the modal syllogistic.

Against this, it is sometimes thought that the requirement of omnitemporality supposedly stated in chapter 1.15 is meant to be restricted only to certain parts of the modal syllogistic.[6] If so, then Aristotle would use two different kinds of a_X-propositions in the modal syllogistic: an omnitemporal one and a nonomnitemporal one. Thus, we would have to attribute to Aristotle a hidden ambiguity in his use of a_X-propositions,

---

4. Philoponus *in APr.* 175.21–4, Hintikka (1973: 136–8), Nortmann (1990: 64, 1996: 45–6), and Barnes (2007: 18).

5. Patterson (1995: 169).

6. Nortmann (1996: 51–4), Thom (1994: 108, 1996: 100), and Striker (2009: 147); similarly, Angelelli (1979: 201).

in much the same way that commentators attribute to him an ambiguity between *de re* and *de dicto* readings of N-propositions. However, there is no independent evidence for such an ambiguity of a_X-propositions in the modal syllogistic. In general, the issue of time and temporality is not discussed in the modal syllogistic, besides the present passage from 1.15. It is therefore preferable to look for a single, uniform interpretation of a_X-propositions throughout the whole modal syllogistic. Given Aristotle's use of a_X-propositions, this cannot be an omnitemporal interpretation.

AN ANALOGY: BEING FAMILIAR WITHOUT QUALIFICATION. If I am correct, the phrase 'without qualification' (ἁπλῶς) in *Prior Analytics* 1.15 should not be interpreted omnitemporally. Rather, it may be interpreted atemporally, in a way that does not appeal to time and the notion of being true at a time.[7] Support for this alternative comes from Aristotle's distinction between what is familiar to particular individuals and what is familiar without qualification:[8]

> Perhaps, also, what is familiar without qualification (ἁπλῶς) is what is familiar, not to all, but to those who are in a sound state of understanding, just as what is healthy without qualification is what is healthy to those in a sound state of body. (*Top.* 6.4 142a9–11)

In this passage, Aristotle characterizes the notion of being familiar without qualification by means of the relative notion of being familiar *to x*. In doing so, he does not identify what is familiar without qualification with what is familiar *to everybody*. Instead, he characterizes it as that which is familiar to certain ideal individuals who are in a sound state of understanding. However, in order to determine what is familiar without qualification, Aristotle does not look for such ideal individuals to see what is familiar to them. Rather, he determines it without reference to particular individuals: for example, he states that genera are more familiar without qualification than their species because knowledge of a species implies knowledge of the genus, but not vice versa (141b22–34).

7. Various atemporal interpretations are given by Alexander *in APr.* 188.20–190.6, Angelelli (1979: 200–1), and Thom (1996: 251–7 and 345–8).

8. *Top.* 6.4 141b15–142a16; see also *APost.* 1.2 71b33–72a5 and *APr.* 2.23 68b35–7.

Likewise, propositions without qualification need not be identified with those that are true *at all times.* Rather, to follow the analogy, they may be characterized as those that are true at certain ideal times (that is, under certain ideal circumstances). However, in order to determine which propositions are without qualification, we should not look for such ideal times to see what is true at them. Rather, as I will show, we may determine this by other, atemporal criteria.

UNNATURAL PREDICATION REVISITED. The contrast between 'at some time' (ποτέ) and 'without qualification' (ἁπλῶς) occurs not only in *Prior Analytics* 1.15 but also in Aristotle's discussion of unnatural predication. When substance terms are predicated of nonsubstance terms, this is an unnatural predication. Aristotle's examples of such predications include the following:

> we sometimes (ποτέ) say that this white is Socrates, and that the approaching is Kallias (*APr.* 1.27 43a35–6)

> we sometimes (ποτ') say that this white is a man (*APost.* 1.19 81b25–6)

Aristotle refers to unnatural predications as predications not without qualification (μὴ ἁπλῶς), and to natural predications as predications without qualification (ἁπλῶς).[9]

I have argued that the modal syllogistic does not allow for unnatural a_X-predications in which a substance term is a_X-predicated of a nonsubstance term (pp. 159–165). As we have seen, this restriction can be used to reject one of Theophrastus's and Eudemus's counterexamples to Barbara NXN, the counterexample that employs the a_X-premise 'Man belongs to all moving'. The same a_X-premise occurs in the apparent counterexample to Barbara XQM in chapter 1.15. Thus, the phrase 'without qualification' in 1.15 might be taken to rule out unnatural a_X-predications in which substance terms are a_X-predicated of nonsubstance terms. This would be in accordance with the fact that Aristotle often assumes the truth of a_X-propositions such as 'Moving belongs to all man'; for these are not unnatural a_X-predications, since the subject is a substance term and the predicate a nonsubstance term.

9. *APost.* 1.22 83a16 and 83a20.

This interpretation of 'without qualification' in terms of unnatural predication may be satisfactory for the case of Barbara XQM. But, as we will see shortly, it does not work well for the case of Celarent XQM.

IS THE PREMISE PAIR OF CELARENT XQM INCONCLUDENT? Aristotle gives two apparent counterexamples that purport to establish the inconcludence of the premise pair of Barbara XQM. Both of them can be carried over to the case of Celarent XQM, provided that 'animal' and 'man' can be taken to be e_X-predicated of 'moving':[10]

'man' is e_X-predicated of 'moving'
'moving' is a_Q-predicated of 'horse'
and 'man' is not i_M-predicated of 'horse'

'animal' is e_X-predicated of 'moving'
'moving' is a_Q-predicated of 'man'
and 'animal' is not o_M-predicated of 'man'

The latter counterexample purports to establish the invalidity of Celarent XQM.[11] In order to defend this mood, Aristotle needs to reject the counterexample. Although he does not explain how he would reject it, it is not unlikely that he would reject it on the grounds that the e_X-premise is not without qualification. This would presuppose that the contrast between 'without qualification' and 'at some time' applies to e_X-propositions as well as to a_X-propositions.[12] In the case of e_X-propositions, however, the condition of being 'without qualification' cannot be understood in terms of unnatural predication. For Aristotle often accepts e_X-predications in which a substance term is e_X-predicated of a nonsubstance term, so that the notion of unnatural predication does not apply to e_X-propositions as it does to a_X-propositions (see p. 165). Hence we should like to find another interpretation of being 'without qualification', applicable to both a_X- and e_X-propositions.

10. Mueller (1999b: 40).

11. For similar counterexamples, see Alexander *in APr.* 232.17–19 and Patterson (1995: 184).

12. The contrast is also taken to apply to e_X-propositions by Thom (1996: 46 and 254–5) and Nortmann (1996: 46 and 194).

PROPRIA WITHOUT QUALIFICATION IN THE *TOPICS*. To this end, let us briefly consider another passage in which the contrast between 'at some time' and 'without qualification' occurs: the characterization of propria in *Topics* 1.5. In this passage, Aristotle defines a proprium as a predicate that is not predicated essentially of the subject, but is counterpredicated of it:[13]

> A proprium is what does not exhibit what it is to be for some subject, but belongs to it alone and counterpredicates with it. For example, it is a proprium of man to be capable of learning grammar: for if something is a man, then it is capable of learning grammar, and if something is capable of learning grammar, it is a man. (*Top.* 1.5 102a18–22)

Aristotle goes on to distinguish propria without qualification from propria at some time, as follows:

> No one would call something a proprium which may possibly belong to something else, e.g., being asleep for man, not even if it happened for a time to belong to it alone. Therefore, if any such thing were to be called a proprium, it will be called not a proprium without qualification (ἁπλῶς), but a proprium at some time (ποτέ) or relative to something.... It is clear that nothing which possibly belongs to something else counterpredicates; for it is not necessary for something to be a man if it is asleep. (*Top.* 1.5 102a22–30)

In this passage, propria without qualification are described as propria for which it is not possible to belong 'to something else'. For example, even if at some time all and only humans are asleep, 'being asleep' is not a proprium without qualification of 'man' because it is possible for it to belong to 'horse', say. On the other hand, 'capable of learning grammar' is a proprium without qualification of 'man' because it is not possible for it to belong to something else than humans. Thus, propria without qualification are propria for which certain possibility propositions are false: A is a proprium without qualification of B only if possibility propositions of the form 'A possibly belongs to C' are false for all C that are incompatible with B.

---

13. See also *Top.* 1.3 103b11–12, and pp. 115–116 above, as well as Barnes (1970: 137) and Primavesi (1996: 93).

A similar account can be given, I suggest, for a_X- and e_X-propositions without qualification in *Prior Analytics* 1.15. To say that propositions like 'A belongs to all (or, no) B' are without qualification may be taken to mean that certain possibility propositions in which A and B occur are false. In the next paragraphs, this suggestion is made more precise.

THE PRINCIPLE OF THE REALIZATION OF Q-PREDICATIONS. In chapter 15 of the second book of the *Prior Analytics,* Aristotle discusses instances of assertoric moods in which the major term is identified with the minor term. For example, he considers the following instance of Ferison in the third figure (2.15 64a27–30):[14]

If Ae_XB, and Ai_XB, then Ao_XA

In this mood, A is both the major and the minor term, while B is the middle term. The conclusion is an o_X-proposition whose predicate is identical with the subject. Aristotle asserts in chapter 2.15 that such o_X-propositions cannot be true.[15]

Likewise, we may consider an instance of Ferio XQM in which the major term is identified with the minor term:

If Ae_XB, and Bi_QA, then Ao_MA

The conclusion of this mood is an o_M-proposition whose predicate is identical with the subject. Can such a proposition be true? In other words, can a term be o_M-predicated of itself? Aristotle's claims of invalidity in the modal syllogistic commit him to the existence of terms that are both e_M- and o_M-predicated of themselves.[16] Aristotle does not

---

14. Strictly speaking, the sentence 'If Ae_XB and Ai_XB then Ao_XA' is not an instance of the mood Ferison, but a conditional whose truth is underwritten by this mood (since a mood is not a single sentence, but an argument consisting of several sentences). For present purposes, however, this point is of minor importance.

15. *APr.* 2.15 64b7–13; see p. 43n23 above.

16. Aristotle holds that Darapti NQX is invalid. Hence he is committed to there being terms A, B, and C such that A is a_N-predicated of B, C is a_Q-predicated of B, and A is not i_X-predicated of C. Given the contradictoriness of i_X- and e_X-propositions, A is e_X-predicated of C. It follows by Camestres NXX (whose validity Aristotle would surely accept) that B is e_X-predicated

specify which terms these are, but it seems clear that they cannot be substance terms. For just as it is implausible that 'animal' should be e_M- or o_M-predicated of 'man', it is also implausible that 'man' should be e_M- or o_M-predicated of itself. On the other hand, it is conceivable that a nonsubstance term like 'being asleep' is e_M- or o_M-predicated of itself, on the grounds that everything asleep will stop sleeping at a later time. But for substance terms this is not conceivable, since they are predicated necessarily of whatever they are predicated of.

Given that substance terms cannot be o_M-predicated of themselves, it follows that the two premises of the above instance of Ferio XQM cannot both be true when A is a substance term. In other words, if A is a substance term and the minor premise is true, the major premise is false:

If A is a substance term, and Bi_QA, then not Ae_XB

Given the contradictoriness of i_X- and e_X-propositions, this entails

If A is a substance term, and Bi_QA, then Ai_XB

Every i_Q-proposition whose subject term is a substance term entails the corresponding i_X-proposition. Now, i_Q-propositions are convertible. Hence any i_Q-proposition that contains a substance term entails the corresponding i_X-proposition. Moreover, every Q-proposition entails the corresponding i_Q-proposition. Consequently, any Q-proposition that contains a substance term entails the corresponding i_X-proposition. In other words, every Q-predication that involves a substance term is realized by the corresponding i_X-predication. We may call this the principle of the (partial) realization of Q-predications.

This principle bears some resemblance to what is known as the principle of plenitude. In its standard version, the principle of plenitude states that every possibility is realized at some time. As far as I can see, there is no evidence for this principle in Aristotle's modal syllogistic. But Aristotle is committed, by his endorsement of Ferio XQM, to an atemporal version of it concerning the realization of Q-predications by their corresponding i_X-predications.

---

of C. At the same time, C is a_Q- and hence also i_Q-predicated of B. Thus, it follows by Celarent XQM and Ferio XQM that B is both e_M- and o_M-predicated of B.

REJECTING THE COUNTEREXAMPLES TO BARBARA XQM AND CELARENT XQM. As we saw above, Aristotle characterizes propria without qualification by the falsity of certain possibility propositions: A is a proprium without qualification of B only if certain possibility propositions of the form 'A possibly belongs to C' are false.

Likewise, an a_X- or e_X-proposition can be taken to be without qualification only if certain possibility propositions are false. More precisely: in order for an a_X- or e_X-proposition to be without qualification, no Q-proposition can be true which would bring it into conflict with the principle of the realization of Q-predications. For example, consider the a_X-proposition 'Man belongs to all moving'. This proposition fails to be without qualification if the i_Q-proposition 'Moving two-sided-possibly belongs to some horse' is true. For given the principle of the realization of Q-predications, the truth of this i_Q-proposition implies that 'moving' is i_X-predicated of the substance term 'horse'. Combining this with the a_X-proposition under consideration, it follows that 'man' is i_X-predicated of 'horse'. But this is impossible. Thus, the a_X-proposition fails to be without qualification because the truth of the i_Q-proposition brings it into conflict with the principle of the realization of Q-predications. This allows us to reject Aristotle's alleged counterexample to Barbara XQM on the grounds that its major premise is not without qualification:

'man' is a_X-predicated of 'moving'
'moving' is a_Q-predicated of 'horse'
but 'man' is not i_M-predicated of 'horse'

As argued above, this counterexample can be rejected because a substance term is a_X-predicated of a nonsubstance term in the minor premise. Given the principle of the realization of Q-predications, it can also be rejected because the major premise is not without qualification—or because the major premise and the minor premise cannot both be true.

Next, consider the e_X-proposition 'Animal belongs to no moving'. This proposition fails to be without qualification if the i_Q-proposition 'Moving two-sided-possibly belongs to some man' is true. For given that 'animal' is necessarily a_X-predicated of 'man', the principle of the realization of Q-predications implies that this i_Q-proposition and the e_X-proposition cannot both be true. This allows us to reject the alleged counterexample to Celarent XQM, on the grounds that its major premise is not without qualification:[17]

'animal' is e_X-predicated of 'moving'
'moving' is a_Q-predicated of 'man'
and 'animal' is not o_M-predicated of 'man'

Thus, the counterexamples to Barbara and Celarent XQM can both be rejected by means of the principle of the realization of Q-predications.

AN OBJECTION TO THE PRINCIPLE OF THE REALIZATION OF Q-PREDICATIONS. Throughout the modal syllogistic, Aristotle gives several examples of e_X-predications in which one term is a substance term and the other a nonsubstance term. For instance, he assumes that 'animal' is e_X-predicated of 'white', and 'motion' of 'animal'.[18] At the same time, he also often assumes that nonsubstance terms like 'white' and 'motion' are a_Q- and i_Q-predicated of substance terms like 'animal'. But given that 'motion' is i_Q-predicated of 'animal', the principle of the realization of Q-predications implies that it is also i_X-predicated of it. So Aristotle should not be able to assume that 'motion' is e_X-predicated of 'animal'.

Similar problems arise for some of Aristotle's examples of a_X-predication. Aristotle frequently assumes that nonsubstance terms like 'motion' are a_X-predicated of substance terms like 'animal'. Now, given that he accepts that 'motion' is i_Q-predicated of 'animal', he should also accept that 'rest' is i_Q-predicated of 'animal'. If so, the principle of the realization of Q-predications implies that 'rest' is i_X-predicated of 'animal'. So Aristotle should not be able to assume that 'motion' is a_X-predicated of 'animal'. Thus, the principle of the realization of Q-predications implies that there is a tension between Aristotle's examples of a_X- and e_X-predications on the one hand and i_Q-predications on

17. Given the principle of the realization of Q-predications, the minor premise implies that 'moving' is i_X-predicated of the substance term 'man'. Combining this with the major premise, it follows that 'animal' is o_X-predicated of 'man'. Since this is impossible, the major and minor premises of the counterexample cannot both be true.

18. For these and the following examples of X-, Q-, and N-predications, see pp. 327–332.

the other. This tension might be taken to show that the principle is not valid, at least not throughout the whole modal syllogistic.

However, the tension can be resolved by observing that the X- and Q-predications in question do not occur within the same counterexample. When Aristotle assumes that 'motion' is e_X-predicated of 'animal' in a given counterexample he must not assume that 'motion' is i_Q-predicated of 'man' within *the same* counterexample. Otherwise, that would be an excellent counterexample to Ferio XQM. But nothing prevents him from assuming that 'motion' is i_Q-predicated of 'man' in *another* counterexample. Aristotle is free to assume the truth in different counterexamples of propositions that cannot be simultaneously true in the same counterexample.[19] For example, he assumes that 'motion' is a_X-predicated of 'animal' in some counterexamples and e_X-predicated in other counterexamples. Similarly, he assumes that 'white' is o_N-predicated of 'man' in one counterexample and a_Q-predicated in others.[20]

Thus, the objection against the principle of the realization of Q-predications can be answered, and the principle can be accepted as valid throughout the whole modal syllogistic. In the remainder of this chapter, I discuss two remarkable consequences of this principle.

WHATEVER IS I_Q-PREDICATED OF AN ATOMIC SUBSTANCE TERM IS ALSO A_X-PREDICATED OF IT. The first consequence I want to discuss concerns atomic terms. A term is atomic just in case everything of which it is a_X-predicated is a_X-predicated of it.[21] Atomic terms have the feature that everything that is i_X-predicated of them is also a_X-predicated of them. That they have this feature is entailed by the following principle, which we may call the principle of i_X-ecthesis:

If Ai_XB, then there is a C such that Ba_XC and Aa_XC

Although Aristotle does not explicitly assert this principle, it is widely thought that he endorses it. The principle is valid in the vast majority

19. See Patterson (1995: 46).

20. Aristotle is committed to the incompatibility of o_N- and a_Q-propositions; cf. pp. 200–201.

21. If mutual a_X-predication were taken to imply identity, this would mean that A is atomic if and only if it is not a_X-predicated of anything except itself.

of semantics for the assertoric syllogistic that have been put forward by commentators.[22] It is also valid in the preorder semantics of the assertoric syllogistic proposed in the present study. Given the principle of i_X-ecthesis, everything i_X-predicated of an atomic term is also a_X-predicated of it.[23]

Now, according to the principle of the realization of Q-predications, everything i_Q-predicated of a substance term is i_X-predicated of it. Hence everything i_Q-predicated of an atomic substance term is a_X-predicated of it. In other words:

If Ai_QB, and B is an atomic substance term, then Aa_XB

Typical examples of atomic substance terms are categorical singular terms such as 'Kallias'.[24] But the preorder semantics also includes models in which other kinds of terms are atomic. For example, we may consider models that do not contain any terms such as 'Kallias' and 'Mikkalos', so that the term 'man' is not a_X-predicated of anything except itself and therefore is atomic. In such a model, everything that is i_Q-predicated of 'man' is also a_X-predicated of it. Consequently, no two terms that are e_X-predicated of each other are both i_Q-predicated of 'man' in this model. For example, 'healthy' and 'ill' are not both i_Q-predicated of 'man'. These consequences may seem implausible, but, given that Aristotle accepts the principle of i_X-ecthesis, he is committed to them by the principle of the realization of Q-predications.

TERMS THAT ARE A_Q-PREDICATED OF THE SAME SUBSTANCE TERM ARE I_X-PREDICATED OF ONE ANOTHER. The second consequence I want to discuss is connected to the first one. It is that no two terms that are e_X-predicated of each other are a_Q-predicated of the same substance

22. Authors who accept the principle of i_X-ecthesis include Łukasiewicz (1957: 61–4), Patzig (1968: 161–4), Rescher & Parks (1971: 685), Rescher (1974: 11), Lejewski (1976: 6), Smith (1983: 226; 1989: xxiii), Detel (1993: 164), and Lagerlund (2000: 8).

23. Suppose Ai_XB and B is atomic. Because of the principle of i_X-ecthesis, we have Ba_XC and Aa_XC for some C. Since B is atomic, Ba_XC implies Ca_XB. Finally, Aa_XC and Ca_XB imply Aa_XB by assertoric Barbara.

24. For categorical singular terms, see pp. 48–49 and 83 above.

term (whether or not it is atomic). For example, given that 'healthy' and 'ill' are e_X-predicated of each other, they are not both a_Q-predicated of the substance term 'man' (even if 'man' is not atomic). In other words:

If Aa_QC, and Ba_QC, and C is a substance term, then Ai_XB

This follows from the principle of the realization of Q-predications, in combination with the principle of i_X-ecthesis and the principle that substance terms are not a_X-predicated of nonsubstance terms.[25] I have argued that Aristotle endorses the last principle (pp. 160–165). If this is correct, then, given that he is committed to the first two principles, he is also committed to the above consequence.

Although the consequence is at least prima facie implausible, it is, as far as I can see, in accordance with Aristotle's treatment of Q-propositions in *Prior Analytics* 1.1–22. It is, however, in conflict with a passage from chapter 1.34. In this passage, Aristotle discusses apparent counterexamples to Darapti QQQ such as the following:[26]

'health' is a_Q-predicated of 'man'
'illness' is a_Q-predicated of 'man'
but 'health' is not i_Q-predicated of 'illness'

Since Aristotle accepts Darapti QQQ as valid, he must reject the counterexample. He agrees that 'health' is not i_Q-predicated of 'illness', so he must deny at least one of the two premises. Now, Aristotle rejects the counterexample by claiming that the terms 'health' and 'illness' should be replaced by 'healthy' and 'ill', respectively (48a24–8). Thus, he seems to deny that 'health' or 'illness' is a_Q-predicated of 'man', but to accept that 'healthy' and 'ill' are a_Q-predicated of 'man', and hence that these

---

25. Suppose Aa_QC, Ba_QC, and C is a substance term. This implies Ai_XC (realization of Q-predications). Hence Aa_XD and Ca_XD for some D (i_X-ecthesis). Consequently, Ba_QD (Barbara QXQ), and D is a substance term (substance terms are not a_X-predicated of nonsubstance terms). Hence Bi_XD (realization of Q-predications). Finally, Aa_XD and Bi_XD imply Ai_XB (assertoric Datisi).

26. *APr.* 1.34 48a18–23; see Alexander *in APr.* 356.1–20, Ross (1949: 403–4), Ebert & Nortmann (2007: 817–18), and Striker (2009: 220).

two terms are i_Q-predicated of each other.[27] Thus, Aristotle seems to accept in chapter 1.34 that two terms that are presumably e_X-predicated of each other, namely, 'healthy' and 'ill', are both a_Q-predicated of the same substance term, namely, 'man'. This conflicts with the consequence stated above, that any two terms that are a_Q-predicated of the same substance term are i_X-predicated of each other. However, given that Aristotle is committed to (i) the principle of the realization of Q-predications, (ii) the principle of i_X-ecthesis, and (iii) the principle that substance terms are not a_X-predicated of nonsubstance terms, he must accept that consequence.

His example from *Prior Analytics* 1.34 suggests that Aristotle was not aware of this consequence. It is even questionable whether he was aware of the fact that he is committed to the principle of the realization of Q-predications through his endorsement of Ferio XQM. Nevertheless, the principle must be taken on board for any adequate semantics of the modal syllogistic. Accordingly, I accept the principle and its various consequences in my interpretation of the modal syllogistic. As a result, this interpretation cannot account for Aristotle's example from *Prior Analytics* 1.34.

In Chapter 17, the principle of the realization of Q-predications will be used to formulate an interpretation of Q-propositions.

27. See Alexander *in APr.* 356.7–19.

17

Two-Sided Possibility Propositions

The purpose of this chapter is to develop an interpretation of Aristotle's two-sided possibility propositions by giving a definition of a_Q- and i_Q-predication. I begin by introducing a formal framework for these definitions. This framework, which I call the predicable semantics, is based on three primitive relations: a_X-predication, a_N-predication, and a strengthened version of a_N-predication which is incompatible with o_M-predication. The three primitive relations will suffice to formulate definitions of a_Q- and i_Q-predication. I discuss some consequences of these definitions and explain how they relate to Aristotle's treatment of Q-propositions and to a distinction he draws in *Prior Analytics* 1.13 between two readings of a_Q-propositions.

AN ALTERNATIVE FORM OF NOTATION. Giving an interpretation of Q-propositions will require a somewhat more technical approach than employed so far. In particular, we will often need to quantify over categorical terms, that is, over argument-terms of categorical propositions. Thus, quantifiers will be applied to variables of the syntactic type of categorical terms. So far, we have represented categorical terms by uppercase letters such as 'A' and 'B'. But since it is typographically more convenient to apply quantifiers to lowercase letters, I will now represent them by lowercase letters such as '*a*' and '*b*'. Thus, categorical propositions will be rewritten as follows:

$\mathbb{X}^{\mathsf{i}}ab$ instead of $\mathrm{Ai_XB}$
$\mathbb{N}^{\mathsf{e}}ab$ instead of $\mathrm{Ae_NB}$
$\mathbb{Q}^{\mathsf{a}}ab$ instead of $\mathrm{Aa_QB}$, and so on

It should be emphasized that this is merely an alternative form of notation; the syntactic structure of Aristotle's categorical propositions is not affected by this notational variant.

THE PREDICABLE SEMANTICS OF THE MODAL SYLLOGISTIC. We have already seen how N-propositions can be interpreted in terms of the primitive relations of a_X- and a_N-predication (Chapters 11 and 12). In what follows, these relations will also be used to give an interpretation of Q- and M-propositions. To this end, it will be useful to indicate the two relations by the letters '**A**' and '**N**' as follows:

$\mathbf{A}ab$	a is a_X-predicated of b
$\mathbf{N}ab$	a is a_N-predicated of b

Aristotle's assertoric *dictum de omni* implies that a_X-predication is both reflexive and transitive (see p. 66). Moreover, the relations of a_X- and a_N-predication validate Barbara NXN and N-X-subordination.[1] Thus, the two relations are governed by the following four theses:

$\mathbf{A}aa$
$\mathbf{A}ab \wedge \mathbf{A}bc \supset \mathbf{A}ac$
$\mathbf{N}ab \wedge \mathbf{A}bc \supset \mathbf{N}ac$
$\mathbf{N}ab \supset \mathbf{A}ab$

Now, a_X- and a_N-predication alone do not suffice to formulate an interpretation of Q- and M-propositions. Part of the reason for this lies in Aristotle's asymmetric treatment of modal opposition. Aristotle is committed to the view that some a_N-predications are compatible with Q-predications. If Q-propositions are to be interpreted by means of a_N-predication, we need to be able to pick out those a_N-predications that are not compatible with Q-predications. Moreover, an interpretation of Q-propositions should incorporate the principle of the realization of Q-predications discussed in the previous section. Since this principle appeals to substance terms, we need to be able to pick out the class of substance terms.

1. See S15 and S16, pp. 129 and 131 above.

These requirements can be met by introducing a third primitive relation which is characterized by the following three conditions: (i) the predicate is a_N-predicated of the subject, (ii) the subject is a substance term, and (iii) the predicate is not o_M-predicated of the subject (and hence, given the principles of Q-M-subordination, not a_Q-predicated of it). This relation may be called strong a_N-predication, and may be indicated by the symbol '$\widehat{\mathbf{N}}$', as follows:

$\widehat{\mathbf{N}}ab$ (i) a is a_N-predicated of b,
(ii) b is a substance term, and
(iii) a is not o_M-predicated of b

Clearly, strong a_N-predication implies a_N-predication:

$$\widehat{\mathbf{N}}ab \supset \mathbf{N}ab$$

Moreover, it can be shown that strong a_N-predication satisfies the pattern of Barbara NXN. In other words, the following thesis is valid:[2]

$$\widehat{\mathbf{N}}ab \wedge \mathbf{A}bc \supset \widehat{\mathbf{N}}ac$$

This and the previous thesis can be added to the four theses governing a_X- and a_N-predication. The result is a system of three primitive relations, namely, a_X-, a_N-, and strong a_N-predication, governed by the above six theses. I will call this system the predicable semantics of the modal syllogistic.

DEFINING THE CLASS OF SUBSTANCE TERMS. The predicable semantics allows us to pick out the class of substance terms as follows. By definition,

---

2. To see this, suppose that a is strongly a_N-predicated of b. This means that a is a_N-predicated of b, b is a substance term, and a is not o_M-predicated of b. Given that b is a_X-predicated of c, it follows by Barbara NXN that a is a_N-predicated of c (S15, p. 129). Further, it follows that c is a substance term, since substance terms are not a_X-predicated of nonsubstance terms (see p. 160). Finally, it follows by a contraposed version of Bocardo MXM (which Aristotle accepts as valid; see pp. 208–209) that a is not o_M-predicated of c. Consequently, a is strongly a_N-predicated of c.

every subject of a strong a_N-predication is a substance term. Conversely, it can be shown that every substance term is the subject of a strong a_N-predication; for, as we saw above, every substance term is a_N-predicated of itself.[3] Moreover, no substance term is o_M-predicated of itself (p. 241). Consequently, every substance term is strongly a_N-predicated of itself. So the class of substance terms includes all and only those terms that are subjects of strong a_N-predications. In other words, a term is a substance term if and only if something is strongly a_N-predicated of it. Thus, the class of substance terms can be characterized by a one-place predicate $\widehat{\mathbf{\Sigma}}$ defined as follows:

$$\widehat{\mathbf{\Sigma}}a =_{df} \exists z \widehat{\mathbf{N}}za$$

The formula $\widehat{\mathbf{\Sigma}}a$ means that a is a substance term. The definition states that a is a substance term if and only if something is $\widehat{\mathbf{N}}$-predicated of a. With this definition at hand, we are now in a position to formulate a definition of a_Q- and i_Q-predication.

DEFINING A_Q- AND I_Q-PREDICATION. Let us begin with i_Q-predication. Aristotle frequently appeals to i_Q-predications in which either the subject or the predicate is a substance term. But he does not employ any i_Q-predications in which both terms are substance terms. In fact, it seems unlikely that there are such i_Q-predications. Neither is 'animal' i_Q-predicated of 'man', nor 'horse' of 'man', nor does this seem to be possible for any other pair of substance terms. Hence it is a necessary condition for i_Q-predication that at least one term be a nonsubstance term.

Moreover, i_Q-predication is incompatible with strong a_N-predication; for the latter excludes o_M-predication, whereas the former implies o_M-predication via Q-M-subordination.[4] Also, Aristotle's rule of i_Q-conversion implies that i_Q-predication is symmetric. Consequently, if

3. See S23, p. 146 (in combination with S10, p. 125).

4. For Q-M-subordination, see p. 215 above. By contrast, some authors define i_Q-predication in such a way that Q-M-subordination is invalid. They assume that o_M-predication, but not i_Q-predication, is incompatible with a_N-predication (see Thom 1996: 157, 160, and 208; Johnson 2004: 289 and 266). This means that i_Q- and hence also o_Q-predication does not imply o_M-predication. However, Aristotle seems to assume that o_Q-predication implies o_M-predication in his proof of Bocardo QXM (see pp. 208–209).

two terms are i_Q-predicated of each other, neither of them should be strongly a_N-predicated of the other.

Finally, a third necessary condition is given by the principle of the realization of Q-predications discussed above: if one of the two terms in an i_Q-predication is a substance term, then the terms are i_X-predicated of each other (p. 241). In other words, if one of them is a substance term, then there is something of which they are both a_X-predicated.

In sum, combining the three necessary conditions yields the following formula:

$$\mathbf{\Pi}ab =_{df} \neg(\widehat{\mathbf{\Sigma}}a \wedge \widehat{\mathbf{\Sigma}}b) \wedge \neg\widehat{\mathbf{N}}ab \wedge \neg\widehat{\mathbf{N}}ba \wedge ((\widehat{\mathbf{\Sigma}}a \vee \widehat{\mathbf{\Sigma}}b) \supset \exists z(\mathbf{A}az \wedge \mathbf{A}bz))$$

For the purposes of the predicable semantics, the three necessary conditions encapsulated in this formula can be taken to be jointly sufficient for i_Q-predication. Thus, the relation of i_Q-predication will be identified with $\mathbf{\Pi}$-predication in the predicable semantics, so that i_Q-predication is defined by means of the above three primitive relations. I should add that this definition is not meant fully to capture the intuitive meaning of i_Q-propositions. Nor do I intend to claim that Aristotle had in mind exactly this definition of i_Q-predication. Rather, the definition captures relevant aspects of Aristotle's explicit and implicit claims about i_Q-predication, and is therefore useful in constructing an adequate semantics for the modal syllogistic.

Let us now turn to a_Q-predication. Aristotle holds that Barbara QXQ is valid. So if a term is a_Q-predicated of another, it must be a_Q- and hence also i_Q-predicated of everything of which the other is a_X-predicated. In order for a to be a_Q-predicated of b, then, the following must hold in the predicable semantics:

$$\forall z(\mathbf{A}bz \supset \mathbf{\Pi}az)$$

It turns out that within the predicable semantics, this formula can be taken to be not only a necessary but also a sufficient condition for a_Q-predication. Thus a_Q-predication will be defined by that formula, with a being the predicate term and b the subject term. Again, this definition is not to be viewed as Aristotle's own, but rather as a tool for reconstructing his claims about a_Q-predication. The definition is in accordance with Aristotle's asymmetric treatment of modal opposition since it ensures that a_Q-predication is compatible with a_N-predication (but not with strong a_N-predication).

NEGATIVE Q-PROPOSITIONS. Aristotle holds that negative Q-propositions are equivalent to affirmative ones: a_Q-propositions are equivalent to e_Q-propositions, and i_Q- to o_Q-propositions. However, he does not take this equivalence for granted, but undertakes to prove it at 1.13 32a29–b1 (see p. 199n6). When he makes use of the equivalence in later chapters, he regards it as being in need of extensive explanation.[5] Sometimes he does not use the equivalence at all where he might be expected to do so, but gives two independent proofs for the inconcludence of premise pairs that differ only in that the one contains an affirmative Q-proposition and the other the corresponding negative Q-proposition.[6]

I do not follow Aristotle's practice in this respect. For our purposes, we may freely use the equivalence, so that the difference between affirmative and negative Q-propositions can be neglected in most cases. Thus affirmative and negative Q-propositions are given the same interpretation in the predicable semantics, as follows:[7]

$$\mathbb{Q}^{a/e}ab \qquad \forall z(\mathbf{A}bz \supset \mathbf{\Pi}az)$$
$$\mathbb{Q}^{i/o}ab \qquad \mathbf{\Pi}ab$$

The rest of this chapter explores some consequences of this interpretation of Q-propositions.

SUBSTANCE TERMS IN Q-PROPOSITIONS. The above interpretation of Q-propositions turns out to entail that no substance term is the predicate

5. See *APr.* 1.14 33a5–20 and 1.15 35a3–20.

6. For instance, the inconcludence of aa-2-QX and aa-2-XQ is proved at 1.18 37b35–8 after that of ea-2-QX and ae-2-XQ at 37b19–23. Likewise, the inconcludence of aa-2-QN and aa-2-NQ is proved at 1.19 38b13–23 after that of ea-2-QN and ae-2-NQ at 38a26–b5.

7. As a result, Aristotle's proof of the equivalence of affirmative and negative Q-propositions cannot be reconstructed within the predicable semantics, since there are not two different interpretations that could be proved to be equivalent.

of a true a_Q- or e_Q-proposition.[8] In other words, substance terms cannot be a_Q- or e_Q-predicated of anything in the predicable semantics. This is in accordance with Aristotle's practice: in all a_Q- and e_Q-predications adduced by him in the modal syllogistic, the predicate is a nonsubstance term; in none of them is the predicate a substance term (see p. 331). When establishing the invalidity of e_Q-conversion, Aristotle assumes that 'white' is e_Q-predicated of 'man', but denies that 'man' is e_Q-predicated of 'white' (1.17 37a4–9). Given that the substance term 'man' is not e_Q-predicated of 'white', it is difficult to see how another substance term might be e_Q-predicated of anything. Thus, there is good evidence that for Aristotle, substance terms cannot be e_Q- or a_Q-predicated of anything.[9]

On the other hand, Aristotle often assumes that substance terms are i_Q-predicated of nonsubstance terms, for instance, 'animal' of 'white' (p. 332). Also, such i_Q-predications follow by Aristotle's conversion rules from his numerous examples of a_Q- and i_Q-predications in which the predicate is a nonsubstance term and the subject is a substance term. This is in accordance with the above definition of i_Q-predication in the predicable semantics; for while **Π**-predication does not allow that both terms are substance terms, it does allow that one of them is a substance term.[10]

THE PERFECT MOODS OF THE PROBLEMATIC SYLLOGISTIC. Let us now consider whether the predicable semantics can account for the validity of the moods Aristotle regards as perfect in the problematic syllogistic.

---

8. This is proved in Fact 14, p. 290.

9. *Pace* Rini (2011: 153–4), who assumes that substance terms can be a_Q-predicated; for example, she takes 'oak' to be a_Q-predicated of 'acorn'.

10. By contrast, some commentators suggest that i_Q-predications whose predicate is a substance term should be rejected on the grounds that substance terms are not two-sided possibly predicated of any individuals; for example, no individual is two-sided possibly an animal (Striker 1994: 45; Nortmann 1996: 286–7, cf. also 378 and 385–6). However, as far as I can see, there is no reason to think that for Aristotle, an i_Q-proposition such as 'Animal two-sided possibly belongs to some white' implies that some individual is two-sided possibly an animal. At any rate, this implication is not valid in the predicable semantics.

There are altogether twelve such perfect moods, namely, the four standard first-figure moods of the form QQQ, QXQ, and QNQ. However, given N-X-subordination and the equivalence of affirmative and negative Q-propositions, we only need to consider four of them, namely, Barbara and Darii of the form QQQ and QXQ.

First, it is obvious that Barbara QXQ is valid in the predicable semantics. This can be seen as follows:

1. $\mathbb{Q}^{\mathbf{a}}ab$	(major premise)
2. $\mathbb{X}^{\mathbf{a}}bc$	(minor premise)
3. $\forall z(\mathbf{A}bz \supset \mathbf{\Pi}az)$	(from 1; by interpretation of a_Q-propositions)
4. $\mathbf{A}bc$	(from 2; by interpretation of a_X-propositions)
5. $\forall z(\mathbf{A}cz \supset \mathbf{\Pi}az)$	(from 3, 4; by transitivity of $\mathbf{A}$)
6. $\mathbb{Q}^{\mathbf{a}}ac$	(from 5; by interpretation of a_Q-propositions)

The steps in lines 3, 4, and 6 of this argument are justified by the interpretation in the predicable semantics of a_Q- and a_X-propositions. The step in line 5 is justified by the transitivity of a_X-predication (which is among the theses of the predicable semantics stated at the beginning of this section). Hence Barbara QXQ is valid in the predicable semantics. It is less obvious whether Darii QXQ is valid in it, but it can be shown that it is. The proof of its validity is not entirely straightforward, and is given in Appendix B.[11]

Likewise, it is not obvious whether Barbara QQQ is valid in the predicable semantics. But again, it can be shown to be valid. This is because the predicable semantics' definition of a_Q-predication turns out to imply that the predicate is $\mathbf{\Pi}$-predicated of everything of which the subject is $\mathbf{\Pi}$-predicated.[12] Given this implication, the validity in the predicable semantics of Barbara QQQ can be justified as follows:

11. Fact 38.2, p. 297.

12. In other words, $\forall z(\mathbf{A}bz \supset \mathbf{\Pi}az)$ implies $\forall z(\mathbf{\Pi}bz \supset \mathbf{\Pi}az)$. For proof, see Fact 23, p. 293.

1. $\mathbb{Q}^{\mathbf{a}}ab$	(major premise)
2. $\mathbb{Q}^{\mathbf{a}}bc$	(minor premise)
3. $\forall z(\mathbf{A}bz \supset \mathbf{\Pi}az)$	(from 1; by interpretation of a_Q-propositions)
4. $\forall z(\mathbf{\Pi}bz \supset \mathbf{\Pi}az)$	(from 3)
5. $\forall z(\mathbf{A}cz \supset \mathbf{\Pi}bz)$	(from 2; by interpretation of a_Q-propositions)
6. $\forall z(\mathbf{A}cz \supset \mathbf{\Pi}az)$	(from 4, 5)
7. $\mathbb{Q}^{\mathbf{a}}ac$	(from 6; by interpretation of a_Q-propositions)

The crucial implication is applied in line 4 of this argument. The validity in the predicable semantics of Darii QQQ can be established by means of the same implication, given that i_Q-predication is identified with $\mathbf{\Pi}$-predication.

Thus, all perfect moods of the problematic syllogistic are valid in the predicable semantics. Next, I discuss how Aristotle justifies the validity of these moods and how his justification relates to their justification within the predicable semantics.

TWO READINGS OF A_Q-PROPOSITIONS. Aristotle's justification of perfect moods in the problematic syllogistic seems to be different from that in the assertoric and apodeictic syllogistic. In the assertoric syllogistic, he justifies the validity of the perfect moods by means of the assertoric *dictum de omni et de nullo,* which is introduced in chapter 1.1 (see p. 37–39 above). In the apodeictic syllogistic, he seems to justify the validity of at least some perfect moods by means of the apodeictic *dictum de omni,* which is mentioned in chapter 1.8 (see p. 107). In the problematic syllogistic, by contrast, Aristotle does not mention a *dictum de omni et de nullo* for Q- or M-propositions. Instead, he introduces a distinction between two readings of a_Q-propositions, as follows:

> Now, the expression 'it is possible for this to belong to that' may be understood in two ways: it may mean either 'to that to which this belongs' or 'to that to which it is possible for this to belong'. For 'of what B, A is possible' signifies one or the other of the following: 'of what B is said' or 'of what it is possible for B to be said'. But 'of what B, A is possible'

is no different from 'it is possible for A to belong to all B'. (*APr.* 1.13 32b25–30)

In the last sentence of this passage, Aristotle states that 'it is possible for A to belong to all B' is equivalent to 'of what B, A is possible'.[13] The former phrase is Aristotle's standard expression for a_Q-propositions. In the second sentence, he states that the latter phrase can be understood in two ways, which implies that a_Q-propositions too can be understood in these two ways. First, a_Q-propositions can be understood to mean that for everything of which the subject is *actually* said, the predicate is possibly said of it: for any Z, if the subject is said of Z, then the predicate is possibly said of Z. Second, they can be taken to mean that for everything of which the subject is *possibly* said, the predicate is possibly said of it: for any Z, if the subject is possibly said of Z, then the predicate is possibly said of Z. In the secondary literature, the latter reading is called the ampliated reading of a_Q-propositions, and the former is called the nonampliated reading.

Aristotle does not explain what he means by the condition of 'being (possibly) said of' in the two readings. Nor does he explain how the two readings are related, or where which of them should be used in the modal syllogistic. However, he makes it clear that the distinction between the ampliated and the nonampliated reading plays a role in justifying the validity of his perfect moods: he refers to this distinction in his justification of the validity of Barbara and Darii QQQ.[14] In connection with

---

13. See Alexander *in APr.* 166.18–19.

14. 1.14 32b40–1 and 33a24–5; see also 33a3–5; cf. Maier (1900a: 144n1), Jenkinson (1928: ad loc.), Ross (1949: 330), Tredennick (1938: 261–3), Patzig (1968: 63), Wieland (1972: 129–33), Smith (1989: 128–9), and Ebert (1995: 239). Aristotle refers to the distinction between the two readings as a ὁρισμός (1.14 32b40 and 33a24). By contrast, the characterization of two-sided possibility as opposed to one-sided possibility at 1.13 32a18–21 is referred to as a διορισμός (1.14 33b23, 1.15 33b28, 33b30, 34b27; cf. Maier 1900a: 154 and 165–6, Jenkinson 1928: ad loc; similarly, the verb διορίζειν at 1.3 25b15 and 1.17 37a27–8). For this difference between ὁρισμός and διορισμός, see Flannery (1987: 468–70) and Thom (1994: 94–5; 1996: 37). I do not agree with the view that the two occurrences of ὁρισμός in chapter 1.14 refer to the characterization of two-sided possibility at 32a18–21 (Hintikka 1973: 184, Seel 1982: 330n77, Waterlow 1982: 16–17, van Rijen 1989: 30, Rini 2011: 119). Nor do

these two moods, he presumably intends to apply the ampliated reading. Aristotle does not explicitly refer to the distinction between the two readings when justifying the validity of the perfect QXQ-moods. Nevertheless, their justification too can be taken to be based on that distinction, and it presumably relies on the nonampliated reading.[15]

Inasmuch as the distinction between the two readings is used to justify the validity of perfect moods, it serves the same function as the assertoric and apodeictic *dictum de omni* in the assertoric and apodeictic syllogistic. Thus, the distinction between the two readings of a_Q-propositions may be regarded as a more complex version of a problematic *dictum de omni* characterizing the semantics of a_Q-propositions.[16]

THE RELATION BETWEEN THE AMPLIATED AND THE NONAMPLIATED READING. Some commentators regard the ampliated and the nonampliated reading as two logically independent kinds of a_Q-predication.[17] On this view, a_Q-propositions would seem to be ambiguous between the two readings in the modal syllogistic, much like Aristotle's N-propositions are often thought to be ambiguous between a *de dicto* and *de re* reading in the apodeictic syllogistic. Given such an ambiguity of a_Q-propositions, Barbara QQQ would be valid only when a_Q-propositions (or at least some of them) are understood in the ampliated sense, whereas Barbara QXQ would be valid only when they are understood in the nonampliated sense.

From the perspective of the predicable semantics, however, there is no need to assume such an ambiguity. For within the predicable semantics, Aristotle's nonampliated reading of a_Q-propositions can be identified with the definition of a_Q-predication given above, according to which the predicate term is **Π**-predicated of everything of which the subject

---

I agree with the view that διορισμός at 33b23–30 refers to the distinction between the ampliated and the nonampliated reading at 32b25–32 (Wieland 1972: 129–33, Ebert 1995: 239n18).

15. See Wieland (1972: 132).

16. Thus, Mueller (1999b: 185n6) attributes to Alexander (*in APr.* 167.10–20) the view that this distinction is just a development of the assertoric *dictum de omni;* cf. also Maier (1900b: 150–1).

17. Becker (1933: 32–7), Thom (1994: 91–109), Patterson (1995: 146 and 218–19), and Rini (2011: 124–31 and 146–7).

term is a_X-predicated. The ampliated reading can be identified with the condition that the predicate term is **Π**-predicated of everything of which the subject term is **Π**-predicated. On this account, the nonampliated reading implies the ampliated one. Thus, there is a single uniform notion of a_Q-predication applied in the entire modal syllogistic, namely, the nonampliated reading; but this reading implies the ampliated reading.[18] When establishing the validity of Barbara QXQ, the major premise can be given the nonampliated reading. When establishing the validity of Barbara QQQ, on the other hand, the major premise can be given the ampliated reading while the minor premise is given the nonampliated reading, as we saw above.[19] Thus, the predicable semantics can help us understand the role that Aristotle's distinction between the two readings plays in the justification of the perfect moods in the problematic syllogistic.

LIMITATIONS OF THE DEFINITIONS OF A_Q- AND I_Q-PREDICATION. It must be admitted that the above definitions of a_Q- and i_Q-predication are not entirely satisfactory, and that they are not in accordance with Aristotle's practice in some respects. In particular, they imply that every nonsubstance term is a_Q-predicated of every nonsubstance term.[20] For example, they imply that 'healthy' is a_Q-predicated not only of 'sleeping' but also of 'ill' and 'being divisible by 3'. Likewise for 'health', 'sleep', and 'divisibility by 3'. But in *Prior Analytics* 1.34, Aristotle

18. Several authors have suggested that the nonampliated reading implies, or is even equivalent to, the ampliated reading (Patzig 1968: 66, Nortmann 1996: 76–80, Thom 1996: 222–3, Johnson 2004: 294–5). Similarly, Wieland (1972: 130–1) assumes that both readings follow from a unique notion of a_Q-predication. Hintikka (1973: 39) defines a_Q-predication as the conjunction of an ampliated and a nonampliated reading. However, since he does not take the nonampliated reading to imply the ampliated one, Hintikka fails to account for the validity of Barbara QXQ (see Thom 1994: 108).

19. In justifying the validity of QQQ-moods, some commentators take both premises to have the ampliated reading (Thom 1994: 91–109, 1996: 42–3, Schmidt 2000: 117–21). However, I follow those commentators who take only the major premise to have the ampliated reading (Becker 1933: 33–5, Patterson 1995: 239, Rini 2011: 129).

20. This is because any two nonsubstance terms are **Π**-predicated of one another; see Fact 140, p. 323.

denies that 'health' is i_Q- or a_Q-predicated of 'illness', while accepting that 'healthy' is i_Q-predicated of 'ill'.[21] Moreover, he would presumably deny that 'healthy' is a_Q- or i_Q-predicated of 'being divisible by 3'.

I will not undertake here to avoid these unwanted consequences by modifying the above definitions of a_Q- and i_Q-predication. These definitions are not meant to provide a fully satisfactory explication of Aristotle's relations of a_Q- and i_Q-predication. Rather, they are intended to capture important aspects of his explicit claims and implicit commitments about these relations. As such, the definitions are useful, and will allow us to construct a semantics that matches Aristotle's claims of validity and invalidity in the modal syllogistic.

---

21. *APr.* 1.34 48a18–28; cf. Alexander *in APr.* 356.1–20. See also pp. 246–247 above.

18

One-Sided Possibility Propositions

The purpose of this final chapter is to formulate an interpretation in the predicable semantics of M-propositions. I also reconsider N-propositions and discuss how they can be interpreted within the predicable semantics.

DEFINING $\mathrm{A_M}$-PREDICATION. If Aristotle accepted the principles of M-N-contradictoriness, the four kinds of M-predication could be simply defined as the contradictories of the four kinds of N-predication. However, as we have seen, Aristotle is committed to denying at least some of these principles (p. 202). Therefore, M-N-contradictoriness cannot serve as a guide in defining the four kinds of M-predication. Instead, the definitions may be guided by the principles of X-M- and Q-M-subordination, which are taken to be valid throughout this study (p. 215).

According to these principles, every $\mathrm{a_X}$-proposition and every $\mathrm{a_Q}$-proposition should imply the corresponding $\mathrm{a_M}$-proposition. In the predicable semantics, $\mathrm{a_X}$-predication amounts to the condition that the predicate be $\mathrm{a_X}$-predicated of everything of which the subject is $\mathrm{a_X}$-predicated. Similarly, $\mathrm{a_Q}$-predication means that the predicate is $\mathbf{\Pi}$-predicated of everything of which the subject is $\mathrm{a_X}$-predicated. Thus, a natural way to define $\mathrm{a_M}$-predication is to require that the predicate be $\mathrm{a_X}$- *or* $\mathbf{\Pi}$-predicated of everything of which the subject is $\mathrm{a_X}$-predicated. In other words, given the following abbreviation,

$$\overline{\mathbf{\Pi}}ab =_{df} \mathbf{\Pi}ab \vee \mathbf{A}ab,$$

a term is $\mathrm{a_M}$-predicated of another if and only if it is $\overline{\mathbf{\Pi}}$-predicated of everything of which the other is $\mathrm{a_X}$-predicated. This will serve as the definition of $\mathrm{a_M}$-predication in the predicable semantics.

DEFINING $\mathrm{I_M}$-PREDICATION. Given the definition of $\mathrm{a_M}$-predication, a natural way to define $\mathrm{i_M}$-predication is to require that the predicate be $\overline{\mathbf{\Pi}}$-predicated of *something* of which the subject is $\mathrm{a_X}$-predicated. This definiens turns out to be equivalent to the negation of the following relation of what we may call $\mathbf{\Gamma}$-incompatibility:[1]

$$\mathbf{\Gamma}ab =_{df} (\widehat{\mathbf{\Sigma}}a \vee \widehat{\mathbf{\Sigma}}b) \wedge \neg\exists z(\mathbf{A}az \wedge \mathbf{A}bz)$$

Hence, an equivalent way to formulate the above definition of $\mathrm{i_M}$-predication is that a term is $\mathrm{i_M}$-predicated of another if and only if it is not $\mathbf{\Gamma}$-incompatible with it, that is, if and only if it is not the case that the two terms are $\mathrm{e_X}$-predicated of each other and at least one of them is a substance term. Thus, $\mathrm{i_M}$-predication amounts to the condition that the two terms be $\mathrm{i_X}$-predicated of each other or that they are both nonsubstance terms. This will serve as the definition of $\mathrm{i_M}$-predication in the predicable semantics.

The definition is in accordance with Aristotle's conversion rules concerning affirmative M-propositions. First of all, $\mathbf{\Gamma}$-incompatibility is symmetric, so that Aristotle's rule of $\mathrm{i_M}$-conversion is valid in the predicable semantics. Second, given the earlier equivalent formulation of the definition of $\mathrm{i_M}$-predication and given the reflexivity of $\mathrm{a_X}$-predication, it is obvious that $\mathrm{i_M}$-predication follows from $\mathrm{a_M}$-predication in the predicable semantics. Hence Aristotle's rule of $\mathrm{a_M}$-conversion is valid in the semantics.

Moreover, the above definition verifies the principles of X-M- and Q-M-subordination for i-propositions. X-M-subordination for i-propositions is verified because $\mathrm{i_X}$-predication implies the negation of $\mathbf{\Gamma}$-incompatibility. Q-M-subordination for i-propositions is verified because $\mathrm{i_Q}$-predication was defined in such a way that it satisfies the principle of the realization of Q-predications: if one of the two terms is a substance term, then they are $\mathrm{i_X}$-predicated of each other. This amounts to a denial of $\mathbf{\Gamma}$-incompatibility.

So the above definitions of $\mathrm{a_M}$- and $\mathrm{i_M}$-predication yield satisfactory results and harmonize with the rest of the predicable semantics.

1. In other words, $\exists z(\mathbf{A}bz \wedge \overline{\mathbf{\Pi}}az)$ is equivalent to $\neg\mathbf{\Gamma}ab$; for proof, see Fact 141, p. 323.

DEFINING $\mathrm{E_M}$-PREDICATION. Now for negative M-propositions. As before, every $\mathrm{e_X}$- and $\mathrm{e_Q}$-proposition should imply the corresponding $\mathrm{e_M}$-proposition. In the predicable semantics, $\mathrm{e_X}$-predication means that the predicate is not $\mathrm{a_X}$-predicated of anything of which the subject is $\mathrm{a_X}$-predicated; $\mathrm{e_Q}$-predication means that the predicate is $\mathbf{\Pi}$-predicated of everything of which the subject is $\mathrm{a_X}$-predicated. Thus, a natural way to define $\mathrm{e_M}$-predication is to require that for everything of which the subject is $\mathrm{a_X}$-predicated, the predicate is $\mathbf{\Pi}$-predicated *or* not $\mathrm{a_X}$-predicated of it.

Now, consider the condition that a term be $\mathbf{\Pi}$-predicated or not $\mathrm{a_X}$-predicated of another. This turns out to be equivalent to the condition that the first term not be $\overline{\mathbf{N}}$-predicated of the other, where $\overline{\mathbf{N}}$-predication is defined as follows:[2]

$$\overline{\mathbf{N}}ab =_{df} \widehat{\mathbf{N}}ab \vee (\widehat{\mathbf{\Sigma}}a \wedge \mathbf{A}ab)$$

$\overline{\mathbf{N}}$-predication means that the predicate is strongly $\mathrm{a_N}$-predicated of the subject, or that it is a substance term that is $\mathrm{a_X}$-predicated of the subject.[3] Given this, the above tentative definition of $\mathrm{e_M}$-predication amounts to the requirement that the predicate term not be $\overline{\mathbf{N}}$-predicated of anything of which the subject term is $\mathrm{a_X}$-predicated.

However, this definition would fail to validate Aristotle's rule of $\mathrm{e_M}$-conversion.[4] Similarly to what we saw above for $\mathrm{i_N}$-predication,

---

2. This equivalence is proved in Fact 143, pp. 323–324.

3. Consider the case that the predicate is a substance term that is $\mathrm{a_X}$-predicated of the subject. It follows that the predicate is $\mathrm{a_N}$-predicated of the subject, and that the subject is a substance term (see p. 160). Further, if it is assumed that substance terms cannot be both $\mathrm{a_N}$- and $\mathrm{o_M}$-predicated of substance terms, it follows that the predicate is strongly $\mathrm{a_N}$-predicated of the subject (see the characterization of strong $\mathrm{a_N}$-predication on p. 250). Thus, given the assumption just mentioned, $\overline{\mathbf{N}}$-predication is simply equivalent to $\widehat{\mathbf{N}}$-predication. The assumption in question is very plausible and can be taken to be true throughout this study. But since the assumption is not important for the proofs performed within the predicable semantics, I will not make it explicit in the predicable semantics. We therefore need to distinguish formally between $\overline{\mathbf{N}}$- and $\widehat{\mathbf{N}}$-predication, although we may think of them as basically the same relation.

4. Aristotle states this rule at *APr.* 1.3 25b3–13 and uses it in the proof of Camestres QXM, at 1.18 37b29.

the problem can be solved by adding to the nonconvertible definiens its converted counterpart, either by conjunction or by disjunction. Since the conjunctive strategy leads to problems with verifying Q-M-subordination for e-propositions, I prefer the disjunctive strategy. Thus, a term is e_M-predicated of another if and only if the first is not $\overline{\mathbf{N}}$-predicated of anything of which the second is a_X-predicated, *or* the second is not $\overline{\mathbf{N}}$-predicated of anything of which the first is a_X-predicated. This, then, is the definition of e_M-predication in the predicable semantics.

DEFINING o_M-PREDICATION. If o_M-propositions were the contradictories of a_N-propositions, o_M-predication could be defined as the negation of a_N-predication. That is, it could be defined as the negation of $\mathbf{N}$-predication. But since o_M-propositions are not contradictory to a_N-propositions in Aristotle's modal syllogistic, this definition does not work. However, o_M-predication can still be defined as the negation of $\overline{\mathbf{N}}$-predication. This turns out to be a viable definition of o_M-predication in the predicable semantics. In particular, it can be shown that on this definition the principle of subalternation is valid for negative M-propositions.[5] So a term is o_M-predicated of another, I suggest, if and only if it is not $\overline{\mathbf{N}}$-predicated of it.

In sum, then, the four kinds of M-propositions are interpreted in the predicable semantics as follows:

$\mathbb{M}^{\mathbf{a}}ab$	$\forall z(\mathbf{A}bz \supset \overline{\mathbf{\Pi}}az)$
$\mathbb{M}^{\mathbf{e}}ab$	$\forall z(\mathbf{A}bz \supset \neg\overline{\mathbf{N}}az) \vee \forall z(\mathbf{A}az \supset \neg\overline{\mathbf{N}}bz)$
$\mathbb{M}^{\mathbf{i}}ab$	$\neg\mathbf{\Gamma}ab$
$\mathbb{M}^{\mathbf{o}}ab$	$\neg\overline{\mathbf{N}}ab$

We now have an interpretation in the predicable semantics of Q- and M-propositions. In order to obtain an interpretation of the whole modal syllogistic, it remains to consider N-propositions. We have already seen how, in the context of the apodeictic syllogistic, i_N- and e_N-predication can be defined by means of a_N- and a_X-predication. In what follows,

5. That is, it can be shown that the negation of $\overline{\mathbf{N}}$-predication follows from the definiens of the above disjunctive definition of e_M-predication; for proof, see Fact 18, pp. 291–292.

I reconsider these definitions to see whether they are suitable for the context of the problematic syllogistic, and if not, how they might be modified appropriately. Also, I propose a definition of o_N-predication.

DEFINING I_N- AND E_N-PREDICATION. In the discussion of the apodeictic syllogistic, I proposed a disjunctive definition of i_N-predication, in which the predicate is required to be a_N-predicated of something of which the subject is a_X-predicated *or* vice versa (p. 179). This definition turns out to be appropriate not only for the apodeictic syllogistic but also for the problematic syllogistic. In particular, its definiens is compatible with the definiens of the above definition of a_Q-predication. This is in line with Aristotle's asymmetric treatment of modal opposition, according to which i_N-predication is compatible with a_Q-predication. Thus, the disjunctive definition of i_N-predication need not be modified and can be used throughout the whole modal syllogistic.

As to e_N-predication, I proposed defining it by the condition that the two terms be essence terms that are e_X-predicated of each other (p. 170). Now, essence terms are exactly those terms that are subjects of a_N-predications (p. 170n2). Thus, the class of essence terms can be characterized by a one-place predicate $\mathbf{\Sigma}$ defined as follows:

$$\mathbf{\Sigma} a =_{df} \exists z \mathbf{N} za$$

The formula $\mathbf{\Sigma} a$ means that a is an essence term. The definition states that a is an essence term if and only if something is a_N-predicated of a. Given this definition, the condition proposed for defining e_N-predication can be written as follows:

$$\mathbf{\Sigma} a \wedge \mathbf{\Sigma} b \wedge \neg \exists z(\mathbf{A} bz \wedge \mathbf{A} az)$$

As far as the apodeictic syllogistic is concerned, this is a perfectly acceptable definition of e_N-predication.[6] However, it leads to difficulties in the problematic syllogistic, and specifically with principles of modal opposition. Aristotle holds that e_N-propositions are incompatible with i_Q-propositions, and some of his indirect proofs are based on this incompatibility (pp. 199–201). But the suggested definiens for e_N-predication is compatible with the above definiens for i_Q-predication; for

6. For further discussion of this point, see p. 312 below.

when two essence terms are $\mathrm{e_X}$-predicated of each other in the predicable semantics, they may still be $\mathrm{i_Q}$-predicated of each other.

In order to address this problem, one might try to identify $\mathrm{e_N}$-predication with the relation of $\mathbf{\Gamma}$-incompatibility introduced above. In this case, $\mathrm{e_N}$-predication would be exactly the contradictory of $\mathrm{i_M}$-predication in the predicable semantics, and hence would be incompatible with $\mathrm{i_Q}$-predication. Unfortunately, however, this definition would imply that the mood Celarent NQN, which is held to be invalid by Aristotle, is valid in the predicable semantics.[7]

These problems can be solved by combining $\mathbf{\Gamma}$-incompatibility with the above tentative definition of $\mathrm{e_N}$-predication. More precisely, they can be solved by taking a term to be $\mathrm{e_N}$-predicated of another if and only if it is $\mathbf{\Gamma}$-incompatible with it and both terms are essence terms. Thus, $\mathrm{e_N}$-predication is identified with the following relation of $\mathbf{K}$-incompatibility:

$$\mathbf{K}ab =_{df} \mathbf{\Sigma}a \wedge \mathbf{\Sigma}b \wedge \mathbf{\Gamma}ab$$

$\mathbf{K}$-incompatibility means that the two terms are essence terms that are $\mathrm{e_X}$-predicated of one another, and that at least one of them is a substance term. As before, I do not claim that this yields a fully satisfactory analysis of Aristotle's notion of $\mathrm{e_N}$-predication. For example, it conflicts with Aristotle's assumption, in *Prior Analytics* 1.34, that the nonsubstance term 'health' is $\mathrm{e_N}$-predicated of the nonsubstance term 'illness'.[8] Such $\mathrm{e_N}$-predications cannot be captured by the relation of $\mathbf{K}$-incompatibility, since this requires that at least one term be a substance term. Nevertheless, this relation yields a formally useful definition of $\mathrm{e_N}$-predication both for the apodeictic and for the problematic syllogistic. It validates Celarent NXN and $\mathrm{e_N}$-conversion in the same way as the old definition of $\mathrm{e_N}$-predication. Moreover, it is incompatible with $\mathrm{i_Q}$-predication and accounts for the invalidity of Celarent NQN. Thus, $\mathrm{e_N}$-predication may be identified with $\mathbf{K}$-incompatibility in the predicable semantics.

---

7. See Fact 142, p. 323. For Aristotle's claim that Celarent NQN is invalid, see p. 154n2. This claim of Aristotle's is often ignored by commentators (e.g., Ross 1949: 286–7, McCall 1963: 84–5, Smith 1989: 234, and Mueller 1999b: 67). In an earlier paper (2006), I have accidentally given a semantics in which Celarent NQN is valid, not noticing that this is in conflict with Aristotle's explicit pronouncements (see Malink 2006: 127n53).

8. *APr.* 1.34 48a4–5; see p. 155n6 above.

DEFINING o_N-PREDICATION. Giving a definition of o_N-predication is not straightforward (which is also why I have not given one so far). The problems mainly stem from Aristotle's claims in the apodeictic syllogistic. Specifically, they stem from his claims that Baroco NNN and Bocardo NNN are valid but Baroco XNN and Bocardo NXN are invalid. As we have seen, the invalidity of these last two moods implies that o_N-predication cannot be defined by the condition that the predicate be e_N-predicated of something of which the subject is a_X-predicated (pp. 181–185). In the predicable semantics, this means that o_N-predication cannot be defined by the following condition (where b is the subject term and a the predicate term):

$$\exists z(\mathbf{A}bz \wedge \mathbf{K}az)$$

Aristotle offers few hints as to what a definition of o_N-predication should look like. In fact, it is doubtful whether he had in mind any such definition when developing his modal syllogistic. I have argued that his treatment of Baroco XNN and Bocardo NXN is not based on a precise interpretation of o_N-propositions within a well-defined semantic framework, but rather on general considerations concerning certain monotonicity properties of o_N-propositions (pp. 185–190). Nevertheless, if we are to give a semantics for the modal syllogistic, we must specify a definition of o_N-predication in it. Given the complications mentioned, we should not expect there to be a simple definition that is in accordance with Aristotle's claims of validity and invalidity. But if some complexity is allowed, there are ways to formulate such a definition. One of them is this:

$$\exists z((\mathbf{A}bz \wedge \mathbf{K}az) \vee (\mathbf{N}bz \wedge \overline{\mathbf{K}}az)),$$

where the new relation of $\overline{\mathbf{K}}$-incompatibility is defined by the following complex condition:

$$\overline{\mathbf{K}}az =_{df} \exists v\mathbf{N}av \wedge \forall u(\mathbf{A}au \wedge \mathbf{\Sigma}u \supset (\widehat{\mathbf{\Sigma}}z \wedge \widehat{\mathbf{\Sigma}}u \wedge \neg\exists s(\mathbf{A}zs \wedge \mathbf{A}us)))$$

On this definition, a term is o_N-predicated of another just in case it is $\mathbf{K}$-incompatible with something of which the other is a_X-predicated, or it is $\overline{\mathbf{K}}$-incompatible with something of which the other is a_N-predicated.[9]

9. It is worth pointing out that this definition is similar in structure to Brenner's (2000: 336) interpretation of o_N-propositions:

Of course, this is a purely technical solution that is far from anything we find in Aristotle. Still, it yields a definition of o_N-predication that is useful in that it verifies Aristotle's claims of validity and invalidity. Specifically, it verifies his treatment of Baroco XNN and Bocardo NXN, and also his endorsement of the principle that o_N-propositions are incompatible with a_Q-propositions.[10] So we may accept this as the definition of o_N-predication in the predicable semantics.

In sum, then, the four kinds of N-propositions are interpreted in the predicable semantics as follows:

$$\begin{array}{ll} \mathbb{N}^{\mathbf{a}}ab & \mathbf{N}ab \\ \mathbb{N}^{\mathbf{e}}ab & \mathbf{K}ab \\ \mathbb{N}^{\mathbf{i}}ab & \exists z((\mathbf{A}bz \wedge \mathbf{N}az) \vee (\mathbf{A}az \wedge \mathbf{N}bz)) \\ \mathbb{N}^{\mathbf{o}}ab & \exists z((\mathbf{A}bz \wedge \mathbf{K}az) \vee (\mathbf{N}bz \wedge \overline{\mathbf{K}}az)) \end{array}$$

CONCLUSION. We have now arrived at an interpretation of N-, Q-, and M-propositions, thereby completing the introduction of the predicable semantics. All details of this semantics are summarized in Appendix B. There it is also established that the predicable semantics is adequate with respect to Aristotle's modal syllogistic. That is, it is established that every mood and conversion rule held to be valid (or invalid) by Aristotle is valid (or invalid) in the predicable semantics, and that every premise pair held to be inconcludent by him is inconcludent in it. Thus one of the main aims of this study is achieved, namely, to prove that the body of Aristotle's claims of validity, invalidity, and inconcludence is consistent.

It must be acknowledged, however, that the predicable semantics has several limitations. One of them is that, while being in accordance with Aristotle's claims of validity, invalidity, and inconcludence, it cannot account for all of his proofs of these claims. The most prominent proofs that cannot properly be reconstructed within the predicable semantics are the following:

$$\exists x(\mathrm{B}x \wedge \forall y(\mathrm{A}y \supset \Box x \neq y)) \;\vee\; \exists x(\Box\mathrm{B}x \wedge \neg\mathrm{A}x \wedge \forall y(\Box\mathrm{A}y \supset \Box x \neq y)).$$

By comparison, his interpretation of e_N-propositions is $\forall x(\mathrm{B}x \supset \forall y(\mathrm{A}y \supset \Box x \neq y))$.

10. For this last principle, see pp. 200–201 above. Within the predicable semantics, the principle is proved in Fact 30, p. 295.

- the proofs of the conversion rules for N-, Q-, and M-propositions in chapter 1.3 (see pp. 172–179)
- the proofs by ecthesis of Baroco NNN and Bocardo NNN in chapter 1.8 (see pp. 181–185)
- the proof of the equivalence of affirmative and negative Q-propositions in chapter 1.13 (see p. 253)
- the proofs of first-figure XQM-moods in chapter 1.15 (which I have not discussed in this study; see pp. 232–233)

Also, some of the counterexamples used by Aristotle to prove his claims of invalidity and inconcludence cannot properly be reconstructed within the predicable semantics.[11]

Since the predicable semantics cannot account for some of Aristotle's proofs, it cannot establish that these proofs and the assumptions on which they rely are consistent with the rest of the modal syllogistic. In this sense, the predicable semantics cannot establish the consistency of the modal syllogistic. In fact, there is reason to think that the modal syllogistic is not consistent in this sense. For example, Aristotle's ecthetic proofs of Baroco NNN and Bocardo NNN seem to be in conflict with his claim that Baroco XNN and Bocardo NXN are invalid (pp. 181–185). Similar problems arise for Aristotle's proofs of XQM-moods in chapter 1.15; there is reason to think that, while these proofs are in themselves well-reasoned and unobjectionable, they are in conflict with some claims that Aristotle makes elsewhere in the modal syllogistic.[12] Moreover, as we saw above, it is not straightforward to interpret Aristotle's proofs of e_M-conversion and of Bocardo QXM in such a way that they are in accordance with his asymmetric treatment of modal opposition (pp. 207–210). Thus, while the set of Aristotle's claims of validity, invalidity, and inconcludence is consistent, this set may not always be entirely consistent with his proofs of these claims.

Let me mention another limitation of the predicable semantics, concerning its explanatory power. It is one thing to match all of Aristotle's

11. For instance, the counterexample used to establish the invalidity of Baroco XNN in *Prior Analytics* 1.10 (see p. 185), and one of the counterexamples used to establish the inconcludence of the premise pair ea-2-QN in chapter 1.19 (see p. 218).

12. This is argued in Malink & Rosen (forthcoming).

claims of validity, invalidity, and inconcludence; it is another to explain why he made the claims he did make. While the first task is completed by the predicable semantics, the second, explanatory task is only partially accomplished. Specifically, the second task has been accomplished for large parts of the apodeictic syllogistic, but not for the problematic syllogistic. As we have seen, the *Topics*' theory of predicables and categories can explain why Aristotle took Barbara NXN to be valid (Chapters 8–10). Based on this theory, the predicable semantics gives definitions of e_N- and i_N-predication that can plausibly be attributed to Aristotle. In this way, it can account for Aristotle's claims concerning e_N- and i_N-predication. For example, it can explain why he endorsed Celarent NXN and the rule of e_N-conversion, without attributing to him an ambiguity in the use of e_N-propositions (Chapter 11).[13] More generally, the predicable semantics can explain all of his claims of validity and invalidity in the apodeictic syllogistic, except his claims about those moods that contain an o_N-premise (Baroco NNN and XNN, Bocardo NNN and NXN). It cannot explain why Aristotle made these last claims because the definition of o_N-predication given on p. 267 is too artificial to be attributed to Aristotle.[14]

Likewise, the interpretations in the predicable semantics of Q- and M-propositions are rather complex and technical. It is therefore unlikely that Aristotle endorsed them and had them in mind when he developed the problematic syllogistic. Consequently, the predicable semantics cannot explain why Aristotle made the claims of validity and invalidity

13. The explanation presented in Chapter 11 is based on the preliminary definition of e_N-predication given on p. 170. But the explanation can be carried over to the strengthened version of e_N-predication given on p. 266 that is used in the predicable semantics.

14. Despite this, the predicable semantics can explain Aristotle's endorsement of moods that have an o_N-conclusion but no o_N-premise (namely, Ferio NXN, Festino NXN, Ferison NXN, and Felapton NXN). For the validity in the predicable semantics of these moods is guaranteed solely by the left disjunct of the above definition of o_N-predication, that is, by $\exists z(\mathbf{A}bz \wedge \mathbf{K}az)$; their validity does not depend in any way on the artificial right disjunct, $\exists z(\mathbf{N}bz \wedge \overline{\mathbf{K}}az)$. Being equivalent to $\exists z(\mathbb{X}^{\mathbf{a}}bz \wedge \mathbb{N}^{\mathbf{e}}az)$, the left disjunct expresses a natural sufficient condition for o_N-predication that can plausibly be attributed to Aristotle. Thus the definition of o_N-predication can explain why Aristotle endorsed those moods.

he made in the problematic syllogistic. Nevertheless, the predicable semantics is, I think, a useful model that can help explain at least some aspects of Aristotle's treatment of Q- and M-propositions. For example, it can help us understand Aristotle's distinction between the two readings of a_Q-propositions in *Prior Analytics* 1.13, and the role this distinction plays in the justification of perfect moods (pp. 254–259). Moreover, it helps us see what the costs are of an interpretation of the modal syllogistic on which the body of Aristotle's claims of (in)validity and inconcludence is consistent. Thus, my hope is that, in addition to proving the consistency of these claims, the predicable semantics can, despite its limitations, improve our understanding of the modal syllogistic.

Appendix A

Aristotle's Claims of Validity, Invalidity, and Inconcludence

This appendix provides a synopsis of Aristotle's claims of validity, invalidity, and inconcludence in *Prior Analytics* 1.1–22. The synopsis consists of seven tables. Table A.1 contains Aristotle's claims about the (in)validity of conversion rules. Tables A.2–A.7 contain his claims about the (in)validity of syllogistic moods and about the inconcludence of premise pairs.

Boldface Bekker page numbers, such as **25a32**, indicate that the conversion rule or mood under consideration is held to be valid by Aristotle. Italic Bekker page numbers, such as *37b12,* indicate that the conversion rule or mood is held to be invalid by Aristotle. If an italic Bekker page number is accompanied by the symbol $\nvdash$, the premise pair under consideration is held to be inconcludent by Aristotle. Bekker page numbers refer to the beginning of the passage in which Aristotle makes the relevant claim of (in)validity or inconcludence. When no Bekker page number is given, the premise pair is not discussed by Aristotle.

Every conversion rule, mood, and premise pair is associated with a number in square brackets, such as [**36**] or [*72*]. This is the number of the Fact in which the conversion rule, mood, or premise pair is discussed in the predicable semantics in Appendix B. Boldface numbers indicate validity in the predicable semantics; italic numbers indicate invalidity in the predicable semantics. If an italic number in square brackets is accompanied by $\nvdash$, the premise pair under consideration is inconcludent in the predicable semantics. Since the predicable semantics is adequate with respect to Aristotle's modal syllogistic, every Bekker page number in the synopsis is boldface (or italic) if and only if the corresponding

(text continues on page 281)

Conversion from	to	
$Aa_{X}B$	$Bi_{X}A$	**25a17 [17]**
$Ai_{X}B$	$Bi_{X}A$	**25a20 [17]**
$Ae_{X}B$	$Be_{X}A$	**25a15 [17]**
$Ao_{X}B$	$Bo_{X}A$	*25a22 [86]*
$Aa_{N}B$	$Bi_{N}A$	**25a32 [17]**
$Ai_{N}B$	$Bi_{N}A$	**25a32 [17]**
$Ae_{N}B$	$Be_{N}A$	**25a29 [17]**
$Ao_{N}B$	$Bo_{N}A$	*25a34 [86]*
$Aa_{M}B$	$Bi_{M}A$	**25a39 [17]**
$Ai_{M}B$	$Bi_{M}A$	**25a39 [17]**
$Ae_{M}B$	$Be_{M}A$	**25b5 [17]**
$Ao_{M}B$	$Bo_{M}A$	*25b13 [86]*
$Aa_{Q}B$	$Bi_{Q}A$	**25a39 [17]**
$Ai_{Q}B$	$Bi_{Q}A$	**25a39 [17]**
$Ae_{Q}B$	$Be_{Q}A$	*25b16, 36b35 [86]*
$Ao_{Q}B$	$Bo_{Q}A$	**25b17 [17]**

Table A.1. Conversion rules

First figure	XXX	NNN	NXN	XNN
AaB BaC AaC	**25b37 [31]**	**29b36 [34]**	**30a17 [33]**	*30a23 [73]*
AeB BaC AeC	**25b40 [31]**	**29b36 [34]**	**30a17 [33]**	*30a32 [88]*
AaB BeC	*26a2*$^{\nvdash}$ *[105]*			
AeB BeC	*26a9*$^{\nvdash}$ *[106]*			
AaB BiC AiC	**26a23 [31]**	**29b36 [34]**	**30a37 [33]**	*30b2 [74]*
AeB BiC AoC	**26a25 [31]**	**29b36 [34]**	**30b1 [33]**	*30b5 [89]*
AiB BaC	*26a30*$^{\nvdash}$ *[118]*			
AoB BaC	*26a30*$^{\nvdash}$ *[107]*			
AiB BeC	*26a30*$^{\nvdash}$ *[116]*			
AoB BeC	*26a30*$^{\nvdash}$ *[117]*			
AaB BoC	*26a39*$^{\nvdash}$ *[116]*			
AeB BoC	*26a39*$^{\nvdash}$ *[117]*			
Ai/oB Bi/oC	*26b21*$^{\nvdash}$ *[119]*			

Table A.2. Assertoric and apodeictic syllogistic: First figure

Second figure	XXX	NNN	NXN	XNN
BeA BaC AeC	**27a5 [44]**	**29b36 [48]**	**30b9 [47]**	*30b18 [88]*
BaA BeC AeC	**27a9 [44]**	**29b36 [48]**	*30b20 [88]*	**30b14 [47]**
BaA BaC	*27a18*$^{\nvdash}$ *[107]*			
BeA BeC	*27a20*$^{\nvdash}$ *[117]*			
BeA BiC AoC	**27a32 [45]**	**29b36 [48]**	**31a5 [47]**	*[89]*
BaA BoC AoC	**27a36 [32]**	**30a6 [35]**	*31a10 [88]*	*31a15 [77]*
BoA BaC	*27b4*$^{\nvdash}$ *[108]*			
BiA BeC	*27b6*$^{\nvdash}$ *[116]*			
BeA BoC	*27b12*$^{\nvdash}$ *[117]*			
BaA BiC	*27b23*$^{\nvdash}$ *[118]*			
BoA BeC	*27b28*$^{\nvdash}$ *[117]*			
BiA BaC	*27b32*$^{\nvdash}$ *[118]*			
Bi/oA Bi/oC	*27b36*$^{\nvdash}$ *[120]*			

Table A.3. Assertoric and apodeictic syllogistic: Second figure

Third figure	XXX	NNN	NXN	XNN
AaB CaB AiC	**28a17 [46]**	**29b36 [50]**	**31a24 [49]**	**31a31 [49]**
AeB CaB AoC	**28a26 [45]**	**29b36 [50]**	**31a33 [49]**	*31a37 [76]*
AaB CeB	*28a30*$^{\nvdash}$ *[116]*			
AeB CeB	*28a33*$^{\nvdash}$ *[117]*			
AiB CaB AiC	**28b7 [46]**	**29b36 [50]**	*31b31 [87]*	**31b12 [49]**
AaB CiB AiC	**28b11 [46]**	**29b36 [50]**	**31b19 [49]**	*31b20 [87]*
AaB CoB	*28b22*$^{\nvdash}$ *[116]*			
AoB CaB AoC	**28b15 [32]**	**30a7 [36]**	*32a4 [78]*	*31b40 [89]*
AeB CiB AoC	**28b31 [45]**	**29b36 [50]**	**31b33 [49]**	*32a1 [89]*
AiB CeB	*28b36*$^{\nvdash}$ *[116]*			
AoB CeB	*28b38*$^{\nvdash}$ *[117]*			
AeB CoB	*28b38*$^{\nvdash}$ *[117]*			
Ai/oB Ci/oB	*29a6*$^{\nvdash}$ *[121]*			

Table A.4. Assertoric and apodeictic syllogistic: Third figure

First figure	QQQ	QXQ	XQM	NQM	QNQ	NQX
AaB BaC AaC	**32b38 [37]**	**33b33 [38]**	**34a34** [40]	**35b38** [41]	**36a2 [39]**	*35b26* [*81*]
AeB BaC AeC	**33a1 [37]**	**33b36 [38]**	**34b19** [40]	**36a15** [41]	**36a17 [39]**	**36a7 [42]**
AaB BeC AaC	**33a5 [37]**	*35a20*$^{\nvdash}$ [*124*]	**35a3** [40]	**36a25** [41]	*36a27*$^{\nvdash}$ [*110*]	*35b28* [*81*]
AeB BeC AeC	**33a12 [37]**	*35a20*$^{\nvdash}$ [*124*]	**35a11** [40]	[41]	*36a28*$^{\nvdash}$ [*110*]	**[42]**
AaB BiC AiC	**33a23 [37]**	**35a30 [38]**	**35a35** [40]	**36a40** [41]	**35b23 [39]**	*36a40* [*90*]
AeB BiC AoC	**33a25 [37]**	**35a30 [38]**	**35a35** [40]	**35b30** [41]	**36a39 [39]**	**36a34 [42]**
AiB BaC	*33a34*$^{\nvdash}$ [*122*]	*35b11*$^{\nvdash}$ [*125*]	*35b11*$^{\nvdash}$ [*127*]	*36b3*$^{\nvdash}$ [*126*]	*36b10*$^{\nvdash}$ [*125*]	*36b3*$^{\nvdash}$ [*126*]
AoB BaC	*33a34*$^{\nvdash}$ [*122*]	*35b11*$^{\nvdash}$ [*125*]	*35b11*$^{\nvdash}$ [*127*]	*36b3*$^{\nvdash}$ [*113*]	*36b10*$^{\nvdash}$ [*125*]	*36b3*$^{\nvdash}$ [*113*]
AiB BeC	*33a34*$^{\nvdash}$ [*122*]	*35b11*$^{\nvdash}$ [*124*]	*35b11*$^{\nvdash}$ [*127*]	*36b3*$^{\nvdash}$ [*126*]	*36b8*$^{\nvdash}$ [*123*]	*36b3*$^{\nvdash}$ [*126*]
AoB BeC	*33a34*$^{\nvdash}$ [*122*]	*35b11*$^{\nvdash}$ [*124*]	*35b11*$^{\nvdash}$ [*127*]	*36b3*$^{\nvdash}$ [*113*]	*36b8*$^{\nvdash}$ [*123*]	*36b3*$^{\nvdash}$ [*113*]
AaB BoC AiC	**33a27 [37]**	*35b8*$^{\nvdash}$ [*124*]	**35b5** [40]	**35b28** [41]	$^{\nvdash}$ [*123*]	*35b28* [*90*]
AeB BoC AoC	**33a21 [37]**	*35b8*$^{\nvdash}$ [*124*]	**35b5** [40]	**35b30** [41]	$^{\nvdash}$ [*123*]	**35b30 [42]**
Ai/oB Bi/oC	*33a37*$^{\nvdash}$ [*122*]	*35b14*$^{\nvdash}$ [*132*]	*35b14*$^{\nvdash}$ [*132*]	*36b12*$^{\nvdash}$ [*131*]	*36b12*$^{\nvdash}$ [*131*]	*36b12*$^{\nvdash}$ [*131*]

Table A.5a. Problematic syllogistic: First figure

First figure	QNX	XQQ	NQQ	XQN	NQN
AaB BaC AaC	*36a2* [*79*]	[*101*]	[*100*]	[*104*]	*35b34* [*92*]
AeB BaC AeC	*36a17* [*80*]	*34b31* [*91*]	*35b30* [*82*]	*34b37* [*95*]	*35b34* [*82*]
AaB BeC AaC	*36a27*$^{\nvdash}$ [*110*]	[*101*]	[*100*]	[*104*]	*35b34* [*92*]
AeB BeC AeC	*36a28*$^{\nvdash}$ [*110*]	[*91*]	[*82*]	[*95*]	*35b34* [*82*]
AaB BiC AiC	*35b26* [*90*]	[*101*]	[*100*]	[*104*]	*35b34* [*92*]
AeB BiC AoC	*36a39* [*93*]	*33b29* [*91*]	*35b30* [*82*]	[*95*]	*35b34* [*94*]
AiB BaC	*36b10*$^{\nvdash}$ [*125*]	*35b11*$^{\nvdash}$ [*127*]	*36b3*$^{\nvdash}$ [*126*]	*35b11*$^{\nvdash}$ [*127*]	*36b3*$^{\nvdash}$ [*126*]
AoB BaC	*36b10*$^{\nvdash}$ [*125*]	*35b11*$^{\nvdash}$ [*127*]	*36b3*$^{\nvdash}$ [*113*]	*35b11*$^{\nvdash}$ [*127*]	*36b3*$^{\nvdash}$ [*113*]
AiB BeC	*36b8*$^{\nvdash}$ [*123*]	*35b11*$^{\nvdash}$ [*127*]	*36b3*$^{\nvdash}$ [*126*]	*35b11*$^{\nvdash}$ [*127*]	*36b3*$^{\nvdash}$ [*126*]
AoB BeC	*36b8*$^{\nvdash}$ [*123*]	*35b11*$^{\nvdash}$ [*127*]	*36b3*$^{\nvdash}$ [*113*]	*35b11*$^{\nvdash}$ [*127*]	*36b3*$^{\nvdash}$ [*113*]
AaB BoC AiC	$^{\nvdash}$ [*123*]	[*101*]	[*100*]	[*104*]	*35b34* [*92*]
AeB BoC AoC	$^{\nvdash}$ [*123*]	[*91*]	[*82*]	[*95*]	*35b34* [*94*]
Ai/oB Bi/oC	*36b12*$^{\nvdash}$ [*131*]	*35b14*$^{\nvdash}$ [*132*]	*36b12*$^{\nvdash}$ [*131*]	*35b14*$^{\nvdash}$ [*132*]	*36b12*$^{\nvdash}$ [*131*]

Table A.5b. Problematic syllogistic: First figure, continued

Second figure	QQQ	QXM	XQM	NQM	QNM	NQX	QNX
BeA BaC AeC	*37a32*$^{\nvdash}$ *[109]*	*37b19*$^{\nvdash}$ *[129]*	**37b24 [51]**	**38a16 [53]**	*38a26*$^{\nvdash}$ *[114]*	**38a21 [52]**	*38a26*$^{\nvdash}$ *[114]*
BaA BeC AeC	*37b10*$^{\nvdash}$ *[109]*	**37b29 [51]**	*37b19*$^{\nvdash}$ *[128]*	*38b4*$^{\nvdash}$ *[115]*	**38a25 [53]**	*38b4*$^{\nvdash}$ *[115]*	**38a25 [52]**
BaA BaC	*37b10*$^{\nvdash}$ *[109]*	*37b35*$^{\nvdash}$ *[129]*	*37b35*$^{\nvdash}$ *[128]*	*38b13*$^{\nvdash}$ *[115]*	*38b13*$^{\nvdash}$ *[114]*	*38b13*$^{\nvdash}$ *[115]*	*38b13*$^{\nvdash}$ *[114]*
BeA BeC AeC	*37b10*$^{\nvdash}$ *[109]*	**37b29 [51]**	**37b29 [51]**	**38b6 [53]**	**38b6 [53]**	**38b6 [52]**	**38b6 [52]**
BeA BiC AoC	*37b13*$^{\nvdash}$ *[122]*	*37b40*$^{\nvdash}$ *[129]*	**38a3 [51]**	**38b25 [53]**	$^{\nvdash}$ *[129]*	**38b25 [52]**	$^{\nvdash}$ *[129]*
BaA BoC AoC	*37b13*$^{\nvdash}$ *[122]*	*38a8*$^{\nvdash}$ *[111]*	*37b40*$^{\nvdash}$ *[128]*	*38b27*$^{\nvdash}$ *[128]*	**[66]**	*38b27*$^{\nvdash}$ *[128]*	**[63]**
BoA BaC	*37b13*$^{\nvdash}$ *[122]*	*37b40*$^{\nvdash}$ *[125]*	*38a8*$^{\nvdash}$ *[130]*	$^{\nvdash}$ *[112]*	*38b27*$^{\nvdash}$ *[125]*	$^{\nvdash}$ *[112]*	*38b27*$^{\nvdash}$ *[125]*
BiA BeC	*37b13*$^{\nvdash}$ *[122]*	$^{\nvdash}$ *[124]*	*37b40*$^{\nvdash}$ *[127]*	$^{\nvdash}$ *[126]*	$^{\nvdash}$ *[123]*	$^{\nvdash}$ *[126]*	$^{\nvdash}$ *[123]*
BeA BoC AoC	*37b13*$^{\nvdash}$ *[122]*	*38a8*$^{\nvdash}$ *[111]*	**38a4 [51]**	**38b31 [53]**	**[66]**	**38b31 [52]**	**[63]**
BaA BiC	*37b13*$^{\nvdash}$ *[122]*	*37b40*$^{\nvdash}$ *[129]*	*37b40*$^{\nvdash}$ *[128]*	*38b29*$^{\nvdash}$ *[128]*	*38b29*$^{\nvdash}$ *[129]*	*38b29*$^{\nvdash}$ *[128]*	*38b29*$^{\nvdash}$ *[129]*
BoA BeC	*37b13*$^{\nvdash}$ *[122]*	$^{\nvdash}$ *[124]*	*38a8*$^{\nvdash}$ *[130]*	$^{\nvdash}$ *[112]*	$^{\nvdash}$ *[123]*	$^{\nvdash}$ *[112]*	$^{\nvdash}$ *[123]*
BiA BaC	*37b13*$^{\nvdash}$ *[122]*	$^{\nvdash}$ *[125]*	$^{\nvdash}$ *[127]*	*38b29*$^{\nvdash}$ *[126]*	*38b29*$^{\nvdash}$ *[125]*	*38b29*$^{\nvdash}$ *[126]*	*38b29*$^{\nvdash}$ *[125]*
Bi/oA Bi/oC	*37b14*$^{\nvdash}$ *[122]*	*38a10*$^{\nvdash}$ *[134]*	*38a10*$^{\nvdash}$ *[134]*	*38b35*$^{\nvdash}$ *[133]*	*38b35*$^{\nvdash}$ *[133]*	*38b35*$^{\nvdash}$ *[133]*	*38b35*$^{\nvdash}$ *[133]*

Table A.6. Problematic syllogistic: Second figure

Third figure	QQQ	QXQ	QXM	XQQ	XQM	QXX	XQX
AaB CaB AiC	**39a14 [54]**	**39b16 [55]**	**[58]**	**[56]**	**39b10 [59]**	*39b8 [83]*	*39b8 [97]*
AeB CaB AoC	**39a19 [54]**	**39b17 [55]**	**[67]**	*[138]*	**39b17 [59]**	*39b8 [84]*	*39b8 [85]*
AaB CeB AiC	**[54]**	$\nvdash$ *[124]*	$\nvdash$ *[124]*	**[56]**	**39b22 [59]**	$\nvdash$ *[124]*	*39b8 [97]*
AeB CeB AoC	**39a23 [54]**	$\nvdash$ *[124]*	$\nvdash$ *[124]*	*[138]*	**39b23** *[59]*	$\nvdash$ *[124]*	*39b8 [85]*
AiB CaB AiC	**39a35 [54]**	*[103]*	**39b26 [58]**	**39b26 [56]**	**[69]**	*39b30 [98]*	*39b30 [99]*
AaB CiB AiC	**39a31 [54]**	**39b26 [55]**	**[68]**	*[103]*	**39b26 [59]**	*39b30 [98]*	*39b30 [99]*
AaB CoB AiC	**[54]**	$\nvdash$ *[124]*	$\nvdash$ *[124]*	*[103]*	**[59]**	$\nvdash$ *[124]*	*39b8 [99]*
AoB CaB AoC	**39a36 [54]**	*[103]*	**39b31 [43]**	*[138]*	**[64]**	*39b8 [98]*	*39b8 [99]*
AeB CiB AoC	**39a36 [54]**	**39b27 [55]**	**[68]**	*[138]*	**39b27 [59]**	*39b30 [98]*	*39b30 [99]*
AiB CeB AiC	**[54]**	$\nvdash$ *[124]*	$\nvdash$ *[124]*	**39b27 [56]**	**[69]**	$\nvdash$ *[124]*	*39b30 [99]*
AoB CeB AoC	**39a38 [54]**	$\nvdash$ *[124]*	$\nvdash$ *[124]*	*[138]*	**[64]**	$\nvdash$ *[124]*	*39b8 [99]*
AeB CoB AoC	**39a38 [54]**	$\nvdash$ *[124]*	$\nvdash$ *[124]*	*[138]*	**[59]**	$\nvdash$ *[124]*	*39b8 [99]*
Ai/oB Ci/oB	*39b2*$^{\nvdash}$ *[122]*	*40a1*$^{\nvdash}$ *[136]*	*40a1*$^{\nvdash}$ *[136]*	*40a1*$^{\nvdash}$ *[136]*	*40a1*$^{\nvdash}$ *[136]*	*40a1*$^{\nvdash}$ *[136]*	*40a1*$^{\nvdash}$ *[136]*

Table A.7a. Problematic syllogistic: Third figure

Third figure	NQQ	NQM	QNQ	QNM	NQX	QNX	NQN
AaB CaB AiC	**[57]**	**40a11 [62]**	**40a16 [57]**	**[61]**	*40a11 [81]*	*40a16 [90]*	*40a9 [96]*
AeB CaB AoC	*[137]*	**40a25 [62]**	**40a18 [57]**	**[70]**	**40a25 [60]**	*40a18 [80]*	*40a9 [82]*
AaB CeB AiC	**[57]**	**40a33 [62]**	*40a35⊬ [123]*	*40a35⊬ [123]*	*40a33 [81]*	*40a35⊬ [123]*	*40a9 [96]*
AeB CeB AoC	*[137]*	**[62]**	⊬ *[123]*	⊬ *[123]*	**[60]**	⊬ *[123]*	*40a9 [82]*
AiB CaB AiC	**40a39 [57]**	**[71]**	*[102]*	**40a39 [61]**	*40a39 [90]*	*40a39 [90]*	*40a9 [96]*
AaB CiB AiC	*[102]*	**40a39 [62]**	**40a39 [57]**	**[71]**	*40a39 [90]*	*40a39 [90]*	*40a9 [96]*
AaB CoB AiC	*[102]*	**[62]**	⊬ *[123]*	⊬ *[123]*	*[90]*	⊬ *[123]*	*40a9 [96]*
AoB CaB AoC	*[137]*	**[72]**	*[102]*	**[70]**	**[65]**	*[93]*	*40a9 [94]*
AeB CiB AoC	*[137]*	**40b3 [62]**	**40b2 [57]**	**[71]**	**40b3 [60]**	*40b2 [93]*	*40a9 [94]*
AiB CeB AiC	**40b8 [57]**	**[71]**	*40b8⊬ [123]*	*40b8⊬ [123]*	*[90]*	*40b8⊬ [123]*	*40a9 [96]*
AoB CeB AoC	*[137]*	**[72]**	⊬ *[123]*	⊬ *[123]*	**[65]**	⊬ *[123]*	*40a9 [94]*
AeB CoB AoC	*[137]*	**[62]**	⊬ *[123]*	⊬ *[123]*	**[60]**	⊬ *[123]*	*40a9 [94]*
Ai/oB Ci/oB	⊬ *[135]*	⊬ *[135]*	⊬ *[135]*	⊬ *[135]*	⊬ *[135]*	⊬ *[135]*	⊬ *[135]*

Table A.7b. Problematic syllogistic: Third figure, continued number in square brackets is boldface (or italic). The symbol $\nvdash$ is often accompanied both by a number in square brackets and by a Bekker page number. In this case, the premise pair under consideration is both inconcludent in the predicable semantics and held to be inconcludent by Aristotle.

Tables A.1–A.7 are intended to provide a comprehensive overview of Aristotle's claims of (in)validity and inconcludence in the assertoric and modal syllogistic (*Prior Analytics* 1.1–22).[1] However, in a number of places it is not straightforward to determine exactly which claims Aristotle makes. It is therefore appropriate to give some additional explanation.

MODAL TYPES. Aristotle's presentation of the syllogistic is arranged according to the three figures and according to the modal type of premise pairs. Examples of modal types of premise pairs are XN, QX, and NQ. Given the fourfold distinction between a-, e-, i-, and o-propositions, there are sixteen premise pairs for every modal type. In four of these sixteen premise pairs, both premises are particular propositions (that is, i- or o-propositions). Aristotle never holds the validity of a mood in which both premises are particular. For most modal types, he states that such purely particular premise pairs are inconcludent. In Tables A.2–A.7, the four purely particular premise pairs are grouped together in the last row. So each of the tables has thirteen rows.

Tables A.2–A.7 differ from each other in the number of columns. Each column corresponds to a modal type of Aristotle's moods, such as XQM, NQQ, and QXQ. The tables take into account only those modal types that are discussed by Aristotle in a given figure. Hence the number of columns varies depending on how many modal types Aristotle discusses in the figure under consideration. For instance, Aristotle discusses the modal type NQN in the first and third figures, but not in the second figure.

In the apodeictic syllogistic, Aristotle focuses on moods of the modal types NNN, NXN, and XNN. He considers only those moods whose corresponding purely assertoric XXX-moods are valid. Accordingly, Tables A.2–A.4 do not take into account moods whose corresponding purely assertoric moods are invalid. When a mood of the modal type NXN or XNN is invalid, Aristotle sometimes states the validity of

1. For similar overviews, see Smith (1989: 230–5) and Mueller (1999b: 59–69).

the corresponding NXX- or XNX-mood.[2] If N-X-subordination is valid, these claims follow from the validity of the corresponding XXX-moods. Although Aristotle never states the principles of N-X-subordination, they are valid in the predicable semantics (see pp. 131–132 and 292). Therefore moods of the modal types XNX and NXX are not taken into account in the above tables. Likewise, the tables do not take into account Aristotle's claim that all third-figure QNN-moods are invalid.[3] For in the predicable semantics, this follows from the invalidity of the corresponding QNX-moods by N-X-subordination. Apart from this, the tables take into account all modal types discussed by Aristotle.

PREMISE SCHEMES. By a premise scheme, I mean a premise pair whose modal type is left unspecified. For example, aa-1 is a premise scheme that gives rise to premise pairs such as aa-1-XX and aa-1-XN, and to moods such as Barbara XXX and Barbara XNN. In Tables A.2–A.7, each row corresponds to a premise scheme (except the last row, which corresponds to four premise schemes). While some premise schemes do not give rise to any valid moods, many premise schemes do. In Tables A.2–A.7, the quantity and quality of the conclusion of these moods is specified in the first column. However, this needs some explanation. Consider, for instance, the premise scheme ee-3. Aristotle asserts the validity of eeo-3-XQM and of eei-3-QQQ.[4] Thus, Aristotle discusses moods that are based on the same premise scheme but that differ in the quality of the conclusion.

In the example just discussed, one of the two conclusions is a Q-proposition. Hence, given the equivalence of affirmative and negative Q-propositions, we can neglect the difference in quality between the two conclusions. As a matter of fact, this is true for all premise schemes discussed by Aristotle. If two moods discussed by Aristotle are based on the same premise scheme, they never differ in the *quantity* of

2. For instance, he states the validity of Baroco NXX (1.10 31a11–14) and Felapton XNX (1.11 31a38–b2). In similar passages elsewhere in chapters 1.9–11, it is less clear whether he is asserting the validity of an NXX- or XNX-mood.

3. *APr.* 1.22 40a9–11; cf. also 1.20 39a8–10.

4. For the latter mood, see 1.20 39a23–8; cf. Smith (1989: 232), Mueller (1999b: 64).

their conclusion. When they differ in the *quality* of their conclusion, at least one conclusion is a Q-proposition. This allows us to neglect the difference in quality. Consequently, the quantity and quality of conclusions can be specified in the first column of the tables and can thus be taken to apply to all moods in a row.

SOME DETAILS OF INTERPRETATION. In some cases, Aristotle's remarks seem to imply the validity of certain moods although commentators generally agree that Aristotle did not intend to state their validity.[5] The reason for their agreement is that there is no obvious way to prove the validity of those moods. The above synopsis follows this view and does not take Aristotle to assert the validity of those moods.

Sometimes Aristotle implies the (in)validity of a mood in general statements summarizing the results of a chapter, although he does not actually state its (in)validity in the chapter.[6] In the above synopsis, these moods are taken to be held to be (in)valid by Aristotle.

Also, it is a moot point whether Aristotle asserts the validity of Bocardo NQX, NQM, and QNM.[7] In the above synopsis, these moods are

---

5. Contrary to appearances, Aristotle does not make any claim of validity concerning the premise pair ie-2-QX or oe-2-QX at 1.18 38a3–7 (Alexander *in APr.* 234.8–11, Maier 1900a: 182n4, Ross 1949: 357, Smith 1989: 233). The same is true of the premise pairs ie-2-QN and oe-2-QN at 1.19 38b25–35 (Maier 1900a: 192n2, Ross 1949: 361, Smith 1989: 235) and of the premise pairs ae-3-QX and ee-3-QX at 1.21 39b22–5 (Alexander *in APr.* 246.13–20, Maier 1900a: 199–200n1, Ross 1949: 365, Smith 1989: 233–4, Ebert & Nortmann 2007: 702). Similarly, Aristotle does not make any claim of validity or invalidity concerning the premise pair ie-3-QX at 1.21 39b26–31 or concerning the premise pairs ao-3-XQ, ao-3-QX, and oa-3-XQ at 39b31–9 (Maier 1900a: 200–1n1 and 201–2n1, Ross 1949: 366, Smith 1989: 234, Ebert & Nortmann 2007: 705).

6. For example, the validity of eoo-1-QQQ is implied at 1.14 33a21–3 (Mueller 1999b: 64). Also, I take it that the invalidity of Barbara NQX, Darii QNX, aea-1-NQX, and aoi-1-NQX is implied at 1.16 35b26–30 (cf. McCall 1963: 85), that the invalidity of Celarent NQQ and Ferio NQQ is implied at 1.16 35b30–4 (cf. McCall 1963: 85), that the invalidity of Felapton QNX is implied at 1.22 40a18–25, and that the invalidity of aei-3-NQX, is implied at 1.22 40a33–5.

7. Some commentators think that Aristotle asserts the validity of Bocardo NQX and NQM at 1.22 40b3–8 (Ross 1949: 286–7 and 368–9, McCall 1963:

not taken to be held to be valid by Aristotle (although they are valid in the predicable semantics).

INDETERMINATE PROPOSITIONS. Indeterminate propositions are propositions that lack quantifying expressions such as 'all', 'some', or 'no'.[8] Aristotle makes some claims of (in)validity and inconcludence concerning moods that contain indeterminate propositions. Nevertheless, indeterminate propositions are not mentioned in the above synopsis, and do not play a role in the predicable semantics. This is because in the *Prior Analytics*, Aristotle seems to treat indeterminate propositions as equivalent to particular ones: he seems to assume that an indeterminate proposition is true just in case the corresponding particular proposition of the same quality and modality is true. Let me explain this in more detail.

85, Nortmann 1996: 310–12, Rini 2011: 212–14). Others think that he does not state the invalidity of these moods there (Maier 1900a: 204–5, Becker 1933: Tafel III, Price 1969: 609–10, Wieland 1975: 81n9, Ebert & Nortmann 2007: 734). Others remain undecided (Smith 1989: 235, Mueller 1999b: 68). The only obvious way to prove Bocardo NQX would be indirectly by means of Barbara XQM (Ross 1949: 286–7, Mueller 1999b: 226). Such a proof would assume that o_N-propositions are incompatible with a_M-propositions. Aristotle does not state or use this principle of incompatibility elsewhere (but it is valid in the predicable semantics; see Fact 29, p. 294). In any case, such an indirect proof would not fit Aristotle's claim that the moods in question are proved in the same way as the corresponding universal moods (40b4–6); for Felapton NQX, the universal mood corresponding to Bocardo NQX, is proved directly by conversion (1.22 40a25–32; see Ross 1949: 369).

Some commentators think that Aristotle asserts the validity of Bocardo QNM at 1.22 40b2–3 (Ross 1949: 286–7, McCall 1963: 85, Wieland 1975: 81, Mueller 1999b: 68), others not (Maier 1900a: 204–5, Becker 1933: Tafel III). Others remain undecided (Smith 1989: 235, Ebert & Nortmann 2007: 733). Bocardo QNM can be proved to be valid in much the same way as Bocardo QXM (cf. p. 208 above). Alexander (*in APr.* 252.27–9, 254.26–9) points out that Aristotle might also be taken to state the validity of oai-3-QNM instead of Bocardo QNM. Others take Aristotle to state the validity of Bocardo QNQ (Nortmann 1996: 313, Schmidt 2000: 208–10, Rini 2011: 212). For further discussion, cf. Mueller (1999b: 226).

8. *APr.* 1.1 24a19–22; see pp. 25–26 above.

Most of what Aristotle says about indeterminate propositions concerns the inconcludence of premise pairs. He often states that a given inconcludent premise pair remains inconcludent when the particular premise(s) in it is (are) replaced by the corresponding indeterminate proposition(s).[9] There are only two claims of validity concerning indeterminate propositions in *Prior Analytics* 1.1–22. Both of them concern valid purely assertoric moods that contain an i_X-premise. Aristotle asserts that when the i_X-premise in them is replaced by an indeterminate affirmative proposition, "there will be the same deduction [συλλογισμός]."[10] His statement that "there will be the same deduction" can be understood in two ways. Either the conclusion of the resulting valid mood remains an (affirmative or negative) particular proposition, or it is replaced by an (affirmative or negative) indeterminate proposition. Alexander prefers the latter option, others the former.[11] For our purposes, it is not necessary to settle this question. What is important for us is that in any case, Aristotle's statements about indeterminate propositions in *Prior Analytics* 1.1–22 can best be explained by the assumption that Aristotle took these propositions to be equivalent to the corresponding particular propositions of the same quality and modality.

This equivalence is generally accepted by commentators, at least for the assertoric syllogistic.[12] The equivalence is also accepted in this study. Consequently, there is no need for a separate treatment of indeterminate propositions, so that they can generally be ignored.

---

9. *APr.* 1.4 26a32, 26a39, 26b23–4, 1.5 27b38, 1.6 29a8, 1.14 33a37, 1.15 35b15, 1.16 36b12, 1.17 37b14, 1.18 38a10–11, 1.19 38b36, 1.20 39b2, 1.21 40a1.

10. *APr.* 1.4 26a28–30; similarly, 1.7 29a27–9.

11. Alexander *in APr.* 51.24–30. The other option is preferred by Whitaker (1996: 86) and Drechsler (2005: 374–6 and 543–5).

12. Alexander *in APr.* 30.29–31, 49.15, 62.24, 111.30–112.2, 267.2, Philoponus *in APr.* 79.4–5, 252.35, Philoponus *in APost.* 296.10–11, Waitz (1844: 369), Kneale & Kneale (1962: 55; 1972: 203), Ackrill (1963: 129), Owen (1965: 86–7), Thom (1981: 19), and Barnes (2007: 141); cf. also Crivelli (2004: 244n19). Some evidence for this equivalence can be found at *Top.* 3.6 120a6–20; see Alexander *in Top.* 288.27–289.4.

Appendix B

The Predicable Semantics of the Modal Syllogistic

In this appendix the predicable semantics is proved to be adequate with respect to Aristotle's modal syllogistic. In other words, it is proved that every mood and conversion rule held to be valid (or invalid) by Aristotle is valid (or invalid) in the predicable semantics, and that every premise pair held to be inconcludent by Aristotle is inconcludent in it. In Appendix A, each of Aristotle's claims of (in)validity and inconcludence was assigned a number in square brackets. This number refers to the Fact in which the claim is proved to be true within the predicable semantics in the present appendix.

An Overview of the Predicable Semantics

First of all, the predicable semantics is formulated in classical first-order logic without identity, based on three primitive binary relations:

$\mathbf{A}ab$	a is $\mathrm{a_X}$-predicated of b	pp. 66–71
$\mathbf{N}ab$	a is $\mathrm{a_N}$-predicated of b	pp. 110–131
$\widehat{\mathbf{N}}ab$	a is strongly $\mathrm{a_N}$-predicated of b	p. 250

These primitive relations are governed by six theses, which are regarded as axioms (see pp. 249–250):[1]

1. The list of axioms could be extended to include the statements S1–25 given on pp. 116–159, but for ease of exposition, I restrict the list to those six theses (see p. 325 below).

(ax_1)	$\mathbf{A}aa$
(ax_2)	$\mathbf{A}ab \wedge \mathbf{A}bc \supset \mathbf{A}ac$
(ax_3)	$\mathbf{N}ab \wedge \mathbf{A}bc \supset \mathbf{N}ac$
(ax_4)	$\widehat{\mathbf{N}}ab \wedge \mathbf{A}bc \supset \widehat{\mathbf{N}}ac$
(ax_5)	$\mathbf{N}ab \supset \mathbf{A}ab$
(ax_6)	$\widehat{\mathbf{N}}ab \supset \mathbf{N}ab$

There are a number of complex notions defined in terms of the three primitive relations. Those required for the interpretation of the apodeictic syllogistic are

$(\mathrm{df}_{\mathbf{\Sigma}})$	$\mathbf{\Sigma}a =_{df} \exists z \mathbf{N}za$	p. 265
$(\mathrm{df}_{\widehat{\mathbf{\Sigma}}})$	$\widehat{\mathbf{\Sigma}}a =_{df} \exists z \widehat{\mathbf{N}}za$	p. 251
$(\mathrm{df}_{\mathbf{\Gamma}})$	$\mathbf{\Gamma}ab =_{df} (\widehat{\mathbf{\Sigma}}a \vee \widehat{\mathbf{\Sigma}}b) \wedge \neg\exists z(\mathbf{A}az \wedge \mathbf{A}bz)$	p. 262
$(\mathrm{df}_{\mathbf{K}})$	$\mathbf{K}ab =_{df} \mathbf{\Sigma}a \wedge \mathbf{\Sigma}b \wedge \mathbf{\Gamma}ab$	p. 266
$(\mathrm{df}_{\widehat{\mathbf{K}}})$	$\widehat{\mathbf{K}}ab =_{df} \widehat{\mathbf{\Sigma}}a \wedge \widehat{\mathbf{\Sigma}}b \wedge \neg\exists z(\mathbf{A}az \wedge \mathbf{A}bz)$	p. 267
$(\mathrm{df}_{\overline{\mathbf{K}}})$	$\overline{\mathbf{K}}ab =_{df} \exists v \mathbf{N}av \wedge \forall u(\mathbf{A}au \wedge \mathbf{\Sigma}u \supset \widehat{\mathbf{K}}bu)$	p. 267

The complex notions additionally required in the problematic syllogistic are

$(\mathrm{df}_{\mathbf{\Pi}})$	$\mathbf{\Pi}ab =_{df} \neg(\widehat{\mathbf{\Sigma}}a \wedge \widehat{\mathbf{\Sigma}}b) \wedge \neg\widehat{\mathbf{N}}ab \wedge \neg\widehat{\mathbf{N}}ba \wedge \neg\mathbf{\Gamma}ab$	p. 252
$(\mathrm{df}_{\overline{\mathbf{\Pi}}})$	$\overline{\mathbf{\Pi}}ab =_{df} \mathbf{\Pi}ab \vee \mathbf{A}ab$	p. 261
$(\mathrm{df}_{\overline{\mathbf{N}}})$	$\overline{\mathbf{N}}ab =_{df} \widehat{\mathbf{N}}ab \vee (\widehat{\mathbf{\Sigma}}a \wedge \mathbf{A}ab)$	p. 263

The interpretation of Aristotle's categorical propositions in the predicable semantics is as follows:

$\mathbb{X}^{\mathbf{a}}ab$	$\mathbf{A}ab$	p. 71
$\mathbb{X}^{\mathbf{e}}ab$	$\forall z(\mathbf{A}bz \supset \neg\mathbf{A}az)$	p. 71
$\mathbb{X}^{\mathbf{i}}ab$	$\exists z(\mathbf{A}bz \wedge \mathbf{A}az)$	p. 71
$\mathbb{X}^{\mathbf{o}}ab$	$\neg\mathbf{A}ab$	p. 71
$\mathbb{N}^{\mathbf{a}}ab$	$\mathbf{N}ab$	p. 249
$\mathbb{N}^{\mathbf{e}}ab$	$\mathbf{K}ab$	p. 266

$\mathbb{N}^{\mathrm{i}}ab$	$\exists z((\mathbf{A}bz \wedge \mathbf{N}az) \vee (\mathbf{A}az \wedge \mathbf{N}bz))$	p. 265
$\mathbb{N}^{\mathrm{o}}ab$	$\exists z((\mathbf{A}bz \wedge \mathbf{K}az) \vee (\mathbf{N}bz \wedge \overline{\mathbf{K}}az))$	p. 267
$\mathbb{M}^{\mathrm{a}}ab$	$\forall z(\mathbf{A}bz \supset \overline{\mathbf{\Pi}}az)$	p. 261
$\mathbb{M}^{\mathrm{e}}ab$	$\forall z(\mathbf{A}bz \supset \neg\overline{\mathbf{N}}az) \vee \forall z(\mathbf{A}az \supset \neg\overline{\mathbf{N}}bz)$	p. 264
$\mathbb{M}^{\mathrm{i}}ab$	$\neg\mathbf{\Gamma}ab$	p. 262
$\mathbb{M}^{\mathrm{o}}ab$	$\neg\overline{\mathbf{N}}ab$	p. 264
$\mathbb{Q}^{\mathrm{a/e}}ab$	$\forall z(\mathbf{A}bz \supset \mathbf{\Pi}az)$	p. 252
$\mathbb{Q}^{\mathrm{i/o}}ab$	$\mathbf{\Pi}ab$	p. 252

We now have to explain when a conversion rule or syllogistic mood is valid in the predicable semantics. A conversion rule allows us to infer a categorical proposition $\mathcal{B}$ from another categorical proposition $\mathcal{A}$; a syllogistic mood allows us to infer $\mathcal{B}$ from a pair of propositions $\mathcal{A}_1, \mathcal{A}_2$. More generally, when a categorical proposition $\mathcal{B}$ is inferred from categorical propositions $\mathcal{A}_1, \ldots, \mathcal{A}_n$, we may call this a *categorical inference*. Let $\mathrm{A}_1, \ldots, \mathrm{A}_n, \mathrm{B}$ be the first-order formulae assigned in the predicable semantics to the categorical propositions $\mathcal{A}_1, \ldots, \mathcal{A}_n, \mathcal{B}$, as determined by the above table. Then a categorical inference from $\mathcal{A}_1, \ldots, \mathcal{A}_n$ to $\mathcal{B}$ is valid in the predicable semantics if and only if the formula $(\mathrm{A}_1 \wedge \cdots \wedge \mathrm{A}_n) \supset \mathrm{B}$ is logically valid in classical first-order logic. Otherwise the categorical inference is invalid in the predicable semantics. We will sometimes write $\mathrm{A} \vdash \mathrm{B}$ to indicate that a first-order formula $\mathrm{A} \supset \mathrm{B}$ is logically valid. Standard conventions for omitting parentheses will be adopted; in particular, implications are taken to have wider scope than conjunctions, and conjunctions are taken to have wider scope than negations.

The rest of this appendix consists of five sections. The first of them establishes some preliminary results, including the validity in the predicable semantics of conversion rules, principles of subalternation, principles of modal subordination, and some principles of modal opposition. In the second section, all moods held to be valid by Aristotle in the modal syllogistic are shown to be valid in the predicable semantics. In the third section, all moods held to be invalid by Aristotle are shown to be invalid in the predicable semantics. In the fourth section, all premise pairs held to be inconcludent by Aristotle are shown to be inconcludent in the predicable semantics. Finally, the fifth section contains some miscellaneous results.

Preliminaries

The following formulae are valid in the predicable semantics:

Fact 1: $\widehat{\mathbf{N}}ab \supset \mathbf{A}ab$

Proof. Follows from (ax_5) and (ax_6). □

Fact 2: $\mathbf{\Sigma}a \wedge \mathbf{A}ab \supset \mathbf{\Sigma}b$

Proof. By ($df_{\mathbf{\Sigma}}$), $\mathbf{\Sigma}a$ implies that there is a z such that $\mathbf{N}za$. By (ax_3), $\mathbf{N}za$ and $\mathbf{A}ab$ imply $\mathbf{N}zb$. Hence $\mathbf{\Sigma}b$ by ($df_{\mathbf{\Sigma}}$). □

The foregoing fact states that no essence term is a_X-predicated of a nonessence term; the next one states that no substance term is a_X-predicated of a nonsubstance term (see p. 160):

Fact 3: $\widehat{\mathbf{\Sigma}}a \wedge \mathbf{A}ab \supset \widehat{\mathbf{\Sigma}}b$

Proof. By ($df_{\widehat{\mathbf{\Sigma}}}$), $\widehat{\mathbf{\Sigma}}a$ implies that there is a z such that $\widehat{\mathbf{N}}za$. By (ax_4), $\widehat{\mathbf{N}}za$ and $\mathbf{A}ab$ imply $\widehat{\mathbf{N}}zb$. Hence $\widehat{\mathbf{\Sigma}}b$ by ($df_{\widehat{\mathbf{\Sigma}}}$). □

Fact 4: $\overline{\mathbf{N}}ab \supset (\widehat{\mathbf{\Sigma}}b \wedge \mathbf{A}ab)$

Proof. By ($df_{\overline{\mathbf{N}}}$), $\overline{\mathbf{N}}ab$ implies $\widehat{\mathbf{N}}ab$ or $\widehat{\mathbf{\Sigma}}a \wedge \mathbf{A}ab$. In the first case, $\widehat{\mathbf{\Sigma}}b \wedge \mathbf{A}ab$ follows by ($df_{\widehat{\mathbf{\Sigma}}}$) and Fact 1. In the latter case, $\widehat{\mathbf{\Sigma}}b$ follows by Fact 3. □

Fact 5: $\widehat{\mathbf{\Sigma}}a \supset \mathbf{\Sigma}a$

Proof. Follows from (ax_6), ($df_{\widehat{\mathbf{\Sigma}}}$), and ($df_{\mathbf{\Sigma}}$). □

Fact 6: $\widehat{\mathbf{K}}ab \supset \mathbf{K}ab$

Proof. By ($df_{\widehat{\mathbf{K}}}$), $\widehat{\mathbf{K}}ab$ implies $\widehat{\mathbf{\Sigma}}a \wedge \widehat{\mathbf{\Sigma}}b \wedge \neg\exists z(\mathbf{A}az \wedge \mathbf{A}bz)$. This implies $\mathbf{\Gamma}ab$ by ($df_{\mathbf{\Gamma}}$). Moreover, $\widehat{\mathbf{\Sigma}}a \wedge \widehat{\mathbf{\Sigma}}b$ implies $\mathbf{\Sigma}a \wedge \mathbf{\Sigma}b$ by Fact 5. So we have $\mathbf{\Sigma}a \wedge \mathbf{\Sigma}b \wedge \mathbf{\Gamma}ab$. This implies $\mathbf{K}ab$ by ($df_{\mathbf{K}}$). □

Fact 7: $\mathbf{K}ab \supset \neg\mathbf{A}ab$

Proof. By ($\text{df}_{\mathbf{K}}$), $\mathbf{K}ab$ implies $\mathbf{\Gamma}ab$. By ($\text{df}_{\mathbf{\Gamma}}$), $\mathbf{\Gamma}ab$ implies $\neg\exists z(\mathbf{A}az \wedge \mathbf{A}bz)$. On the other hand, $\mathbf{A}ab$ implies $\mathbf{A}ab \wedge \mathbf{A}bb$ by (ax_1), hence $\exists z(\mathbf{A}az \wedge \mathbf{A}bz)$. □

Fact 8: $\widehat{\mathbf{K}}ab \supset \neg\mathbf{A}ab$

Proof. Follows from Facts 6 and 7. □

Fact 9: $\neg\mathbf{K}aa$

Proof. Follows from (ax_1) and Fact 7. □

Fact 10: $\neg\widehat{\mathbf{K}}aa$

Proof. Follows from Facts 6 and 9. □

Fact 11: $\mathbf{K}ab \leftrightarrow \mathbf{K}ba$

Proof. The definition of $\mathbf{\Gamma}$ in ($\text{df}_{\mathbf{\Gamma}}$) is symmetric. So the definition of $\mathbf{K}$ in ($\text{df}_{\mathbf{K}}$) is also symmetric. □

Fact 12: $\mathbf{\Pi}ab \leftrightarrow \mathbf{\Pi}ba$

Proof. The definition of $\mathbf{\Gamma}$ in ($\text{df}_{\mathbf{\Gamma}}$) is symmetric. So the definition of $\mathbf{\Pi}$ in ($\text{df}_{\mathbf{\Pi}}$) is also symmetric. □

Fact 13: $\mathbf{\Pi}ab \wedge (\widehat{\mathbf{\Sigma}}a \vee \widehat{\mathbf{\Sigma}}b) \supset \exists z(\mathbf{A}az \wedge \mathbf{A}bz)$

Proof. By ($\text{df}_{\mathbf{\Pi}}$), $\mathbf{\Pi}ab$ implies $\neg\mathbf{\Gamma}ab$. By ($\text{df}_{\mathbf{\Gamma}}$), $\neg\mathbf{\Gamma}ab$ and $\widehat{\mathbf{\Sigma}}a \vee \widehat{\mathbf{\Sigma}}b$ imply $\exists z(\mathbf{A}az \wedge \mathbf{A}bz)$. □

Fact 14: $\forall z(\mathbf{A}bz \supset \mathbf{\Pi}az) \supset \neg\widehat{\mathbf{\Sigma}}a$ (see pp. 253–254)

Proof. By (ax_1), $\forall z(\mathbf{A}bz \supset \mathbf{\Pi}az)$ implies $\mathbf{\Pi}ab$. For *reductio*, assume $\widehat{\mathbf{\Sigma}}a$. By Fact 13, $\mathbf{\Pi}ab$ and $\widehat{\mathbf{\Sigma}}a$ imply $\mathbf{A}az \wedge \mathbf{A}bz$, hence $\widehat{\mathbf{\Sigma}}z$ by Fact 3. Also, $\mathbf{A}bz$ and $\forall z(\mathbf{A}bz \supset \mathbf{\Pi}az)$ imply $\mathbf{\Pi}az$, contradicting $\widehat{\mathbf{\Sigma}}a \wedge \widehat{\mathbf{\Sigma}}z$ and ($\text{df}_{\mathbf{\Pi}}$). □

Fact 15: $\mathbf{\Gamma}ab \wedge \mathbf{A}bc \supset \mathbf{\Gamma}ac$

Proof. By ($\text{df}_{\mathbf{\Gamma}}$), $\mathbf{\Gamma}ab$ is tantamount to $(\widehat{\mathbf{\Sigma}}a \vee \widehat{\mathbf{\Sigma}}b) \wedge \neg\exists z(\mathbf{A}az \wedge \mathbf{A}bz)$. $\widehat{\mathbf{\Sigma}}a \vee \widehat{\mathbf{\Sigma}}b$ and $\mathbf{A}bc$ imply $\widehat{\mathbf{\Sigma}}a \vee \widehat{\mathbf{\Sigma}}c$ by Fact 3. Moreover, $\neg\exists z(\mathbf{A}az \wedge \mathbf{A}bz)$ and $\mathbf{A}bc$ imply $\neg\exists z(\mathbf{A}az \wedge \mathbf{A}cz)$ by (ax_2). So we have $\mathbf{\Gamma}ac$. □

Fact 16: $\mathbf{\Gamma}ab \wedge \mathbf{A}ac \supset \mathbf{\Gamma}cb$

Proof. Follows from Fact 15 and the symmetry of $\mathbf{\Gamma}$, which is guaranteed by ($\mathrm{df}_{\mathbf{\Gamma}}$). □

We are now in a position to establish the validity in the predicable semantics of Aristotle's conversion rules, full-blown subalternation, and full-blown modal subordination.

Fact 17: Aristotle's conversion rules are valid in the predicable semantics: (see p. 274)

(17.1) $\mathbb{X}^{\mathbf{i}}ab \supset \mathbb{X}^{\mathbf{i}}ba$	(17.7) $\mathbb{M}^{\mathbf{e}}ab \supset \mathbb{M}^{\mathbf{e}}ba$
(17.2) $\mathbb{X}^{\mathbf{e}}ab \supset \mathbb{X}^{\mathbf{e}}ba$	(17.8) $\mathbb{M}^{\mathbf{i}}ab \supset \mathbb{M}^{\mathbf{i}}ba$
(17.3) $\mathbb{N}^{\mathbf{i}}ab \supset \mathbb{N}^{\mathbf{i}}ba$	(17.9) $\mathbb{X}^{\mathbf{a}}ab \supset \mathbb{X}^{\mathbf{i}}ba$
(17.4) $\mathbb{N}^{\mathbf{e}}ab \supset \mathbb{N}^{\mathbf{e}}ba$	(17.10) $\mathbb{N}^{\mathbf{a}}ab \supset \mathbb{N}^{\mathbf{i}}ba$
(17.5) $\mathbb{Q}^{\mathbf{i}}ab \supset \mathbb{Q}^{\mathbf{i}}ba$	(17.11) $\mathbb{M}^{\mathbf{a}}ab \supset \mathbb{M}^{\mathbf{i}}ba$
(17.6) $\mathbb{Q}^{\mathbf{o}}ab \supset \mathbb{Q}^{\mathbf{o}}ba$	(17.12) $\mathbb{Q}^{\mathbf{a}}ab \supset \mathbb{Q}^{\mathbf{i}}ba$

Proof. 17.1–17.8 follow from the fact that the interpretation in the predicable semantics of the relevant categorical propositions is symmetric (for the case of $\mathbb{N}^{\mathbf{e}}$, $\mathbb{Q}^{\mathbf{i}}$, and $\mathbb{Q}^{\mathbf{o}}$, see Facts 11 and 12; in the other cases, the symmetry is obvious). As to 17.9, $\mathbf{A}ab$ implies $\exists z(\mathbf{A}az \wedge \mathbf{A}bz)$ by (ax_1); and similarly for 17.10. As to 17.11, we have to show that $\forall z(\mathbf{A}bz \supset \overline{\mathbf{\Pi}}az)$ implies $\neg\mathbf{\Gamma}ba$. By (ax_1), $\forall z(\mathbf{A}bz \supset \overline{\mathbf{\Pi}}az)$ implies $\overline{\mathbf{\Pi}}ab$. By ($\mathrm{df}_{\overline{\mathbf{\Pi}}}$), $\overline{\mathbf{\Pi}}ab$ implies $\mathbf{\Pi}ab$ or $\mathbf{A}ab$. In the first case, $\mathbf{\Pi}ab$ implies $\neg\mathbf{\Gamma}ab$ by ($\mathrm{df}_{\mathbf{\Pi}}$). In the second case, $\mathbf{A}ab$ implies $\neg\mathbf{\Gamma}ab$ by ($\mathrm{df}_{\mathbf{\Gamma}}$) and ($\mathrm{ax}_1$). So in either case, we have $\neg\mathbf{\Gamma}ab$, and hence $\neg\mathbf{\Gamma}ba$ by the symmetry of $\mathbf{\Gamma}$. As to 17.12, $\forall z(\mathbf{A}bz \supset \mathbf{\Pi}az)$ implies $\mathbf{\Pi}ba$ by (ax_1) and Fact 12. □

Fact 18: For all four modalities $\mathbb{H} \in \{\mathbb{X}, \mathbb{N}, \mathbb{M}, \mathbb{Q}\}$, affirmative and negative subalternation is valid in the predicable semantics:(see p. 214)

(18.1) $\mathbb{H}^{\mathbf{a}}ab \supset \mathbb{H}^{\mathbf{i}}ab$ (18.2) $\mathbb{H}^{\mathbf{e}}ab \supset \mathbb{H}^{\mathbf{o}}ab$

Proof. 18.1 follows from Fact 17. 18.2 is identical to 18.1 in the case of $\mathbb{Q}$, and follows from (ax_1) in the case of $\mathbb{X}$ and $\mathbb{N}$. As to the case of $\mathbb{M}$, we have to show that $\forall z(\mathbf{A}bz \supset \neg\overline{\mathbf{N}}az) \vee \forall z(\mathbf{A}az \supset \neg\overline{\mathbf{N}}bz)$ implies $\neg\overline{\mathbf{N}}ab$. First, assume $\forall z(\mathbf{A}bz \supset \neg\overline{\mathbf{N}}az)$. By ($\mathrm{ax}_1$), this implies $\neg\overline{\mathbf{N}}ab$. Second, assume $\forall z(\mathbf{A}az \supset \neg\overline{\mathbf{N}}bz)$ and $\overline{\mathbf{N}}ab$. $\overline{\mathbf{N}}ab$ implies $\hat{\mathbf{\Sigma}}b \wedge \mathbf{A}ab$ by Fact 4.

$\mathbf{A}ab$ and $\forall z(\mathbf{A}az \supset \neg\overline{\mathbf{N}}bz)$ imply $\neg\overline{\mathbf{N}}bb$. This implies $\neg(\widehat{\mathbf{\Sigma}}b \wedge \mathbf{A}bb)$ by ($\mathrm{df}_{\overline{\mathbf{N}}}$), contradicting $\widehat{\mathbf{\Sigma}}b$ and (ax_1). □

Next we observe the validity in the predicable semantics of N-X-, X-M-, and Q-M-subordination:

Fact 19: N-X-subordination is valid in the predicable semantics. For all $* \in \{\mathbf{a}, \mathbf{e}, \mathbf{i}, \mathbf{o}\}$: $\mathbb{N}^*ab \supset \mathbb{X}^*ab$. (see pp. 131–132)

Proof. As to a- and i-propositions, $\mathbf{N}ab$ implies $\mathbf{A}ab$ by (ax_5). As to e-propositions, $\mathbf{K}ab$ implies $\forall z(\mathbf{A}bz \supset \neg\mathbf{A}az)$ by ($\mathrm{df}_{\mathbf{K}}$) and ($\mathrm{df}_{\mathbf{\Gamma}}$). As to o-propositions, we have to show that $\exists z((\mathbf{A}bz \wedge \mathbf{K}az) \vee (\mathbf{N}bz \wedge \overline{\mathbf{K}}az))$ implies $\neg\mathbf{A}ab$. First, assume $\mathbf{A}bz \wedge \mathbf{K}az$ and $\mathbf{A}ab$. $\mathbf{A}ab$ and $\mathbf{A}bz$ imply $\mathbf{A}az$ by (ax_2), contradicting $\mathbf{K}az$ and Fact 7. Second, assume $\mathbf{N}bz \wedge \overline{\mathbf{K}}az$ and $\mathbf{A}ab$. By ($\mathrm{df}_{\overline{\mathbf{K}}}$), $\overline{\mathbf{K}}az$ implies $\mathbf{N}av \wedge \forall u(\mathbf{A}au \wedge \mathbf{\Sigma}u \supset \widehat{\mathbf{K}}zu)$. $\mathbf{N}bz$ implies $\mathbf{A}bz$ and $\mathbf{\Sigma}z$ by (ax_5) and ($\mathrm{df}_{\mathbf{\Sigma}}$). $\mathbf{A}ab$ and $\mathbf{A}bz$ imply $\mathbf{A}az$ by (ax_2). Finally, $\mathbf{A}az$, $\mathbf{\Sigma}z$, and $\forall u(\mathbf{A}au \wedge \mathbf{\Sigma}u \supset \widehat{\mathbf{K}}zu)$ imply $\widehat{\mathbf{K}}zz$, contradicting Fact 10. □

Fact 20: X-M-subordination is valid in the predicable semantics. For all $* \in \{\mathbf{a}, \mathbf{e}, \mathbf{i}, \mathbf{o}\}$: $\mathbb{X}^*ab \supset \mathbb{M}^*ab$. (see p. 215)

Proof. As to a-propositions, $\mathbf{A}ab$ implies $\forall z(\mathbf{A}bz \supset \mathbf{A}az)$ by (ax_2), and hence $\forall z(\mathbf{A}bz \supset \overline{\mathbf{\Pi}}az)$ by ($\mathrm{df}_{\overline{\mathbf{\Pi}}}$). As to i-propositions, $\exists z(\mathbf{A}bz \wedge \mathbf{A}az)$ implies $\neg\mathbf{\Gamma}ab$ by ($\mathrm{df}_{\mathbf{\Gamma}}$). As to e-propositions, $\forall z(\mathbf{A}bz \supset \neg\mathbf{A}az)$ implies $\forall z(\mathbf{A}bz \supset \neg\overline{\mathbf{N}}az)$ by Fact 4; and similarly for o-propositions. □

Fact 21: Q-M-subordination is valid in the predicable semantics. For all $* \in \{\mathbf{a}, \mathbf{e}, \mathbf{i}, \mathbf{o}\}$: $\mathbb{Q}^*ab \supset \mathbb{M}^*ab$. (see pp. 208–209)

Proof. As to a-propositions, $\forall z(\mathbf{A}bz \supset \mathbf{\Pi}az)$ implies $\forall z(\mathbf{A}bz \supset \overline{\mathbf{\Pi}}az)$ by ($\mathrm{df}_{\overline{\mathbf{\Pi}}}$). As to i-propositions, $\mathbf{\Pi}ab$ implies $\neg\mathbf{\Gamma}ab$ by ($\mathrm{df}_{\mathbf{\Pi}}$). As to e-propositions, it suffices to show that $\forall z(\mathbf{A}bz \supset \mathbf{\Pi}az)$ implies $\forall z(\mathbf{A}bz \supset \neg\overline{\mathbf{N}}az)$. For *reductio,* assume $\mathbf{A}bz \wedge \overline{\mathbf{N}}az$. $\mathbf{A}bz$ and $\forall z(\mathbf{A}bz \supset \mathbf{\Pi}az)$ imply $\mathbf{\Pi}az$. By ($\mathrm{df}_{\overline{\mathbf{N}}}$), $\overline{\mathbf{N}}az$ implies $\widehat{\mathbf{N}}az$ or $\widehat{\mathbf{\Sigma}}a$. The first case contradicts $\mathbf{\Pi}az$ and ($\mathrm{df}_{\mathbf{\Pi}}$); the latter case contradicts $\forall z(\mathbf{A}bz \supset \mathbf{\Pi}az)$ and Fact 14. As to o-propositions, we have to show that $\mathbf{\Pi}ab$ implies $\neg\overline{\mathbf{N}}ab$. For *reductio,* assume $\overline{\mathbf{N}}ab$. By ($\mathrm{df}_{\overline{\mathbf{N}}}$), this implies $\widehat{\mathbf{N}}ab$ or $\widehat{\mathbf{\Sigma}}a \wedge \mathbf{A}ab$. The first case contradicts $\mathbf{\Pi}ab$ and ($\mathrm{df}_{\mathbf{\Pi}}$). In the latter case, we have $\widehat{\mathbf{\Sigma}}b$ by Fact 3, hence $\widehat{\mathbf{\Sigma}}a \wedge \widehat{\mathbf{\Sigma}}b$. This contradicts $\mathbf{\Pi}ab$ and (df$_{\mathbf{\Pi}}$). □

It follows from the principles of subalternation and modal subordination that every affirmative (or negative) categorical proposition implies the corresponding affirmative (or negative) particular M-proposition:

Fact 22: For all four modalities $\mathbb{H} \in \{\mathbb{X}, \mathbb{N}, \mathbb{M}, \mathbb{Q}\}$, the following implications are valid in the predicable semantics: (see pp. 213–214)

(22.1) $\mathbb{H}^{\mathbf{a}}ab \supset \mathbb{M}^{\mathbf{i}}ab$ (22.3) $\mathbb{H}^{\mathbf{e}}ab \supset \mathbb{M}^{\mathbf{o}}ab$
(22.2) $\mathbb{H}^{\mathbf{i}}ab \supset \mathbb{M}^{\mathbf{i}}ab$ (22.4) $\mathbb{H}^{\mathbf{o}}ab \supset \mathbb{M}^{\mathbf{o}}ab$

Proof. Follows from Facts 18, 19, 20, and 21. □

The next result will be useful for establishing the validity in the predicable semantics of Aristotle's perfect QQQ-moods:

Fact 23: $\forall z(\mathbf{A}bz \supset \mathbf{\Pi}az) \supset \forall z(\mathbf{\Pi}bz \supset \mathbf{\Pi}az)$ (see p. 255)

Proof. Suppose $\forall z(\mathbf{A}bz \supset \mathbf{\Pi}az)$ and $\mathbf{\Pi}bz$. Our aim is to establish $\mathbf{\Pi}az$. By Fact 14, $\forall z(\mathbf{A}bz \supset \mathbf{\Pi}az)$ implies $\neg\widehat{\mathbf{\Sigma}}a$. Hence $\neg(\widehat{\mathbf{\Sigma}}a \wedge \widehat{\mathbf{\Sigma}}z)$ and $\neg\widehat{\mathbf{N}}za$ by (df$_{\widehat{\mathbf{\Sigma}}}$). In order to establish $\mathbf{\Pi}az$, it remains to show $\neg\widehat{\mathbf{N}}az$ and $\neg\mathbf{\Gamma}az$. First, assume $\widehat{\mathbf{N}}az$. This implies $\widehat{\mathbf{\Sigma}}z$ by (df$_{\widehat{\mathbf{\Sigma}}}$). By Fact 13, $\widehat{\mathbf{\Sigma}}z$ and $\mathbf{\Pi}bz$ imply $\mathbf{A}bu \wedge \mathbf{A}zu$. $\mathbf{A}zu$ and $\widehat{\mathbf{N}}az$ imply $\widehat{\mathbf{N}}au$ by (ax$_4$). But $\mathbf{A}bu$ and $\forall z(\mathbf{A}bz \supset \mathbf{\Pi}az)$ imply $\mathbf{\Pi}au$, contradicting $\widehat{\mathbf{N}}au$ and (df$_{\mathbf{\Pi}}$). Second, assume $\mathbf{\Gamma}az$. By (df$_{\mathbf{\Gamma}}$), this implies $\widehat{\mathbf{\Sigma}}a \vee \widehat{\mathbf{\Sigma}}z$ and $\neg\exists u(\mathbf{A}au \wedge \mathbf{A}zu)$. $\widehat{\mathbf{\Sigma}}a \vee \widehat{\mathbf{\Sigma}}z$ and $\neg\widehat{\mathbf{\Sigma}}a$ imply $\widehat{\mathbf{\Sigma}}z$. By Fact 13, $\mathbf{\Pi}bz$ and $\widehat{\mathbf{\Sigma}}z$ imply $\mathbf{A}bu \wedge \mathbf{A}zu$. $\mathbf{A}zu$ and $\widehat{\mathbf{\Sigma}}z$ imply $\widehat{\mathbf{\Sigma}}u$ by Fact 3. Moreover, $\mathbf{A}bu$ and $\forall z(\mathbf{A}bz \supset \mathbf{\Pi}az)$ imply $\mathbf{\Pi}au$. By Fact 13, $\widehat{\mathbf{\Sigma}}u$ and $\mathbf{\Pi}au$ imply $\mathbf{A}aw \wedge \mathbf{A}uw$. Now, $\mathbf{A}uw$ and $\mathbf{A}zu$ imply $\mathbf{A}zw$ by (ax$_2$), hence $\mathbf{A}aw \wedge \mathbf{A}zw$. This contradicts $\neg\exists u(\mathbf{A}au \wedge \mathbf{A}zu)$. □

Next, we may state some principles of modal opposition between possibility propositions and negative N-propositions.

Fact 24: $\mathbf{K}ab \supset \neg\mathbf{\Pi}ab$

Proof. $\mathbf{K}ab$ implies $\mathbf{\Gamma}ab$ by (df$_{\mathbf{K}}$), whereas $\mathbf{\Pi}ab$ implies $\neg\mathbf{\Gamma}ab$ by (df$_{\mathbf{\Pi}}$). □

Fact 25: $\mathbf{K}ab \wedge \mathbf{A}bc \supset \mathbf{K}ac$

Proof. By ($\mathrm{df}_{\mathbf{K}}$), $\mathbf{K}ab$ is tantamount to $\mathbf{\Sigma}a \wedge \mathbf{\Sigma}b \wedge \mathbf{\Gamma}ab$. $\mathbf{\Sigma}b$ and $\mathbf{A}bc$ imply $\mathbf{\Sigma}c$ by Fact 2. $\mathbf{\Gamma}ab$ and $\mathbf{A}bc$ imply $\mathbf{\Gamma}ac$ by Fact 15. So we have $\mathbf{K}ac$. □

Fact 26: $\mathbf{K}ab \wedge \mathbf{A}ac \supset \mathbf{K}cb$

Proof. Follows from Facts 11 and 25. □

Fact 27: $\mathbf{K}ab \supset \mathbf{\Gamma}ab$

Proof. Follows from ($\mathrm{df}_{\mathbf{K}}$). □

Fact 28: $\forall z(\mathbf{A}bz \supset \overline{\mathbf{\Pi}}az) \supset \neg\exists z((\mathbf{A}bz \wedge \mathbf{K}az) \vee (\mathbf{N}bz \wedge \overline{\mathbf{K}}az))$

Proof. For *reductio,* assume first $\mathbf{A}bz \wedge \mathbf{K}az$. $\mathbf{A}bz$ and $\forall z(\mathbf{A}bz \supset \overline{\mathbf{\Pi}}az)$ imply $\overline{\mathbf{\Pi}}az$, hence $\mathbf{A}az$ or $\mathbf{\Pi}az$ by ($\mathrm{df}_{\overline{\mathbf{\Pi}}}$). Either case contradicts $\mathbf{K}az$, by Fact 7 and Fact 24. Second, assume $\mathbf{N}bz \wedge \overline{\mathbf{K}}az$, that is, $\mathbf{N}bz \wedge \mathbf{N}av \wedge \forall u(\mathbf{A}au \wedge \mathbf{\Sigma}u \supset \widehat{\mathbf{K}}zu)$. By ($\mathrm{ax}_5$) and ($\mathrm{df}_{\mathbf{\Sigma}}$), $\mathbf{N}av$ implies $\mathbf{A}av \wedge \mathbf{\Sigma}v$. Thus, $\forall u(\mathbf{A}au \wedge \mathbf{\Sigma}u \supset \widehat{\mathbf{K}}zu)$ implies $\widehat{\mathbf{K}}zv$. By ($\mathrm{df}_{\widehat{\mathbf{K}}}$), $\widehat{\mathbf{K}}zv$ implies $\widehat{\mathbf{\Sigma}}z$, hence $\mathbf{\Sigma}z$ by Fact 5. Now, $\mathbf{N}bz$ implies $\mathbf{A}bz$ by (ax_5). $\mathbf{A}bz$ and $\forall z(\mathbf{A}bz \supset \overline{\mathbf{\Pi}}az)$ imply $\mathbf{A}az$ or $\mathbf{\Pi}az$ by ($\mathrm{df}_{\overline{\mathbf{\Pi}}}$). In the first case, we have $\mathbf{A}az \wedge \mathbf{\Sigma}z$ so that $\forall u(\mathbf{A}au \wedge \mathbf{\Sigma}u \supset \widehat{\mathbf{K}}zu)$ implies $\widehat{\mathbf{K}}zz$, contradicting Fact 10. In the second case, $\mathbf{\Pi}az$ and $\widehat{\mathbf{\Sigma}}z$ yield $\mathbf{A}as \wedge \mathbf{A}zs$ by Fact 13. By Fact 3, $\widehat{\mathbf{\Sigma}}z$ and $\mathbf{A}zs$ imply $\widehat{\mathbf{\Sigma}}s$, hence $\mathbf{\Sigma}s$ by Fact 5. So we have $\mathbf{A}as \wedge \mathbf{\Sigma}s$ so that $\forall u(\mathbf{A}au \wedge \mathbf{\Sigma}u \supset \widehat{\mathbf{K}}zu)$ implies $\widehat{\mathbf{K}}zs$, contradicting $\mathbf{A}zs$ and Fact 8. □

Fact 29: The following principles of modal opposition are valid in the predicable semantics:

(29.1) $\mathbb{N}^{\mathbf{e}}ab \supset \neg\mathbb{M}^{\mathbf{i}}ab$ (29.2) $\mathbb{N}^{\mathbf{o}}ab \supset \neg\mathbb{M}^{\mathbf{a}}ab$

Proof. Given the interpretations of categorical propositions listed on pp. 287–288, this is a reformulation of Facts 27 and 28. □

The following three principles of Q-N-incompatibility are used by Aristotle in his proofs of validity (see pp. 200–201).

Fact 30: The following principles of modal opposition are valid in the predicable semantics: (see pp. 226–227 and 268)

(30.1) $\mathbb{N}^{\mathsf{e}}ab \supset \neg\mathbb{Q}^{\mathsf{i}}ab$ (30.3) $\mathbb{N}^{\mathsf{o}}ab \supset \neg\mathbb{Q}^{\mathsf{a}}ab$
(30.2) $\mathbb{N}^{\mathsf{e}}ab \supset \neg\mathbb{Q}^{\mathsf{a}}ab$

Proof. Follows from Fact 29 by subalternation and Q-M-subordination (Facts 18 and 21). □

Valid Moods

In this section, all moods held to be valid by Aristotle are proved to be valid in the predicable semantics (Facts 31–62). In proving them so, it will be helpful to make use of the deductive system given in Table 15.1 on p. 225. This is a system of direct deductions based on three kinds of deduction rules, namely, on Aristotle's conversion rules, the equivalence of affirmative and negative Q-propositions (qualitative conversion), and several syllogistic moods. The moods that figure as deduction rules may be called the *primitive moods* of the system.

Now, Aristotle's conversion rules are valid in the predicable semantics (Fact 17). Qualitative conversion is also valid in it, and trivially so, because affirmative and negative Q-propositions are assigned the same interpretation. In what follows, we may therefore neglect the difference between affirmative and negative Q-propositions. In order to show that all moods held to be valid by Aristotle are valid in the predicable semantics, then, it remains to establish two theses: first, that all primitive moods of the deductive system are valid in the predicable semantics, and second, that all moods held to be valid by Aristotle are deducible in the deductive system. The first thesis will be established in Facts 31–43, and the second in Facts 44–62.

Fact 31: The following moods are valid in the predicable semantics:

(31.1) **aaa-1-XXX**: $\mathbf{A}ab \wedge \mathbf{A}bc \vdash \mathbf{A}ac$
(31.2) **eae-1-XXX**: $\forall z(\mathbf{A}bz \supset \neg\mathbf{A}az) \wedge \mathbf{A}bc \vdash \forall z(\mathbf{A}cz \supset \neg\mathbf{A}az)$
(31.3) **aii-1-XXX**: $\mathbf{A}ab \wedge \exists z(\mathbf{A}cz \wedge \mathbf{A}bz) \vdash \exists z(\mathbf{A}cz \wedge \mathbf{A}az)$
(31.4) **eio-1-XXX**: $\forall z(\mathbf{A}bz \supset \neg\mathbf{A}az) \wedge \exists z(\mathbf{A}cz \wedge \mathbf{A}bz) \vdash \neg\mathbf{A}ac$

Proof. Follows by (ax_2). □

Fact 32: The following moods are valid in the predicable semantics:

(32.1) **aoo-2-XXX**: $\mathbf{A}ba \wedge \neg\mathbf{A}bc \vdash \neg\mathbf{A}ac$
(32.2) **oao-3-XXX**: $\neg\mathbf{A}ab \wedge \mathbf{A}cb \vdash \neg\mathbf{A}ac$

Proof. Follows by (ax_2). □

Fact 33: The following moods are valid in the predicable semantics:

(33.1) **aaa-1-NXN**: $\mathbf{N}ab \wedge \mathbf{A}bc \vdash \mathbf{N}ac$
(33.2) **eae-1-NXN**: $\mathbf{K}ab \wedge \mathbf{A}bc \vdash \mathbf{K}ac$
(33.3) **aii-1-NXN**: $\mathbf{N}ab \wedge \exists z(\mathbf{A}cz \wedge \mathbf{A}bz) \vdash \exists z((\mathbf{A}cz \wedge \mathbf{N}az) \vee (\mathbf{A}az \wedge \mathbf{N}cz))$
(33.4) **eio-1-NXN**: $\mathbf{K}ab \wedge \exists z(\mathbf{A}cz \wedge \mathbf{A}bz) \vdash \exists z((\mathbf{A}cz \wedge \mathbf{K}az) \vee (\mathbf{N}cz \wedge \overline{\mathbf{K}}az))$

Proof. 33.1 is (ax_3). 33.2 is Fact 25. As to 33.3, $\mathbf{N}ab$ and $\mathbf{A}cz \wedge \mathbf{A}bz$ imply $\mathbf{A}cz \wedge \mathbf{N}az$ by (ax_3). As to 33.4, $\mathbf{K}ab$ and $\mathbf{A}cz \wedge \mathbf{A}bz$ imply $\mathbf{A}cz \wedge \mathbf{K}az$ by Fact 25. □

Fact 34: The moods **aaa-1-NNN**, **eae-1-NNN**, **aii-1-NNN**, and **eio-1-NNN** are valid in the predicable semantics.

Proof. Follows from Fact 33 by N-X-subordination (Fact 19). □

Fact 35: The mood **aoo-2-NNN** is valid in the predicable semantics: $\mathbf{N}ba \wedge \exists z((\mathbf{A}cz \wedge \mathbf{K}bz) \vee (\mathbf{N}cz \wedge \overline{\mathbf{K}}bz)) \vdash \exists z((\mathbf{A}cz \wedge \mathbf{K}az) \vee (\mathbf{N}cz \wedge \overline{\mathbf{K}}az))$

Proof. First, assume $\mathbf{N}ba$ and $\mathbf{A}cz \wedge \mathbf{K}bz$. This implies $\mathbf{A}cz \wedge \mathbf{K}az$ by (ax_5) and Fact 26. Second, assume $\mathbf{N}ba$ and $\mathbf{N}cz \wedge \overline{\mathbf{K}}bz$. By ($\text{df}_{\overline{\mathbf{K}}}$), we have $\mathbf{N}cz \wedge \mathbf{N}bv \wedge \forall u(\mathbf{A}bu \wedge \mathbf{\Sigma}u \supset \widehat{\mathbf{K}}zu)$. $\mathbf{N}ba$ implies $\mathbf{\Sigma}a$ by ($\text{df}_{\mathbf{\Sigma}}$), and $\mathbf{A}ba$ by (ax_5). Now, $\mathbf{\Sigma}a$, $\mathbf{A}ba$, and $\forall u(\mathbf{A}bu \wedge \mathbf{\Sigma}u \supset \widehat{\mathbf{K}}zu)$ imply $\widehat{\mathbf{K}}za$. Hence $\widehat{\mathbf{K}}az$ by ($\text{df}_{\widehat{\mathbf{K}}}$), and $\mathbf{K}az$ by Fact 6. $\mathbf{N}cz$ implies $\mathbf{A}cz$ by (ax_5). So we have $\mathbf{A}cz \wedge \mathbf{K}az$. □

Fact 36: The mood **oao-3-NNN** is valid in the predicable semantics: $\exists z((\mathbf{A}bz \wedge \mathbf{K}az) \vee (\mathbf{N}bz \wedge \overline{\mathbf{K}}az)) \wedge \mathbf{N}cb \vdash \exists z((\mathbf{A}cz \wedge \mathbf{K}az) \vee (\mathbf{N}cz \wedge \overline{\mathbf{K}}az))$

Proof. First, assume $\mathbf{A}bz \wedge \mathbf{K}az$ and $\mathbf{N}cb$. This implies $\mathbf{A}cz \wedge \mathbf{K}az$ by (ax_5) and (ax_2). Second, assume $\mathbf{N}bz \wedge \overline{\mathbf{K}}az$ and $\mathbf{N}cb$. $\mathbf{N}bz$ and $\mathbf{N}cb$ imply $\mathbf{N}cz$ by (ax_5) and (ax_3). So we have $\mathbf{N}cz \wedge \overline{\mathbf{K}}az$. □

Fact 37: The following moods are valid in the predicable semantics:

(37.1) **aaa,eae,aea,eee-1-QQQ**:
$\forall z(\mathbf{A}bz \supset \mathbf{\Pi}az) \wedge \forall z(\mathbf{A}cz \supset \mathbf{\Pi}bz) \vdash \forall z(\mathbf{A}cz \supset \mathbf{\Pi}az)$

(37.2) **aii,eio,aoi,eoo-1-QQQ**: $\forall z(\mathbf{A}bz \supset \mathbf{\Pi}az) \wedge \mathbf{\Pi}bc \vdash \mathbf{\Pi}ac$

Proof. In both moods, the major premise $\forall z(\mathbf{A}bz \supset \mathbf{\Pi}az)$ implies $\forall z(\mathbf{\Pi}bz \supset \mathbf{\Pi}az)$ by Fact 23. This yields the conclusion. □

Fact 38: The following moods are valid in the predicable semantics:

(38.1) **aaa,eae-1-QXQ**: $\forall z(\mathbf{A}bz \supset \mathbf{\Pi}az) \wedge \mathbf{A}bc \vdash \forall z(\mathbf{A}cz \supset \mathbf{\Pi}az)$
(38.2) **aii,eio-1-QXQ**: $\forall z(\mathbf{A}bz \supset \mathbf{\Pi}az) \wedge \exists z(\mathbf{A}cz \wedge \mathbf{A}bz) \vdash \mathbf{\Pi}ac$

Proof. 38.1 follows from (ax_2). As to 38.2, assume $\mathbf{A}cz \wedge \mathbf{A}bz$. $\mathbf{A}bz$ and $\forall z(\mathbf{A}bz \supset \mathbf{\Pi}az)$ imply $\mathbf{\Pi}az$. First, we have $\neg(\widehat{\mathbf{\Sigma}}a \wedge \widehat{\mathbf{\Sigma}}c)$; otherwise, $\mathbf{A}cz$ would imply $\widehat{\mathbf{\Sigma}}a \wedge \widehat{\mathbf{\Sigma}}z$ by Fact 3, contradicting $\mathbf{\Pi}az$ and ($df_{\mathbf{\Pi}}$). Second, we have $\neg\widehat{\mathbf{N}}ac$; otherwise, $\mathbf{A}cz$ would imply $\widehat{\mathbf{N}}az$ by (ax_4), contradicting $\mathbf{\Pi}az$ and ($df_{\mathbf{\Pi}}$). Third, we have $\neg\widehat{\mathbf{N}}ca$; otherwise, we would have $\widehat{\mathbf{\Sigma}}a$ by ($df_{\widehat{\mathbf{\Sigma}}}$), contradicting $\forall z(\mathbf{A}bz \supset \mathbf{\Pi}az)$ and Fact 14. Fourth, we have $\neg\mathbf{\Gamma}ac$; otherwise, $\mathbf{A}cz$ would imply $\mathbf{\Gamma}az$ by Fact 15, contradicting $\mathbf{\Pi}az$ and ($df_{\mathbf{\Pi}}$). These four consequences together imply $\mathbf{\Pi}ac$ by ($df_{\mathbf{\Pi}}$). □

Fact 39: The moods **aaa,eae-1-QNQ** and **aii,eio-1-QNQ** are valid in the predicable semantics.

Proof. Follows from Fact 38 by N-X-subordination (Fact 19). □

Fact 40: The following moods are valid in the predicable semantics:

(40.1) **aaa,aea-1-XQM**: $\mathbf{A}ab \wedge \forall z(\mathbf{A}cz \supset \mathbf{\Pi}bz) \vdash \forall z(\mathbf{A}cz \supset \overline{\mathbf{\Pi}}az)$
(40.2) **eae,eee-1-XQM**: $\forall z(\mathbf{A}bz \supset \neg\mathbf{A}az) \wedge \forall z(\mathbf{A}cz \supset \mathbf{\Pi}bz) \vdash$
$\forall z(\mathbf{A}cz \supset \neg\overline{\mathbf{N}}az) \vee \forall z(\mathbf{A}az \supset \neg\overline{\mathbf{N}}cz)$

(40.3) **aii,aoi-1-XQM**: $\mathbf{A}ab \wedge \mathbf{\Pi}bc \vdash \neg\mathbf{\Gamma}ac$
(40.4) **eio,eoo-1-XQM**: $\forall z(\mathbf{A}bz \supset \neg\mathbf{A}az) \wedge \mathbf{\Pi}bc \vdash \neg\overline{\mathbf{N}}ac$

Proof. 40.1: For *reductio,* assume $\mathbf{A}cz$ and $\neg\overline{\mathbf{\Pi}}az$. By ($\text{df}_{\overline{\Pi}}$), $\neg\overline{\mathbf{\Pi}}az$ implies $\neg\mathbf{A}az$ and $\neg\mathbf{\Pi}az$. $\neg\mathbf{A}az$ implies $\neg\widehat{\mathbf{N}}az$ by Fact 1. Moreover, we have $\neg\widehat{\mathbf{\Sigma}}a$. Otherwise, $\mathbf{A}ab$ would imply $\widehat{\mathbf{\Sigma}}b$ by Fact 3, contradicting $\forall z(\mathbf{A}cz \supset \mathbf{\Pi}bz)$ and Fact 14. Also, $\neg\widehat{\mathbf{\Sigma}}a$ implies $\neg\widehat{\mathbf{N}}za$ by ($\text{df}_{\widehat{\Sigma}}$). So we have $\neg(\widehat{\mathbf{\Sigma}}a \wedge \widehat{\mathbf{\Sigma}}z) \wedge \neg\widehat{\mathbf{N}}az \wedge \neg\widehat{\mathbf{N}}za$. This and $\neg\mathbf{\Pi}az$ imply $\mathbf{\Gamma}az$ by (df_{Π}). By (df_{Γ}), $\mathbf{\Gamma}az$ implies $\widehat{\mathbf{\Sigma}}a \vee \widehat{\mathbf{\Sigma}}z$ and $\neg\exists u(\mathbf{A}au \wedge \mathbf{A}zu)$. $\widehat{\mathbf{\Sigma}}a \vee \widehat{\mathbf{\Sigma}}z$ and $\neg\widehat{\mathbf{\Sigma}}a$ imply $\widehat{\mathbf{\Sigma}}z$. $\mathbf{A}cz$ and $\forall z(\mathbf{A}cz \supset \mathbf{\Pi}bz)$ imply $\mathbf{\Pi}bz$. $\mathbf{\Pi}bz$ and $\widehat{\mathbf{\Sigma}}z$ imply $\exists u(\mathbf{A}bu \wedge \mathbf{A}zu)$ by Fact 13. But $\mathbf{A}ab$ and $\neg\exists u(\mathbf{A}au \wedge \mathbf{A}zu)$ imply $\neg\exists u(\mathbf{A}bu \wedge \mathbf{A}zu)$ by (ax_2).

40.2: For *reductio,* assume $\mathbf{A}cz \wedge \overline{\mathbf{N}}az$. $\overline{\mathbf{N}}az$ implies $\widehat{\mathbf{\Sigma}}z \wedge \mathbf{A}az$ by Fact 4. $\mathbf{A}cz$ and $\forall z(\mathbf{A}cz \supset \mathbf{\Pi}bz)$ imply $\mathbf{\Pi}bz$. $\widehat{\mathbf{\Sigma}}z$ and $\mathbf{\Pi}bz$ imply $\exists u(\mathbf{A}bu \wedge \mathbf{A}zu)$ by Fact 13. $\mathbf{A}az$ and $\exists u(\mathbf{A}bu \wedge \mathbf{A}zu)$ imply $\exists u(\mathbf{A}bu \wedge \mathbf{A}au)$ by (ax_2). This contradicts $\forall z(\mathbf{A}bz \supset \neg\mathbf{A}az)$.

40.3: $\mathbf{\Gamma}ac$ and $\mathbf{A}ab$ imply $\mathbf{\Gamma}bc$ by Fact 16. $\mathbf{\Gamma}bc$ contradicts $\mathbf{\Pi}bc$ and (df_{Π}).

40.4: For *reductio,* assume $\overline{\mathbf{N}}ac$. This implies $\widehat{\mathbf{\Sigma}}c \wedge \mathbf{A}ac$ by Fact 4. $\widehat{\mathbf{\Sigma}}c$ and $\mathbf{\Pi}bc$ imply $\exists u(\mathbf{A}bu \wedge \mathbf{A}cu)$ by Fact 13. This and $\mathbf{A}ac$ imply $\neg\forall z(\mathbf{A}bz \supset \neg\mathbf{A}az)$ by (ax_2). □

Fact 41: The moods **aaa,aea-1-NQM**, **eae,eee-1-NQM**, **aii,aoi-1-NQM**, and **eio,eoo-1-NQM** are valid in the predicable semantics.

Proof. Follows from Fact 40 by N-X-subordination (Fact 19). □

Fact 42: The following moods are valid in the predicable semantics:

(42.1) **eae,eee-1-NQX**: $\mathbf{K}ab \wedge \forall z(\mathbf{A}cz \supset \mathbf{\Pi}bz) \vdash \forall z(\mathbf{A}cz \supset \neg\mathbf{A}az)$

(42.2) **eio,eoo-1-NQX**: $\mathbf{K}ab \wedge \mathbf{\Pi}bc \vdash \neg\mathbf{A}ac$

Proof. 42.1: For *reductio,* assume $\mathbf{A}cz \wedge \mathbf{A}az$. By Fact 26, $\mathbf{K}ab$ and $\mathbf{A}az$ imply $\mathbf{K}zb$, hence $\mathbf{K}bz$ by Fact 11. $\mathbf{A}cz$ and $\forall z(\mathbf{A}cz \supset \mathbf{\Pi}bz)$ imply $\mathbf{\Pi}bz$, contradicting $\mathbf{K}bz$ and Fact 24. 42.2: For *reductio,* assume $\mathbf{A}ac$. $\mathbf{K}ab$ and $\mathbf{A}ac$ imply $\mathbf{K}cb$ by Fact 26, hence $\mathbf{K}bc$ by Fact 11. This contradicts $\mathbf{\Pi}bc$ and Fact 24. □

Fact 43: The mood **oao-3-QXM** is valid in the predicable semantics: $\mathbf{\Pi}ab \wedge \mathbf{A}cb \vdash \neg\overline{\mathbf{N}}ac$

Proof. For *reductio,* assume $\overline{\mathbf{N}}ac$. By $(\mathrm{df}_{\overline{\mathbf{N}}})$, this implies $\widehat{\mathbf{N}}ac$ or $\widehat{\mathbf{\Sigma}}a \wedge \mathbf{A}ac$. In the first case, $\widehat{\mathbf{N}}ac$ and $\mathbf{A}cb$ imply $\widehat{\mathbf{N}}ab$ by (ax_4), contradicting $\mathbf{\Pi}ab$ and (df_{Π}). In the latter case, $\mathbf{A}ac$ and $\mathbf{A}cb$ imply $\mathbf{A}ab$ by (ax_2). $\mathbf{A}ab$ and $\widehat{\mathbf{\Sigma}}a$ imply $\widehat{\mathbf{\Sigma}}a \wedge \widehat{\mathbf{\Sigma}}b$ by Fact 3, contradicting $\mathbf{\Pi}ab$ and (df_{Π}). □

The moods listed in Facts 31–43 are precisely the primitive moods used as deduction rules in the deductive system given in Table 15.1 on p. 225. Hence all deduction rules of this system are valid in the predicable semantics, with the result that every mood deducible in the system is valid in the predicable semantics.

We now go on to show that all moods held to be valid by Aristotle are deducible in the deductive system (Facts 44–62). In other words, it will be shown that all these moods are deducible by means of direct deductions using only the primitive moods (Facts 31–43) and Aristotle's conversion rules (Fact 17). Given this, it will follow that all moods held to be valid by Aristotle are valid in the predicable semantics.

In the proofs of Facts 44–62, the term "deducible" is employed to mean "deducible by means of direct deductions (without using *reductio ad absurdum*)."

Fact 44: The moods **eae-2-XXX** and **aee-2-XXX** are valid in the predicable semantics.

Proof. Deducible by eae-1-XXX (Fact 31) using $\mathrm{e_X}$-conversion (Fact 17). □

Fact 45: The moods **eio-2-XXX**, **eao-3-XXX**, and **eio-3-XXX** are valid in the predicable semantics.

Proof. Deducible by eio-1-XXX (Fact 31) using $\mathrm{e_X}$-, $\mathrm{a_X}$-, and $\mathrm{i_X}$-conversion (Fact 17). □

Fact 46: The moods **aai-3-XXX**, **aii-3-XXX**, and **iai-3-XXX** are valid in the predicable semantics.

Proof. Deducible by aii-1-XXX (Fact 31) using a_X- and i_X-conversion (Fact 17). □

Fact 47: The moods **eae-2-NXN**, **aee-2-XNN**, and **eio-2-NXN** are valid in the predicable semantics.

Proof. Deducible by eae-1-NXN and eio-1-NXN (Fact 33) using e_N-conversion (Fact 17). □

Fact 48: The moods **eae-2-NNN**, **aee-2-NNN**, and **eio-2-NNN** are valid in the predicable semantics.

Proof. Deducible by eae-1-NNN and eio-1-NNN (Fact 34) using e_N-conversion (Fact 17). □

Fact 49: The moods **aai-3-XNN**, **aai-3-NXN**, **eao-3-NXN**, **iai-3-XNN**, **aii-3-NXN**, and **eio-3-NXN** are valid in the predicable semantics.

Proof. Deducible by aii-1-NXN and eio-1-NXN (Fact 33) using a_X-, i_X-, and i_N-conversion (Fact 17). □

Fact 50: The moods **aai-3-NNN**, **eao-3-NNN**, **iai-3-NNN**, **aii-3-NNN**, and **eio-3-NNN** are valid in the predicable semantics.

Proof. Deducible by aii-1-NNN and eio-1-NNN (Fact 34) using a_N- and i_N-conversion (Fact 17). □

Fact 51: The moods **eae,eee-2-XQM**, **aee,eee-2-QXM**, and **eio,eoo-2-XQM** are valid in the predicable semantics.

Proof. Deducible by eae-1-XQM and eio-1-XQM (Fact 40) using e_X- and e_M-conversion (Fact 17). □

Fact 52: The moods **eae,eee-2-NQX**, **aee,eee-2-QNX**, and **eio,eoo-2-NQX** are valid in the predicable semantics.

Proof. Deducible by eae-1-NQX and eio-1-NQX (Fact 42) using e_N- and e_X-conversion (Fact 17). □

Fact 53: The moods **eae,eee-2-NQM**, **aee,eee-2-QNM**, and **eio,eoo-2-NQM** are valid in the predicable semantics.

Proof. Deducible by eae-1-NQM and eio-1-NQM (Fact 41) using e_N- and e_M-conversion (Fact 17). □

Fact 54: The moods **aai,eao,aei,eeo-3-QQQ**, **aii,aoi,eio,eoo-3-QQQ**, and **iai,oao,iei,oeo-3-QQQ** are valid in the predicable semantics.

Proof. Deducible by aii-1-QQQ (Fact 37) using a_Q- and i_Q-conversion (Fact 17). □

Fact 55: The moods **aai,eao-3-QXQ** and **aii,eio-3-QXQ** are valid in the predicable semantics.

Proof. Deducible by aii-1-QXQ (Fact 38) using a_X- and i_X-conversion (Fact 17). □

Fact 56: The moods **aai,aei-3-XQQ** and **iai,iei-3-XQQ** are valid in the predicable semantics.

Proof. Deducible by aii-1-QXQ (Fact 38) using a_X-, i_X-, and i_Q-conversion (Fact 17). □

Fact 57: The moods **aai,eao-3-QNQ**, **aii,eio-3-QNQ**, **aai,aei-3-NQQ**, and **iai,iei-3-NQQ** are valid in the predicable semantics.

Proof. Deducible by aii-1-QNQ (Fact 39) using a_N-, i_N-, and i_Q-conversion (Fact 17). □

Fact 58: The moods **aai-3-QXM** and **iai-3-QXM** are valid in the predicable semantics.

Proof. Deducible by aii-1-XQM (Fact 40) using a_Q-, i_Q-, and i_M-conversion (Fact 17). □

Fact 59: The moods **aai,aei-3-XQM**, **eao,eeo-3-XQM**, **aii,aoi-3-XQM**, and **eio,eoo-3-XQM** are valid in the predicable semantics.

Proof. Deducible by aii-1-XQM and eio-1-XQM (Fact 40) using a_Q- and i_Q-conversion (Fact 17). □

Fact 60: The moods **eao,eeo-3-NQX** and **eio,eoo-3-NQX** are valid in the predicable semantics.

Proof. Deducible by eio-1-NQX (Fact 42) using a_Q- and i_Q-conversion (Fact 17). □

Fact 61: The moods **aai-3-QNM** and **iai-3-QNM** are valid in the predicable semantics.

Proof. Deducible by aii-1-NQM (Fact 41) using a_Q-, i_Q-, and i_M-conversion (Fact 17). □

Fact 62: The moods **aai,aei-3-NQM**, **eao,eeo-3-NQM**, **aii,aoi-3-NQM**, and **eio,eoo-3-NQM** are valid in the predicable semantics.

Proof. Deducible by aii-1-NQM and eio-1-NQM (Fact 41) using a_Q- and i_Q-conversion (Fact 17). □

The moods listed in Facts 31–62 include all moods whose validity Aristotle asserts in *Prior Analytics* 1.1–22. This can be checked by means of the synopsis in Appendix A. Thus, it has been proved that all moods held to be valid by Aristotle are valid in the predicable semantics.

Moreover, it has been shown that all moods held to be valid by Aristotle are deducible in the deductive system given in Table 15.1 on p. 225. The moods in Facts 31–43 are the primitive moods used as deduction rules in this system. The moods in Facts 44–62 are deducible in the system by means of direct deductions using only the primitive moods and Aristotle's conversion rules.

Next, we list some moods that are not mentioned by Aristotle but are valid in the predicable semantics (Facts 63–72). The moods mentioned in Facts 65, 70, and 72 are special in that it is not entirely clear whether or not Aristotle asserts their validity (see pp. 283–284n7 above).

Fact 63: The mood **aoo,eoo-2-QNX** is valid in the predicable semantics: $\forall z(\mathbf{A}az \supset \mathbf{\Pi}bz) \wedge \exists z((\mathbf{A}cz \wedge \mathbf{K}bz) \vee (\mathbf{N}cz \wedge \overline{\mathbf{K}}bz)) \vdash \neg\mathbf{A}ac$

Proof. For *reductio,* assume $\mathbf{A}ac$. By (ax$_2$), $\mathbf{A}ac$ and the major premise imply $\forall z(\mathbf{A}cz \supset \mathbf{\Pi}bz)$. This contradicts the minor premise and Fact 30.3. □

Fact 64: The mood **oao,oeo-3-XQM** is valid in the predicable semantics: $\neg\mathbf{A}ab \wedge \forall z(\mathbf{A}bz \supset \mathbf{\Pi}cz) \vdash \neg\overline{\mathbf{N}}ac$

Proof. For *reductio,* assume $\overline{\mathbf{N}}ac$. This implies $\widehat{\mathbf{\Sigma}}c$ by Fact 4, contradicting $\forall z(\mathbf{A}bz \supset \mathbf{\Pi}cz)$ and Fact 14. □

Fact 65: The mood **oao,oeo-3-NQX** is valid in the predicable semantics: $\exists z((\mathbf{A}bz \wedge \mathbf{K}az) \vee (\mathbf{N}bz \wedge \overline{\mathbf{K}}az)) \wedge \forall z(\mathbf{A}bz \supset \mathbf{\Pi}cz) \vdash \neg\mathbf{A}ac$

Proof. For *reductio,* assume $\mathbf{A}ac$. By Fact 40.1, $\mathbf{A}ac$ and the minor premise imply $\forall z(\mathbf{A}bz \supset \overline{\mathbf{\Pi}}az)$. This contradicts the major premise and Fact 28. □

Fact 66: The mood **aoo,eoo-2-QNM** is valid in the predicable semantics.

Proof. Follows from aoo-2-QNX (Fact 63) by X-M-subordination (Fact 20). □

Fact 67: The mood **eao-3-QXM** is valid in the predicable semantics.

Proof. Follows from oao-3-QXM (Fact 43) by Q-subalternation, that is, by $\mathrm{a_Q}$- and $\mathrm{i_Q}$-conversion (Fact 17). □

Fact 68: The moods **aii-3-QXM** and **eio-3-QXM** are valid in the predicable semantics.

Proof. Follows from aii-3-QXQ and eio-3-QXQ (Fact 55) by Q-M-subordination (Fact 21). □

Fact 69: The mood **iai,iei-3-XQM** is valid in the predicable semantics.

Proof. Follows from iai-3-XQQ (Fact 56) by Q-M-subordination (Fact 21). □

Fact 70: The moods **eao-3-QNM** and **oao-3-QNM** are valid in the predicable semantics.

Proof. Follows from eao-3-QXM (Fact 67) and oao-3-QXM (Fact 43) by N-X-subordination (Fact 19). □

Fact 71: The moods **aii-3-QNM**, **eio-3-QNM**, and **iai,iei-3-NQM** are valid in the predicable semantics.

Proof. Follows from aii-3-QNQ and iai-3-NQQ (Fact 57) by Q-M-subordination (Fact 21). □

Fact 72: The mood **oao,oeo-3-NQM** is valid in the predicable semantics.

Proof. Follows from oao-3-NQX (Fact 65) by X-M-subordination (Fact 20). □

Invalid Moods

In this section, all moods and conversion rules held to be invalid by Aristotle are proved to be invalid in the predicable semantics. To this end, it will be helpful to use the list of primitive claims of invalidity given in Table 15.3 on p. 230. We will proceed in two steps. First, each of the claims on the list is shown to be true in the predicable semantics. That is, every mood and conversion rule included in the list is proved to be invalid in the predicable semantics. This is done in Facts 73–86. As a second step, the list of primitive claims is shown to capture all of Aristotle's claims of invalidity. Specifically, all of his claims of invalidity that are not included in the list are shown to follow from one of the claims on the list by means of the rules of conversion, subalternation, and modal subordination (that is, by means of Facts 17–21). This is done in Facts 87–99. Taken together, the two steps imply that Aristotle's claims about the invalidity of moods and conversion rules are true in the predicable semantics. (His claims of inconcludence will be treated in the next section.)

In the first step, the moods and conversion rules on the list are shown to be invalid in the predicable semantics. This is achieved by means of first-order models in which the premises of the moods and conversion rules are true and the conclusion is false. Of course, the characteristic axioms of the predicable semantics, (ax_1)–(ax_6), will be true in these models. In order to make the models easier to grasp, they are represented by diagrams. The relation of a_X-predication, **A**, is indicated by downward dotted lines (·······), with the upper term being the predicate term. Likewise, the relation of a_N-predication, **N**, is indicated by downward dashed lines (‒‒‒). Strong a_N-predication, $\widehat{\mathbf{N}}$, is indicated by downward solid lines (——), sometimes also by horizontal solid lines. When an a_X-predication coincides with an a_N-predication, the dotted lines indicating the former are omitted. And when an a_N-predication coincides with a strong a_N-predication, the dashed lines are omitted. Whenever possible, a_N-predication is taken to coincide with strong a_N-predication, in order to reduce the use of troublesome a_N-predications that coincide with an o_M-predication.[2]

Self-a_X-predications and self-a_N-predications are not explicitly indicated in the diagrams. Substance terms ($\widehat{\mathbf{\Sigma}}$) are indicated by solid circles (•), nonsubstance terms by open circles (∘). In all models, every term is taken to be its own semantic value; for example, the semantic value of the term 'a' is taken to be this term itself (see pp. 73–74). The abbreviation 'Id(A)' stands for the identity relation on a set A. Otherwise, the notation used to specify the models and diagrams is, I hope, self-explanatory.

Fact 73: The mood **aaa-1-XNN** is invalid in the predicable semantics.

Proof. Let $\mathfrak{A}$ be a model with $A = \{a, b, c\}$, $\mathbf{N}^{\mathfrak{A}} = \widehat{\mathbf{N}}^{\mathfrak{A}} = \{bc, bb, cc\}$, and $\mathbf{A}^{\mathfrak{A}} = \{ab, ac\} \cup \widehat{\mathbf{N}}^{\mathfrak{A}} \cup \mathrm{Id}(A)$. We have $\mathfrak{A} \models \mathbf{A}ab \wedge \mathbf{N}bc$, and therefore $\mathfrak{A} \models \mathbb{X}^{\mathsf{a}}ab \wedge \mathbb{N}^{\mathsf{a}}bc$; but also $\mathfrak{A} \not\models \mathbf{N}ac$, hence $\mathfrak{A} \not\models \mathbb{N}^{\mathsf{a}}ac$. □

2. Models in which an a_N-predication coincides not with a strong a_N-predication, but with an o_M-predication are used only in the proofs of Facts 81, 82, 100, 114, and 115. Three of these Facts (namely, 81, 114, and 115) establish claims of invalidity and inconcludence which commit Aristotle to denying principles of modal opposition for affirmative N-propositions (see pp. 201–203 above). For Fact 100, see also p. 325 below.

Fact 74: The mood **aii-1-XNN** is invalid in the predicable semantics.

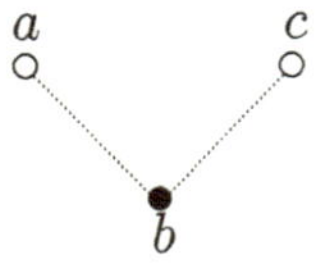

Proof. Let $\mathfrak{A}$ be a model with $\mathrm{A} = \{a, b, c\}$, $\mathbf{N}^{\mathfrak{A}} = \widehat{\mathbf{N}}^{\mathfrak{A}} = \{bb\}$, and $\mathbf{A}^{\mathfrak{A}} = \{ab, cb\} \cup \mathrm{Id(A)}$. We have $\mathfrak{A} \models \mathbf{A}ab$, and therefore $\mathfrak{A} \models \mathbb{X}^{\mathbf{a}}ab$. Moreover, we have $\mathfrak{A} \models \mathbf{A}cb \wedge \mathbf{N}bb$, hence $\mathfrak{A} \models \mathbb{N}^{\mathbf{i}}bc$. On the other hand, we have $\mathfrak{A} \not\models \exists z((\mathbf{A}cz \wedge \mathbf{N}az) \vee (\mathbf{A}az \wedge \mathbf{N}cz))$, and therefore $\mathfrak{A} \not\models \mathbb{N}^{\mathbf{i}}ac$. □

Fact 75: The mood **aeo-2-NXN** is invalid in the predicable semantics.

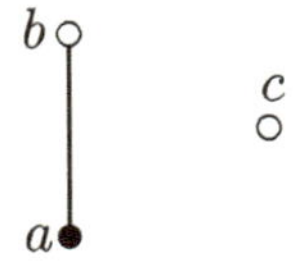

Proof. Let $\mathfrak{A}$ be a model with $\mathrm{A} = \{a, b, c\}$, $\mathbf{N}^{\mathfrak{A}} = \widehat{\mathbf{N}}^{\mathfrak{A}} = \{ba, aa\}$, and $\mathbf{A}^{\mathfrak{A}} = \mathbf{N}^{\mathfrak{A}} \cup \mathrm{Id(A)}$. We have $\mathfrak{A} \models \mathbf{N}ba$, and therefore $\mathfrak{A} \models \mathbb{N}^{\mathbf{a}}ba$. Moreover, we have $\mathfrak{A} \models \forall z(\mathbf{A}cz \supset \neg\mathbf{A}bz)$, and therefore $\mathfrak{A} \models \mathbb{X}^{\mathbf{e}}bc$. On the other hand, we have $\mathfrak{A} \not\models \mathbf{\Sigma}c$, hence $\mathfrak{A} \not\models \mathbf{K}ac$ and $\mathfrak{A} \not\models \exists z(\mathbf{A}cz \wedge \mathbf{K}az)$. Also, we have $\mathfrak{A} \not\models \exists z\mathbf{N}cz$, hence $\mathfrak{A} \not\models \exists z(\mathbf{N}cz \wedge \overline{\mathbf{K}}az)$. So we have $\mathfrak{A} \not\models \mathbb{N}^{\mathbf{o}}ac$. □

Fact 76: The mood **eao-3-XNN** is invalid in the predicable semantics.

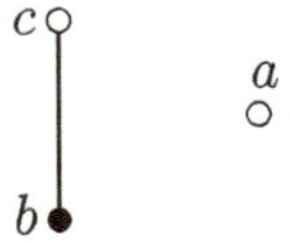

Proof. Let $\mathfrak{A}$ be a model with $\mathrm{A} = \{a, b, c\}$, $\mathbf{N}^{\mathfrak{A}} = \widehat{\mathbf{N}}^{\mathfrak{A}} = \{cb, bb\}$, and $\mathbf{A}^{\mathfrak{A}} = \mathbf{N}^{\mathfrak{A}} \cup \mathrm{Id(A)}$. We have $\mathfrak{A} \models \forall z(\mathbf{A}bz \supset \neg\mathbf{A}az)$, and therefore $\mathfrak{A} \models \mathbb{X}^{\mathbf{e}}ab$. Moreover, we have $\mathfrak{A} \models \mathbf{N}cb$, and therefore $\mathfrak{A} \models \mathbb{N}^{\mathbf{a}}cb$. On the other hand, we have $\mathfrak{A} \not\models \mathbf{\Sigma}a$, hence $\mathfrak{A} \not\models \exists z\mathbf{K}az$ and $\mathfrak{A} \not\models \exists z(\mathbf{A}cz \wedge \mathbf{K}az)$. Also, we have $\mathfrak{A} \not\models \exists v\mathbf{N}av$, hence $\mathfrak{A} \not\models \exists z\overline{\mathbf{K}}az$ and $\mathfrak{A} \not\models \exists z(\mathbf{N}cz \wedge \overline{\mathbf{K}}az)$. So we have $\mathfrak{A} \not\models \mathbb{N}^{\mathbf{o}}ac$. □

Fact 77: The mood **aoo-2-XNN** is invalid in the predicable semantics.

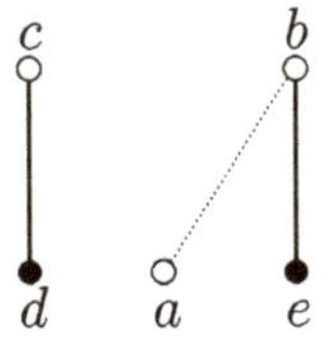

Proof. Let $\mathfrak{A}$ be a model with $\mathrm{A} = \{a, b, c, d, e\}$, $\mathbf{N}^{\mathfrak{A}} = \widehat{\mathbf{N}}^{\mathfrak{A}} = \{cd, dd, be, ee\}$, and $\mathbf{A}^{\mathfrak{A}} = \{ba\} \cup \mathbf{N}^{\mathfrak{A}} \cup \mathrm{Id(A)}$. We have $\mathfrak{A} \models \mathbf{A}ba$, and therefore $\mathfrak{A} \models \mathbb{X}^{\mathbf{a}}ba$. Moreover, we have $\mathfrak{A} \models \mathbf{N}cd \wedge \mathbf{N}be \wedge \forall u(\mathbf{A}bu \wedge \mathbf{\Sigma}u \supset \widehat{\mathbf{K}}du)$, hence $\mathfrak{A} \models \mathbb{N}^{\mathbf{o}}bc$. On the other hand, we have $\mathfrak{A} \not\models \mathbf{\Sigma}a$, hence $\mathfrak{A} \not\models \exists z\mathbf{K}az$ and $\mathfrak{A} \not\models \exists z(\mathbf{A}cz \wedge \mathbf{K}az)$. Also, we have $\mathfrak{A} \not\models \exists v\mathbf{N}av$, hence $\mathfrak{A} \not\models \exists z\overline{\mathbf{K}}az$ and $\mathfrak{A} \not\models \exists z(\mathbf{N}cz \wedge \overline{\mathbf{K}}az)$. So we have $\mathfrak{A} \not\models \mathbb{N}^{\mathbf{o}}ac$. □

Fact 78: The mood **oao-3-NXN** is invalid in the predicable semantics.

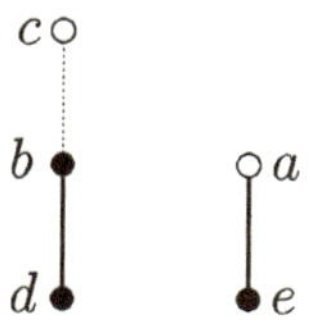

Proof. Let $\mathfrak{A}$ be a model with $\mathrm{A} = \{a, b, c, d, e\}$, $\mathbf{N}^{\mathfrak{A}} = \widehat{\mathbf{N}}^{\mathfrak{A}} = \{bd, bb, dd, ae, ee\}$, and $\mathbf{A}^{\mathfrak{A}} = \{cb, cd\} \cup \mathbf{N}^{\mathfrak{A}} \cup \mathrm{Id}(\mathrm{A})$. We have $\mathfrak{A} \models \mathbf{N}bd \wedge \mathbf{N}ae \wedge \forall u(\mathbf{A}au \wedge \mathbf{\Sigma}u \supset \widehat{\mathbf{K}}du)$, hence $\mathfrak{A} \models \mathbb{N}^{\mathsf{o}}ab$. Moreover, we have $\mathfrak{A} \models \mathbf{A}cb$, and therefore $\mathfrak{A} \models \mathbb{X}^{\mathsf{a}}cb$. On the other hand, we have $\mathfrak{A} \not\models \mathbf{\Sigma}a$, hence $\mathfrak{A} \not\models \exists z\mathbf{K}az$ and $\mathfrak{A} \not\models \exists z(\mathbf{A}cz \wedge \mathbf{K}az)$. Also, we have $\mathfrak{A} \not\models \exists z\mathbf{N}cz$, hence $\mathfrak{A} \not\models \exists z(\mathbf{N}cz \wedge \overline{\mathbf{K}}az)$. So we have $\mathfrak{A} \not\models \mathbb{N}^{\mathsf{o}}ac$. □

Fact 79: The mood **aaa-1-QNX** is invalid in the predicable semantics.

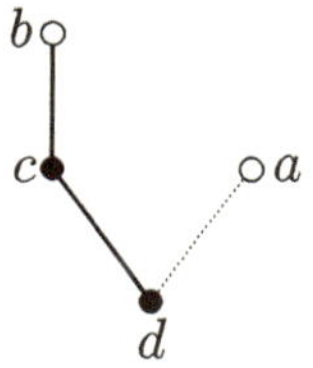

Proof. Let $\mathfrak{A}$ be a model with $\mathrm{A} = \{a, b, c, d\}$, $\mathbf{N}^{\mathfrak{A}} = \widehat{\mathbf{N}}^{\mathfrak{A}} = \{bc, cd, bd, cc, dd\}$, and $\mathbf{A}^{\mathfrak{A}} = \{ad\} \cup \mathbf{N}^{\mathfrak{A}} \cup \mathrm{Id}(\mathrm{A})$. We have $\mathfrak{A} \models \forall z(\mathbf{A}bz \supset \mathbf{\Pi}az)$, and therefore $\mathfrak{A} \models \mathbb{Q}^{\mathsf{a}}ab$. Moreover, we have $\mathfrak{A} \models \mathbf{N}bc$, and therefore $\mathfrak{A} \models \mathbb{N}^{\mathsf{a}}bc$. On the other hand, we have $\mathfrak{A} \not\models \mathbf{A}ac$, and therefore $\mathfrak{A} \not\models \mathbb{X}^{\mathsf{a}}ac$. □

Fact 80: The moods **eae-1-QNX** and **eao-3-QNX** are invalid in the predicable semantics.

Proof. Let $\mathfrak{A}$ be a model with $\mathrm{A} = \{a, b, c\}$, $\mathbf{N}^{\mathfrak{A}} = \widehat{\mathbf{N}}^{\mathfrak{A}} = \{bc, cb, bb, cc\}$, and $\mathbf{A}^{\mathfrak{A}} = \{ab, ac\} \cup \mathbf{N}^{\mathfrak{A}} \cup \mathrm{Id}(\mathrm{A})$. We have $\mathfrak{A} \models \forall z(\mathbf{A}bz \supset \mathbf{\Pi}az)$, and therefore $\mathfrak{A} \models \mathbb{Q}^{\mathsf{e}}ab$. Moreover, we have $\mathfrak{A} \models \mathbf{N}bc$, and therefore $\mathfrak{A} \models \mathbb{N}^{\mathsf{a}}bc$. Also, we have $\mathfrak{A} \models \mathbf{N}cb$, and therefore $\mathfrak{A} \models \mathbb{N}^{\mathsf{a}}cb$. On the other hand, we have $\mathfrak{A} \models \mathbf{A}ac$, hence $\mathfrak{A} \not\models \mathbb{X}^{\mathsf{e}}ac$ and $\mathfrak{A} \not\models \mathbb{X}^{\mathsf{o}}ac$. □

Fact 81: The moods **aaa,aea-1-NQX** and **aai,aei-3-NQX** are invalid in the predicable semantics.

Proof. Let $\mathfrak{A}$ be a model with $\mathrm{A} = \{a, b, c\}$, $\mathbf{N}^{\mathfrak{A}} = \{ab, bb\}$, and $\mathbf{A}^{\mathfrak{A}} = \mathbf{N}^{\mathfrak{A}} \cup \mathrm{Id}(\mathrm{A})$. We have $\mathfrak{A} \models \mathbf{N}ab$, hence $\mathfrak{A} \models \mathbb{N}^{\mathsf{a}}ab$. Moreover, we have $\mathfrak{A} \models \forall z(\mathbf{A}cz \supset \mathbf{\Pi}bz)$, and therefore $\mathfrak{A} \models \mathbb{Q}^{\mathsf{a}}bc$. Also, we have $\mathfrak{A} \models \forall z(\mathbf{A}bz \supset \mathbf{\Pi}cz)$, and therefore $\mathfrak{A} \models \mathbb{Q}^{\mathsf{a}}cb$. On the other hand, we have $\mathfrak{A} \models \forall z(\mathbf{A}cz \supset \neg\mathbf{A}az)$, hence $\mathfrak{A} \not\models \mathbb{X}^{\mathsf{a}}ac$ and $\mathfrak{A} \not\models \mathbb{X}^{\mathsf{i}}ac$. □

Fact 82: The moods **eae,eee-1-NQQ**, **eio,eoo-1-NQQ**, **eae,eee-1-NQN**, and **eao,eeo-3-NQN** are invalid in the predicable semantics.

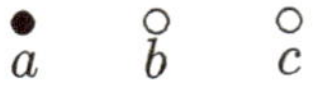

Proof. Let $\mathfrak{A}$ be a model with $\mathrm{A} = \{a, b, c\}$, $\mathbf{N}^{\mathfrak{A}} = \{aa, bb\}$, $\widehat{\mathbf{N}}^{\mathfrak{A}} = \{aa\}$, and $\mathbf{A}^{\mathfrak{A}} = \mathrm{Id}(\mathrm{A})$. We have $\mathfrak{A} \models \mathbf{K}ab$, and therefore $\mathfrak{A} \models \mathbb{N}^{\mathbf{e}}ab$. Moreover, we have $\mathfrak{A} \models \forall z(\mathbf{A}cz \supset \mathbf{\Pi}bz)$, and therefore $\mathfrak{A} \models \mathbb{Q}^{\mathbf{a}}bc$, hence also $\mathfrak{A} \models \mathbb{Q}^{\mathbf{i}}bc$. Also, we have $\mathfrak{A} \models \forall z(\mathbf{A}bz \supset \mathbf{\Pi}cz)$, and therefore $\mathfrak{A} \models \mathbb{Q}^{\mathbf{a}}cb$. On the other hand, we have $\mathfrak{A} \models \mathbf{\Gamma}ac$, hence $\mathfrak{A} \not\models \mathbf{\Pi}ac$. So we have $\mathfrak{A} \not\models \mathbb{Q}^{\mathbf{o}}ac$, hence also $\mathfrak{A} \not\models \mathbb{Q}^{\mathbf{e}}ac$. Moreover, we have $\mathfrak{A} \not\models \mathbf{\Sigma}c$, hence $\mathfrak{A} \not\models \mathbf{K}ac$, and therefore $\mathfrak{A} \not\models \mathbb{N}^{\mathbf{e}}ac$. Also, we have $\mathfrak{A} \not\models \exists z(\mathbf{A}cz \wedge \mathbf{K}az)$ and $\mathfrak{A} \not\models \exists z\mathbf{N}cz$, hence $\mathfrak{A} \not\models \exists z(\mathbf{N}cz \wedge \overline{\mathbf{K}}az)$. So we have $\mathfrak{A} \not\models \mathbb{N}^{\mathbf{o}}ac$. □

Fact 83: The mood **aai-3-QXX** is invalid in the predicable semantics.

Proof. Let $\mathfrak{A}$ be a model with $\mathrm{A} = \{a, b, c\}$ and $\mathbf{A}^{\mathfrak{A}} = \{cb\} \cup \mathrm{Id}(\mathrm{A})$. We have $\mathfrak{A} \models \forall z(\mathbf{A}bz \supset \mathbf{\Pi}az)$, and therefore $\mathfrak{A} \models \mathbb{Q}^{\mathbf{a}}ab$. Moreover, we have $\mathfrak{A} \models \mathbf{A}cb$, and therefore $\mathfrak{A} \models \mathbb{X}^{\mathbf{a}}cb$. On the other hand, we have $\mathfrak{A} \not\models \exists z(\mathbf{A}cz \wedge \mathbf{A}az)$, hence $\mathfrak{A} \not\models \mathbb{X}^{\mathbf{i}}ac$. □

Fact 84: The mood **eao-3-QXX** is invalid in the predicable semantics.

Proof. Let $\mathfrak{A}$ be a model with $\mathrm{A} = \{a, b, c\}$ and $\mathbf{A}^{\mathfrak{A}} = \{ac, ab, cb\} \cup \mathrm{Id}(\mathrm{A})$. We have $\mathfrak{A} \models \forall z(\mathbf{A}bz \supset \mathbf{\Pi}az)$, and therefore $\mathfrak{A} \models \mathbb{Q}^{\mathbf{e}}ab$. Moreover, we have $\mathfrak{A} \models \mathbf{A}cb$, and therefore $\mathfrak{A} \models \mathbb{X}^{\mathbf{a}}cb$. On the other hand, we have $\mathfrak{A} \not\models \neg\mathbf{A}ac$, hence $\mathfrak{A} \not\models \mathbb{X}^{\mathbf{o}}ac$. □

Fact 85: The mood **eao,eeo-3-XQX** is invalid in the predicable semantics.

Proof. Let $\mathfrak{A}$ be a model with $\mathrm{A} = \{a, b, c\}$ and $\mathbf{A}^{\mathfrak{A}} = \{ac\} \cup \mathrm{Id}(\mathrm{A})$. We have $\mathfrak{A} \models \forall z(\mathbf{A}bz \supset \neg\mathbf{A}az)$, and therefore $\mathfrak{A} \models \mathbb{X}^{\mathbf{e}}ab$. Moreover, we have $\mathfrak{A} \models \forall z(\mathbf{A}bz \supset \mathbf{\Pi}cz)$, and therefore $\mathfrak{A} \models \mathbb{Q}^{\mathbf{a}}cb$. On the other hand, we have $\mathfrak{A} \not\models \neg\mathbf{A}ac$, hence $\mathfrak{A} \not\models \mathbb{X}^{\mathbf{o}}ac$. □

Fact 86: The following conversion rules are invalid in the predicable semantics:

(86.1) $\mathbb{X}^{\mathrm{o}}ab \supset \mathbb{X}^{\mathrm{o}}ba$	(86.3) $\mathbb{M}^{\mathrm{o}}ab \supset \mathbb{M}^{\mathrm{o}}ba$
(86.2) $\mathbb{N}^{\mathrm{o}}ab \supset \mathbb{N}^{\mathrm{o}}ba$	(86.4) $\mathbb{Q}^{\mathrm{e}}ab \supset \mathbb{Q}^{\mathrm{e}}ba$

Proof. Let $\mathfrak{A}$ be a model with $\mathrm{A} = \{a, b, c, d\}$, $\mathbf{N}^{\mathfrak{A}} = \widehat{\mathbf{N}}^{\mathfrak{A}} = \{ba, aa, dd\}$, and $\mathbf{A}^{\mathfrak{A}} = \{cd\} \cup \mathbf{N}^{\mathfrak{A}} \cup \mathrm{Id}(\mathrm{A})$. 86.1: We have $\mathfrak{A} \models \neg\mathbf{A}ab$, and therefore $\mathfrak{A} \models \mathbb{X}^{\mathrm{o}}ab$. On the other hand, we have $\mathfrak{A} \models \mathbf{A}ba$, and therefore $\mathfrak{A} \not\models \mathbb{X}^{\mathrm{o}}ba$. 86.2: We have $\mathfrak{A} \models \mathbf{A}cd \wedge \mathbf{K}ad$, hence $\mathfrak{A} \models \mathbb{N}^{\mathrm{o}}ac$. On the other hand, we have $\mathfrak{A} \not\models \mathbf{\Sigma}c$, hence $\mathfrak{A} \not\models \exists z\mathbf{K}cz$ and $\mathfrak{A} \not\models \exists z(\mathbf{A}az \wedge \mathbf{K}cz)$. Moreover, we have $\mathfrak{A} \not\models \exists z\mathbf{N}cz$, hence $\mathfrak{A} \not\models \exists z\overline{\mathbf{K}}cz$ and $\mathfrak{A} \not\models \exists z(\mathbf{N}az \wedge \overline{\mathbf{K}}cz)$. So we have $\mathfrak{A} \not\models \mathbb{N}^{\mathrm{o}}ca$. 86.3: We have $\mathfrak{A} \models \neg\mathbf{A}ab$, hence $\mathfrak{A} \models \mathbb{M}^{\mathrm{o}}ab$. On the other hand, we have $\mathfrak{A} \models \widehat{\mathbf{N}}ba$, hence $\mathfrak{A} \not\models \mathbb{M}^{\mathrm{o}}ba$. 86.4: We have $\mathfrak{A} \models \mathbf{\Pi}cd$ and $\mathfrak{A} \models \forall z(\mathbf{A}dz \supset \mathbf{\Pi}cz)$, and therefore $\mathfrak{A} \models \mathbb{Q}^{\mathrm{e}}cd$. On the other hand, we have $\mathfrak{A} \models \widehat{\mathbf{\Sigma}}d$. By Fact 14, $\mathfrak{A} \models \widehat{\mathbf{\Sigma}}d$ implies $\mathfrak{A} \not\models \forall z(\mathbf{A}cz \supset \mathbf{\Pi}dz)$, and therefore $\mathfrak{A} \not\models \mathbb{Q}^{\mathrm{e}}dc$. □

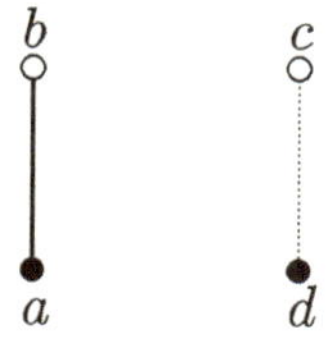

The moods and conversion rules listed in Facts 73–86 are precisely those whose invalidity is taken as primitive in Table 15.3 on p. 230. All of them have been shown to be invalid in the predicable semantics. As a second step, we will now consider all other moods and conversion rules held to be invalid by Aristotle. Their invalidity will be shown to follow from the foregoing primitive invalidities by means of conversion rules, principles of subalternation, and principles of modal subordination (which are valid in the predicable semantics; see Facts 17–21). This is done in Facts 87–99:

Fact 87: The moods **aii-3-XNN** and **iai-3-NXN** are invalid in the predicable semantics.

Proof. Follows from the invalidity of aii-1-XNN (Fact 74) by $\mathrm{i_N}$-conversion (Fact 17). □

Fact 88: The moods **aoo-2-NXN**, **aee-2-NXN**, **eae-2-XNN**, and **eae-1-XNN** are invalid in the predicable semantics.

Proof. Follows from the invalidity of aeo-2-NXN (Fact 75) by negative N-subalternation and negative X-subalternation (Fact 18), and by e_N- and e_X-conversion (Fact 17). □

Fact 89: The moods **eio-1-XNN**, **eio-2-XNN**, **eio-3-XNN**, and **oao-3-XNN** are invalid in the predicable semantics.

Proof. Follows from the invalidity of eao-3-XNN (Fact 76) by a_N-, i_N-, and e_X-conversion (Fact 17), and by negative X-subalternation (Fact 18). □

Fact 90: The moods **aii,aoi-1-NQX**, **iai,iei-3-NQX**, **aii,aoi-3-NQX**, **aai-3-QNX**, **aii-1-QNX**, **aii-3-QNX**, and **iai-3-QNX** are invalid in the predicable semantics.

Proof. Follows from the invalidity of aai-3-NQX (Fact 81) by a_Q-, i_Q-, i_X-, a_N-, and i_N-conversion (Fact 17). □

Fact 91: The moods **eae,eee-1-XQQ** and **eio,eoo-1-XQQ** are invalid in the predicable semantics.

Proof. Follows from the invalidity of the corresponding NQQ-moods (Fact 82) by N-X-subordination (Fact 19). □

Fact 92: The moods **aaa,aea-1-NQN** and **aii,aoi-1-NQN** are invalid in the predicable semantics.

Proof. Follows from the invalidity of the corresponding NQX-moods (Facts 81 and 90) by N-X-subordination (Fact 19). □

Fact 93: The moods **eio-1-QNX**, **oao-3-QNX**, and **eio-3-QNX** are invalid in the predicable semantics.

Proof. Follows from the invalidity of eao-3-QNX (Fact 80) by a_N-, i_N-, a_Q-, and i_Q-conversion (Fact 17). □

Fact 94: The moods **eio,eoo-1-NQN**, **oao,oeo-3-NQN**, and **eio,eoo-3-NQN** are invalid in the predicable semantics.

Proof. Follows from the invalidity of eao-3-NQN (Fact 82) by a_Q- and i_Q-conversion (Fact 17), and by negative N-subalternation (Fact 18). □

Fact 95: The moods **eae,eee-1-XQN** and **eio,eoo-1-XQN** are invalid in the predicable semantics.

Proof. Follows from the invalidity of the corresponding NQN-moods (Facts 82 and 94) by N-X-subordination (Fact 19). □

Fact 96: The moods **aai,aei-3-NQN**, **iai,iei-3-NQN**, and **aii,aoi-3-NQN** are invalid in the predicable semantics.

Proof. Follows from the invalidity of the corresponding NQX-moods (Facts 81 and 90) by N-X-subordination (Fact 19). □

Fact 97: The mood **aai,aei-3-XQX** is invalid in the predicable semantics.

Proof. Follows from the invalidity of aai-3-QXX (Fact 83) by i_X-conversion (Fact 17). □

Fact 98: The moods **iai-3-QXX**, **aii-3-QXX**, **oao-3-QXX**, and **eio-3-QXX** are invalid in the predicable semantics.

Proof. Follows from the invalidity of aai-3-QXX (Fact 83) and eao-3-QXX (Fact 84) by subalternation (Fact 18). □

Fact 99: The moods **iai,iei-3-XQX**, **aii,aoi-3-XQX**, **oao,oeo-3-XQX**, and **eio,eoo-3-XQX** are invalid in the predicable semantics.

Proof. Follows from the invalidity of aai-3-XQX (Fact 97) and eao-3-XQX (Fact 85) by subalternation (Fact 18). □

Facts 73–99 include all moods and conversion rules whose invalidity Aristotle asserts in *Prior Analytics* 1.1–22. This can be checked by means of the synopsis in Appendix A. Thus it has been proved that all moods and conversion rules held to be invalid by Aristotle are invalid in the predicable semantics.

Moreover, it has been shown in Facts 87–99 that, given the rules of conversion, subalternation, and modal subordination, all of Aristotle's claims about the invalidity of moods and conversion rules follow from the primitive claims of invalidity listed in Table 15.3 on p. 230 (Facts 73–86).

Consider for a moment the apodeictic syllogistic (*Prior Analytics* 1.3 and 1.8–12). All of Aristotle's claims of invalidity in the apodeictic syllogistic are included in Facts 73–78 and 86–89 (he does not make any claims of inconcludence there). None of the proofs of these facts relies on the distinction between a_N-predication (**N**) and strong a_N-predication ($\widehat{\mathbf{N}}$); the two relations coincide completely in all models used to establish these facts. This means, in effect, that the distinction between the two relations is not relevant for the interpretation of the apodeictic syllogistic in the predicable semantics. As far as the apodeictic syllogistic is concerned, there is no need to introduce strong a_N-predication as a third primitive relation; the two relations of a_X- and a_N-predication suffice. Thus, all occurrences of the symbol $\widehat{}$ could be deleted in the interpretation of N-propositions in the apodeictic syllogistic. Strong a_N-predication is needed only for the interpretation of the problematic syllogistic and for a uniform interpretation of the whole modal syllogistic (see p. 305n2).

Next, we list some moods not mentioned by Aristotle that are invalid in the predicable semantics:

Fact 100: The moods **aaa,aea-1-NQQ** and **aii,aoi-1-NQQ** are invalid in the predicable semantics.

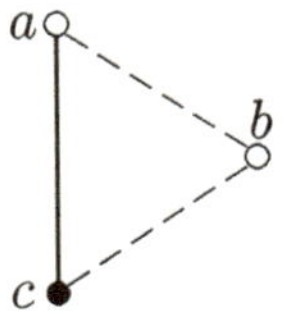

Proof. Let $\mathfrak{A}$ be a model with $\mathrm{A} = \{a, b, c\}$, $\widehat{\mathbf{N}}^{\mathfrak{A}} = \{ac, cc\}$, $\mathbf{N}^{\mathfrak{A}} = \{ab, bc, bb\} \cup \widehat{\mathbf{N}}^{\mathfrak{A}}$, and $\mathbf{A}^{\mathfrak{A}} = \mathbf{N}^{\mathfrak{A}} \cup \mathrm{Id}(\mathrm{A})$.[3] We have $\mathfrak{A} \models \mathbf{N}ab$, and therefore $\mathfrak{A} \models \mathbb{N}^{\mathbf{a}}ab$. Moreover, we have $\mathfrak{A} \models \forall z(\mathbf{A}cz \supset \mathbf{\Pi}bz)$, hence $\mathfrak{A} \models \mathbb{Q}^{\mathbf{a}}bc$ and $\mathfrak{A} \models \mathbb{Q}^{\mathbf{i}}bc$. On the other hand, we have $\mathfrak{A} \models \widehat{\mathbf{N}}ac$, hence $\mathfrak{A} \not\models \mathbf{\Pi}ac$. So we have $\mathfrak{A} \not\models \mathbb{Q}^{\mathbf{i}}ac$ and $\mathfrak{A} \not\models \mathbb{Q}^{\mathbf{a}}ac$. □

Fact 101: The moods **aaa,aea-1-XQQ** and **aii,aoi-1-XQQ** are invalid in the predicable semantics.

Proof. Follows from the invalidity of the corresponding NQQ-moods (Fact 100) by N-X-subordination (Fact 19). □

---

3. It is worth noting that this model violates statement S24 on p. 151 above (since the nonsubstance term b is a_X-predicated of a substance term, c, while being the subject of an a_N-predication). For further discussion of this point, see p. 325 below.

Fact 102: The moods **iai,oao-3-QNQ** and **aii,aoi-3-NQQ** are invalid in the predicable semantics.

Proof. Follows from the invalidity of aii-1-NQQ (Fact 100) by i_Q-conversion (Fact 17). □

Fact 103: The moods **iai,oao-3-QXQ** and **aii,aoi-3-XQQ** are invalid in the predicable semantics.

Proof. Follows from the invalidity of iai-3-QNQ and aii-3-NQQ (Fact 102) by N-X-subordination (Fact 19). □

Fact 104: The moods **aaa,aea-1-XQN** and **aii,aoi-1-XQN** are invalid in the predicable semantics.

Proof. Follows from the invalidity of the corresponding NQN-moods (Fact 92) by N-X-subordination (Fact 19). □

Inconcludent Premise Pairs

In this section, all premise pairs held to be inconcludent by Aristotle are proved to be inconcludent in the predicable semantics. For a premise pair to be inconcludent means that no conclusion follows from it in the figure to which it belongs. In other words, no categorical proposition is the conclusion of a valid mood with this premise pair. Now, there are sixteen kinds of categorical propositions in the modal syllogistic: N-, X-, M-, and Q-propositions, each of the type a, e, i, or o. In the predicable semantics, every affirmative (or negative) categorical proposition implies the corresponding affirmative (or negative) particular M-proposition (Fact 22). Hence if any conclusion follows from a premise pair in the predicable semantics, then an i_M- or o_M-proposition also follows from it. The inconcludence of a premise pair can therefore be established by showing that neither an i_M- nor an o_M-proposition follows from it in its figure (see pp. 211–214).

As before, the argument of this section will proceed in two steps. First, we establish the inconcludence in the predicable semantics of those premise pairs whose inconcludence is taken as primitive in Table 15.3 on p. 230. This is done in Facts 105–115. As a second step, the inconcludence of all other premise pairs held to be inconcludent by Aristotle is shown to follow from these primitive inconcludences by means of conversion, subalternation, and modal subordination (that is, by means of Facts 17–21). This is done in Facts 116–136.

Aristotle's way of establishing inconcludence in the assertoric syllogistic differs from that in the modal syllogistic. In the assertoric syllogistic, he only shows that no *assertoric* conclusion follows from the premise pair, not taking into account modalized propositions. We shall follow his strategy in Facts 105–108 and Facts 116–121, although it would be easy to prove that none of the assertoric premise pairs mentioned in these Facts yields an i_{M}- or o_{M}-conclusion in the predicable semantics.

Fact 105: The premise pair **ae-1-XX** is inconcludent in the predicable semantics.

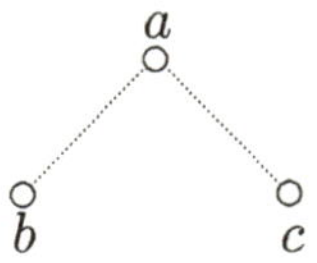

Proof. Let $\mathfrak{A}$ be a model with $\mathrm{A} = \{a, b, c\}$ and $\mathbf{A}^{\mathfrak{A}} = \{ab, ac\} \cup \mathrm{Id}(\mathrm{A})$. We have $\mathfrak{A} \models \mathbf{A}ab$, and therefore $\mathfrak{A} \models \mathbb{X}^{\mathsf{a}}ab$. Moreover, we have $\mathfrak{A} \models \forall z(\mathbf{A}cz \supset \neg\mathbf{A}bz)$, and therefore $\mathfrak{A} \models \mathbb{X}^{\mathsf{e}}bc$. On the other hand, we have $\mathfrak{A} \models \mathbf{A}ac$, hence $\mathfrak{A} \not\models \mathbb{X}^{\mathsf{o}}ac$.

Let $\mathfrak{B}$ be a model with $\mathrm{B} = \{a, b, c\}$ and $\mathbf{A}^{\mathfrak{B}} = \{ab\} \cup \mathrm{Id}(\mathrm{B})$. We have $\mathfrak{B} \models \mathbf{A}ab$, and therefore $\mathfrak{B} \models \mathbb{X}^{\mathsf{a}}ab$. Moreover, we have $\mathfrak{B} \models \forall z(\mathbf{A}cz \supset \neg\mathbf{A}bz)$, and therefore $\mathfrak{B} \models \mathbb{X}^{\mathsf{e}}bc$. On the other hand, we have $\mathfrak{B} \models \forall z(\mathbf{A}cz \supset \neg\mathbf{A}az)$, hence $\mathfrak{B} \not\models \mathbb{X}^{\mathsf{i}}ac$. □

Fact 106: The premise pair **ee-1-XX** is inconcludent in the predicable semantics.

Proof. Let $\mathfrak{A}$ be a model with $\mathrm{A} = \{a, b, c\}$ and $\mathbf{A}^{\mathfrak{A}} = \{ac\} \cup \mathrm{Id}(\mathrm{A})$. We have $\mathfrak{A} \models \forall z(\mathbf{A}bz \supset \neg\mathbf{A}az)$, and therefore $\mathfrak{A} \models \mathbb{X}^{\mathsf{e}}ab$. Moreover, we have $\mathfrak{A} \models \forall z(\mathbf{A}cz \supset \neg\mathbf{A}bz)$, and therefore $\mathfrak{A} \models \mathbb{X}^{\mathsf{e}}bc$. On the other hand, we have $\mathfrak{A} \models \mathbf{A}ac$, hence $\mathfrak{A} \not\models \mathbb{X}^{\mathsf{o}}ac$.

Let $\mathfrak{B}$ be a model with $\mathrm{B} = \{a, b, c\}$ and $\mathbf{A}^{\mathfrak{B}} = \mathrm{Id}(\mathrm{B})$. We have $\mathfrak{B} \models \forall z(\mathbf{A}bz \supset \neg\mathbf{A}az)$, and therefore $\mathfrak{B} \models \mathbb{X}^{\mathsf{e}}ab$. Moreover, we have $\mathfrak{B} \models \forall z(\mathbf{A}cz \supset \neg\mathbf{A}bz)$, and therefore $\mathfrak{B} \models \mathbb{X}^{\mathsf{e}}bc$. On the other hand, we have $\mathfrak{B} \models \forall z(\mathbf{A}cz \supset \neg\mathbf{A}az)$, hence $\mathfrak{B} \not\models \mathbb{X}^{\mathsf{i}}ac$. □

Fact 107: The premise pairs **oa-1-XX** and **aa-2-XX** are inconcludent in the predicable semantics.

Proof. Let $\mathfrak{A}$ be a model with $\mathrm{A} = \{a, b, c\}$ and $\mathbf{A}^{\mathfrak{A}} = \{ba, bc, ac\} \cup \mathrm{Id}(\mathrm{A})$. We have $\mathfrak{A} \models \neg\mathbf{A}ab$ and $\mathfrak{A} \models \mathbf{A}ba$, and therefore $\mathfrak{A} \models \mathbb{X}^{\mathbf{o}}ab$ and $\mathfrak{A} \models \mathbb{X}^{\mathbf{a}}ba$. Moreover, we have $\mathfrak{A} \models \mathbf{A}bc$, and therefore $\mathfrak{A} \models \mathbb{X}^{\mathbf{a}}bc$. On the other hand, we have $\mathfrak{A} \models \mathbf{A}ac$, hence $\mathfrak{A} \not\models \mathbb{X}^{\mathbf{o}}ac$.

Let $\mathfrak{B}$ be a model with $\mathrm{B} = \{a, b, c\}$ and $\mathbf{A}^{\mathfrak{B}} = \{ba, bc\} \cup \mathrm{Id}(\mathrm{B})$. We have $\mathfrak{B} \models \neg\mathbf{A}ab$ and $\mathfrak{B} \models \mathbf{A}ba$, and therefore $\mathfrak{B} \models \mathbb{X}^{\mathbf{o}}ab$ and $\mathfrak{B} \models \mathbb{X}^{\mathbf{a}}ba$. Moreover, we have $\mathfrak{B} \models \mathbf{A}bc$, and therefore $\mathfrak{B} \models \mathbb{X}^{\mathbf{a}}bc$. On the other hand, we have $\mathfrak{B} \models \forall z(\mathbf{A}cz \supset \neg\mathbf{A}az)$, hence $\mathfrak{B} \not\models \mathbb{X}^{\mathbf{i}}ac$. □

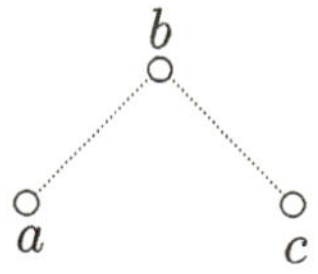

Fact 108: The premise pair **oa-2-XX** is inconcludent in the predicable semantics.

Proof. Let $\mathfrak{A}$ be a model with $\mathrm{A} = \{a, b, c\}$ and $\mathbf{A}^{\mathfrak{A}} = \{ac, bc\} \cup \mathrm{Id}(\mathrm{A})$. We have $\mathfrak{A} \models \neg\mathbf{A}ba$, and therefore $\mathfrak{A} \models \mathbb{X}^{\mathbf{o}}ba$. Moreover, we have $\mathfrak{A} \models \mathbf{A}bc$, and therefore $\mathfrak{A} \models \mathbb{X}^{\mathbf{a}}bc$. On the other hand, we have $\mathfrak{A} \models \mathbf{A}ac$, hence $\mathfrak{A} \not\models \mathbb{X}^{\mathbf{o}}ac$.

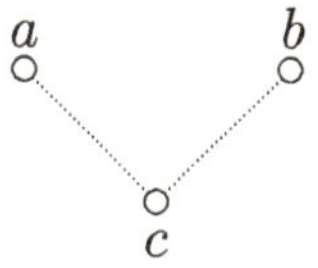

Let $\mathfrak{B}$ be a model with $\mathrm{B} = \{a, b, c\}$ and $\mathbf{A}^{\mathfrak{B}} = \{bc\} \cup \mathrm{Id}(\mathrm{B})$. We have $\mathfrak{B} \models \neg\mathbf{A}ba$, and therefore $\mathfrak{B} \models \mathbb{X}^{\mathbf{o}}ba$. Moreover, we have $\mathfrak{B} \models \mathbf{A}bc$, and therefore $\mathfrak{B} \models \mathbb{X}^{\mathbf{a}}bc$. On the other hand, we have $\mathfrak{B} \models \forall z(\mathbf{A}cz \supset \neg\mathbf{A}az)$, hence $\mathfrak{B} \not\models \mathbb{X}^{\mathbf{i}}ac$. □

Fact 109: The premise pair **ea,ae,aa,ee-2-QQ** is inconcludent in the predicable semantics.

Proof. Let $\mathfrak{A}$ be a model with $\mathrm{A} = \{a, b, c\}$, $\mathbf{N}^{\mathfrak{A}} = \widehat{\mathbf{N}}^{\mathfrak{A}} = \{ac, aa, cc\}$, and $\mathbf{A}^{\mathfrak{A}} = \{ba, bc\} \cup \mathbf{N}^{\mathfrak{A}} \cup \mathrm{Id}(\mathrm{A})$. We have $\mathfrak{A} \models \forall z\mathbf{\Pi}bz$, hence $\mathfrak{A} \models \mathbb{Q}^{\mathbf{a}}ba$ and $\mathfrak{A} \models \mathbb{Q}^{\mathbf{a}}bc$. On the other hand, we have $\mathfrak{A} \models \widehat{\mathbf{N}}ac$, hence $\mathfrak{A} \not\models \mathbb{M}^{\mathbf{o}}ac$.

Let $\mathfrak{B}$ be a model with $\mathrm{B} = \{a, b, c\}$, $\mathbf{N}^{\mathfrak{B}} = \widehat{\mathbf{N}}^{\mathfrak{B}} = \{aa, cc\}$, and $\mathbf{A}^{\mathfrak{B}} = \{ba, bc\} \cup \mathrm{Id}(\mathrm{B})$. We have $\mathfrak{B} \models \forall z\mathbf{\Pi}bz$, hence $\mathfrak{B} \models \mathbb{Q}^{\mathbf{a}}ba$ and $\mathfrak{B} \models \mathbb{Q}^{\mathbf{a}}bc$. On the other hand, we have $\mathfrak{B} \models \mathbf{\Gamma}ac$, hence $\mathfrak{B} \not\models \mathbb{M}^{\mathbf{i}}ac$. □

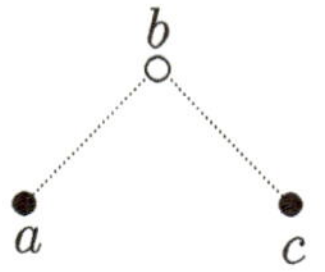

Fact 110: The premise pair **ae,ee-1-QN** is inconcludent in the predicable semantics.

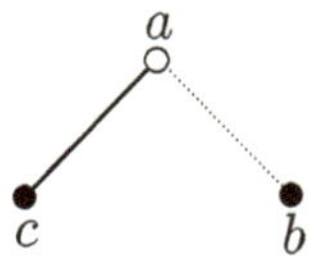

Proof. Let $\mathfrak{A}$ be a model with $\mathrm{A} = \{a, b, c\}$, $\mathbf{N}^{\mathfrak{A}} = \widehat{\mathbf{N}}^{\mathfrak{A}} = \{ac, cc, bb\}$, and $\mathbf{A}^{\mathfrak{A}} = \{ab\} \cup \mathbf{N}^{\mathfrak{A}} \cup \mathrm{Id(A)}$. We have $\mathfrak{A} \models \boldsymbol{\Pi}ab$ and $\mathfrak{A} \models \forall z(\mathbf{A}bz \supset \boldsymbol{\Pi}az)$, hence $\mathfrak{A} \models \mathbb{Q}^{\mathbf{a}}ab$. Moreover, we have $\mathfrak{A} \models \mathbf{K}bc$, and therefore $\mathfrak{A} \models \mathbb{N}^{\mathbf{e}}bc$. On the other hand, we have $\mathfrak{A} \models \widehat{\mathbf{N}}ac$, hence $\mathfrak{A} \not\models \mathbb{M}^{\mathbf{o}}ac$.

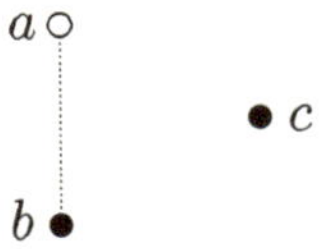

Let $\mathfrak{B}$ be a model with $\mathrm{B} = \{a, b, c\}$, $\mathbf{N}^{\mathfrak{B}} = \widehat{\mathbf{N}}^{\mathfrak{B}} = \{bb, cc\}$, and $\mathbf{A}^{\mathfrak{B}} = \{ab\} \cup \mathrm{Id(B)}$. We have $\mathfrak{B} \models \boldsymbol{\Pi}ab$ and $\mathfrak{B} \models \forall z(\mathbf{A}bz \supset \boldsymbol{\Pi}az)$, and therefore $\mathfrak{B} \models \mathbb{Q}^{\mathbf{a}}ab$. Moreover, we have $\mathfrak{B} \models \mathbf{K}bc$, and therefore $\mathfrak{B} \models \mathbb{N}^{\mathbf{e}}bc$. On the other hand, we have $\mathfrak{B} \models \boldsymbol{\Gamma}ac$, and therefore $\mathfrak{B} \not\models \mathbb{M}^{\mathbf{i}}ac$. □

Fact 111: The premise pair **ao,eo-2-QX** is inconcludent in the predicable semantics.

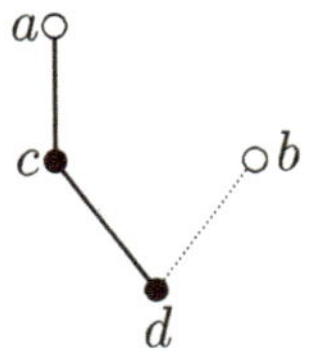

Proof. Let $\mathfrak{A}$ be a model with $\mathrm{A} = \{a, b, c, d\}$, $\mathbf{N}^{\mathfrak{A}} = \widehat{\mathbf{N}}^{\mathfrak{A}} = \{ac, cd, ad, cc, dd\}$, and $\mathbf{A}^{\mathfrak{A}} = \{bd\} \cup \mathbf{N}^{\mathfrak{A}} \cup \mathrm{Id(A)}$. We have $\mathfrak{A} \models \forall z(\mathbf{A}az \supset \boldsymbol{\Pi}bz)$, and therefore $\mathfrak{A} \models \mathbb{Q}^{\mathbf{a}}ba$. Moreover, we have $\mathfrak{A} \models \neg\mathbf{A}bc$, and therefore $\mathfrak{A} \models \mathbb{X}^{\mathbf{o}}bc$. On the other hand, we have $\mathfrak{A} \models \widehat{\mathbf{N}}ac$, hence $\mathfrak{A} \not\models \mathbb{M}^{\mathbf{o}}ac$.

Let $\mathfrak{B}$ be a model with $\mathrm{B} = \{a, b, c\}$, $\mathbf{N}^{\mathfrak{B}} = \widehat{\mathbf{N}}^{\mathfrak{B}} = \{aa, cc\}$, and $\mathbf{A}^{\mathfrak{B}} = \{ba\} \cup \mathrm{Id(B)}$. We have $\mathfrak{B} \models \boldsymbol{\Pi}ba$ and $\mathfrak{B} \models \forall z(\mathbf{A}az \supset \boldsymbol{\Pi}bz)$, and therefore $\mathfrak{B} \models \mathbb{Q}^{\mathbf{a}}ba$. Moreover, we have $\mathfrak{B} \models \neg\mathbf{A}bc$, and therefore $\mathfrak{B} \models \mathbb{X}^{\mathbf{o}}bc$. On the other hand, we have $\mathfrak{B} \models \boldsymbol{\Gamma}ac$, hence $\mathfrak{B} \not\models \mathbb{M}^{\mathbf{i}}ac$. □

Fact 112: The premise pair **oa,oe-2-NQ** is inconcludent in the predicable semantics.

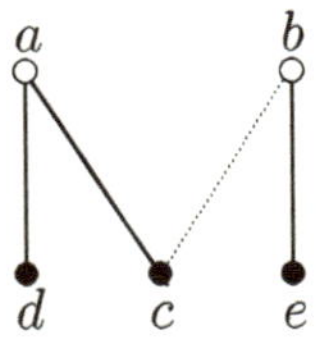

Proof. Let $\mathfrak{A}$ be a model with $\mathrm{A} = \{a, b, c, d, e\}$, $\mathbf{N}^{\mathfrak{A}} = \widehat{\mathbf{N}}^{\mathfrak{A}} = \{ad, dd, ac, cc, be, ee\}$, and $\mathbf{A}^{\mathfrak{A}} = \{bc\} \cup \mathbf{N}^{\mathfrak{A}} \cup \mathrm{Id(A)}$. We have $\mathfrak{A} \models \mathbf{N}ad \wedge \mathbf{N}be \wedge \forall u(\mathbf{A}bu \wedge \boldsymbol{\Sigma}u \supset \widehat{\mathbf{K}}du)$, hence $\mathfrak{A} \models \mathbb{N}^{\mathbf{o}}ba$. Moreover, we have $\mathfrak{A} \models \forall z(\mathbf{A}cz \supset \boldsymbol{\Pi}bz)$, and therefore $\mathfrak{A} \models \mathbb{Q}^{\mathbf{a}}bc$. On the other hand, we have $\mathfrak{A} \models \widehat{\mathbf{N}}ac$, hence $\mathfrak{A} \not\models \mathbb{M}^{\mathbf{o}}ac$.

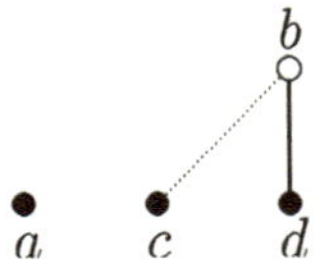

Let $\mathfrak{B}$ be a model with $\mathrm{B} = \{a, b, c, d\}$, $\mathbf{N}^{\mathfrak{B}} = \widehat{\mathbf{N}}^{\mathfrak{B}} = \{bd, aa, cc, dd\}$, and $\mathbf{A}^{\mathfrak{B}} = \{bc\} \cup \mathbf{N}^{\mathfrak{B}} \cup \mathrm{Id(B)}$. We have $\mathfrak{B} \models \mathbf{N}aa \wedge \mathbf{N}bd \wedge \forall u(\mathbf{A}bu \wedge \boldsymbol{\Sigma}u \supset \widehat{\mathbf{K}}au)$, hence $\mathfrak{B} \models \mathbb{N}^{\mathbf{o}}ba$. Moreover, we have $\mathfrak{B} \models \forall z(\mathbf{A}cz \supset \boldsymbol{\Pi}bz)$, and therefore $\mathfrak{B} \models \mathbb{Q}^{\mathbf{a}}bc$. On the other hand, we have $\mathfrak{B} \models \boldsymbol{\Gamma}ac$, hence $\mathfrak{B} \not\models \mathbb{M}^{\mathbf{i}}ac$. □

Fact 113: The premise pair **oa,oe-1-NQ** is inconcludent in the predicable semantics.

Proof. Let $\mathfrak{A}$ be a model with $\mathrm{A} = \{a, b, c, d\}$, $\mathbf{N}^{\mathfrak{A}} = \widehat{\mathbf{N}}^{\mathfrak{A}} = \{aa, ac, cc, dd\}$, and $\mathbf{A}^{\mathfrak{A}} = \{bc, bd\} \cup \mathbf{N}^{\mathfrak{A}} \cup \mathrm{Id(A)}$. We have $\mathfrak{A} \models \mathbf{A}bd \wedge \mathbf{K}ad$, hence $\mathfrak{A} \models \mathbb{N}^{\mathrm{o}}ab$. Moreover, we have $\mathfrak{A} \models \forall z(\mathbf{A}cz \supset \mathbf{\Pi}bz)$, and therefore $\mathfrak{A} \models \mathbb{Q}^{\mathrm{a}}bc$. On the other hand, we have $\mathfrak{A} \models \widehat{\mathbf{N}}ac$, hence $\mathfrak{A} \not\models \mathbb{M}^{\mathrm{o}}ac$.

Let $\mathfrak{B}$ be a model with $\mathrm{B} = \{a, b, c\}$, $\mathbf{N}^{\mathfrak{B}} = \widehat{\mathbf{N}}^{\mathfrak{B}} = \{aa, cc\}$, and $\mathbf{A}^{\mathfrak{B}} = \{ba, bc\} \cup \mathrm{Id(B)}$. We have $\mathfrak{B} \models \mathbf{A}bc \wedge \mathbf{K}ac$, hence $\mathfrak{B} \models \mathbb{N}^{\mathrm{o}}ab$. Moreover, we have $\mathfrak{B} \models \forall z(\mathbf{A}cz \supset \mathbf{\Pi}bz)$, and therefore $\mathfrak{B} \models \mathbb{Q}^{\mathrm{a}}bc$. On the other hand, we have $\mathfrak{B} \models \mathbf{\Gamma}ac$, hence $\mathfrak{B} \not\models \mathbb{M}^{\mathrm{i}}ac$. □

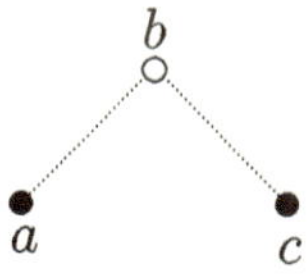

Fact 114: The premise pair **aa,ea-2-QN** is inconcludent in the predicable semantics.

Proof. Let $\mathfrak{A}$ be a model with $\mathrm{A} = \{a, b, c\}$, $\widehat{\mathbf{N}}^{\mathfrak{A}} = \{ac, cc\}$, $\mathbf{N}^{\mathfrak{A}} = \{bc\} \cup \widehat{\mathbf{N}}^{\mathfrak{A}}$, and $\mathbf{A}^{\mathfrak{A}} = \mathbf{N}^{\mathfrak{A}} \cup \mathrm{Id(A)}$. We have $\mathfrak{A} \models \mathbf{\Pi}ba$ and $\mathfrak{A} \models \mathbf{\Pi}bc$, hence $\mathfrak{A} \models \forall z(\mathbf{A}az \supset \mathbf{\Pi}bz)$ and $\mathfrak{A} \models \mathbb{Q}^{\mathrm{a}}ba$. Moreover, we have $\mathfrak{A} \models \mathbf{N}bc$, and therefore $\mathfrak{A} \models \mathbb{N}^{\mathrm{a}}bc$. On the other hand, we have $\mathfrak{A} \models \widehat{\mathbf{N}}ac$, hence $\mathfrak{A} \not\models \mathbb{M}^{\mathrm{o}}ac$.

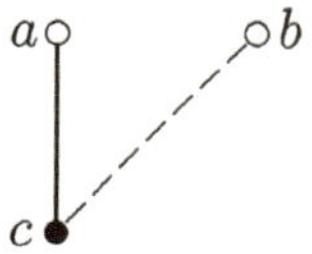

Let $\mathfrak{B}$ be a model with $\mathrm{B} = \{a, b, c\}$, $\mathbf{N}^{\mathfrak{B}} = \widehat{\mathbf{N}}^{\mathfrak{B}} = \{bc, cc, aa\}$, and $\mathbf{A}^{\mathfrak{B}} = \{ba\} \cup \mathbf{N}^{\mathfrak{B}} \cup \mathrm{Id(B)}$. We have $\mathfrak{B} \models \mathbf{\Pi}ba$ and $\mathfrak{B} \models \forall z(\mathbf{A}az \supset \mathbf{\Pi}bz)$, and therefore $\mathfrak{B} \models \mathbb{Q}^{\mathrm{a}}ba$. Moreover, we have $\mathfrak{B} \models \mathbf{N}bc$, and therefore $\mathfrak{B} \models \mathbb{N}^{\mathrm{a}}bc$. On the other hand, we have $\mathfrak{B} \models \mathbf{\Gamma}ac$, hence $\mathfrak{B} \not\models \mathbb{M}^{\mathrm{i}}ac$. □

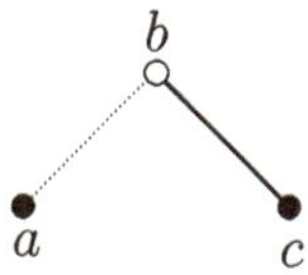

Fact 115: The premise pair **aa,ae-2-NQ** is inconcludent in the predicable semantics. pp. 219–220

Proof. Let $\mathfrak{A}$ be a model with $\mathrm{A} = \{a, b, c\}$, $\widehat{\mathbf{N}}^{\mathfrak{A}} = \{aa, ac, cc\}$, and $\mathbf{N}^{\mathfrak{A}} = \{ba, bc\} \cup \widehat{\mathbf{N}}^{\mathfrak{A}}$, $\mathbf{A}^{\mathfrak{A}} = \mathbf{N}^{\mathfrak{A}} \cup \mathrm{Id(A)}$. We have $\mathfrak{A} \models \mathbf{N}ba$, and therefore $\mathfrak{A} \models \mathbb{N}^{\mathrm{a}}ba$. Moreover, we have $\mathfrak{A} \models \mathbf{\Pi}bc$ and $\mathfrak{A} \models \forall z(\mathbf{A}cz \supset \mathbf{\Pi}bz)$, and therefore $\mathfrak{A} \models \mathbb{Q}^{\mathrm{a}}bc$. On the other hand, we have $\mathfrak{A} \models \widehat{\mathbf{N}}ac$, hence $\mathfrak{A} \not\models \mathbb{M}^{\mathrm{o}}ac$.

Let $\mathfrak{B}$ be a model with $\mathrm{B} = \{a, b, c\}$, $\mathbf{N}^{\mathfrak{B}} = \widehat{\mathbf{N}}^{\mathfrak{B}} = \{ba, aa, cc\}$, and $\mathbf{A}^{\mathfrak{B}} = \{bc\} \cup \mathbf{N}^{\mathfrak{B}} \cup \mathrm{Id(B)}$. We have $\mathfrak{B} \models \mathbf{N}ba$, and therefore $\mathfrak{B} \models \mathbb{N}^{\mathrm{a}}ba$. Moreover, we have $\mathfrak{B} \models \mathbf{\Pi}bc$ and $\mathfrak{B} \models \forall z(\mathbf{A}cz \supset \mathbf{\Pi}bz)$, and therefore $\mathfrak{B} \models \mathbb{Q}^{\mathrm{a}}bc$. On the other hand, we have $\mathfrak{B} \models \mathbf{\Gamma}ac$, hence $\mathfrak{B} \not\models \mathbb{M}^{\mathrm{i}}ac$. □

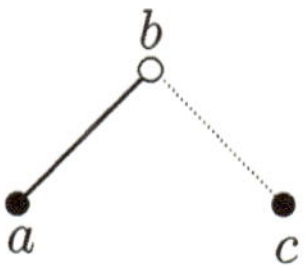

The premise pairs listed in Facts 105–115 are precisely those whose inconcludence is taken as primitive in Table 15.3 on p. 230. All of them have been shown to be inconcludent in the predicable semantics. As a second step, we will now consider all other premise pairs held to be invalid by Aristotle. Their inconcludence will be shown to follow from the foregoing primitive inconcludences by means of conversion, subalternation, and modal subordination (that is, by means of Facts 17–21). This is done in Facts 116–136:

Fact 116: The premise pairs **ie-1-XX**, **ao-1-XX**, **ie-2-XX**, **ae-3-XX**, **ao-3-XX**, and **ie-3-XX** are inconcludent in the predicable semantics.

Proof. Follows from the inconcludence of ae-1-XX (Fact 105) by a_X-, i_X-, and e_X-conversion (Fact 17) and by negative X-subalternation (Fact 18). □

Fact 117: The premise pairs **oe-1-XX**, **eo-1-XX**, **ee-2-XX**, **eo-2-XX**, **oe-2-XX**, **ee-3-XX**, **eo-3-XX**, and **oe-3-XX** are inconcludent in the predicable semantics.

Proof. Follows from the inconcludence of ee-1-XX (Fact 106) by e_X-conversion (Fact 17) and negative X-subalternation (Fact 18). □

Fact 118: The premise pairs **ia-1-XX**, **ai-2-XX**, and **ia-2-XX** are inconcludent in the predicable semantics.

Proof. Follows from the inconcludence of aa-2-XX (Fact 107) by a_X- and i_X-conversion (Fact 17). □

Fact 119: The premise pairs **ii-1-XX**, **io-1-XX**, **oi-1-XX**, and **oo-1-XX** are inconcludent in the predicable semantics.

Proof. Follows from the inconcludence of ia-1-XX (Fact 118), ie-1-XX (Fact 116), oa-1-XX (Fact 107), and oe-1-XX (Fact 117) by subalternation (Fact 18). □

Fact 120: The premise pairs **ii-2-XX**, **io-2-XX**, **oi-2-XX**, and **oo-2-XX** are inconcludent in the predicable semantics.

Proof. Follows from the inconcludence of ia-2-XX (Fact 118), ie-2-XX (Fact 116), oa-2-XX (Fact 108), and oe-2-XX (Fact 117) by subalternation (Fact 18). □

Fact 121: The premise pairs **ii-3-XX**, **io-3-XX**, **oi-3-XX**, and **oo-3-XX** are inconcludent in the predicable semantics.

Proof. The inconcludence of ii-3-XX follows from that of aa-2-XX (Fact 107) by a_X-conversion (Fact 17). The inconcludence of io-3-XX follows from that of ao-3-XX (Fact 116) by subalternation (Fact 18). The inconcludence of oi-3-XX follows from that of oi-1-XX (Fact 119) by i_X-conversion (Fact 17). The inconcludence of oo-3-XX follows from that of oe-3-XX (Fact 117) by subalternation (Fact 18). □

Fact 122: The premise pairs **ia,oa,ie,oe-1-QQ**, **ei,ao,eo,ai-2-QQ**, **oa,ie,oe,ia-2-QQ**, **ii,io,oi,oo-1-QQ**, **ii,io,oi,oo-2-QQ**, and **ii,io,oi,oo-3-QQ** are inconcludent in the predicable semantics.

Proof. Follows from the inconcludence of aa-2-QQ (Fact 109) by a_Q- and i_Q-conversion (Fact 17). □

Fact 123: The premise pairs **ao,eo-1-QN**, **ie,oe-1-QN**, **ie,oe-2-QN**, **ae,ee-3-QN**, **ie,oe-3-QN**, and **ao,eo-3-QN** are inconcludent in the predicable semantics.

Proof. Follows from the inconcludence of ae-1-QN (Fact 110) by a_Q-, i_Q-, and e_N-conversion (Fact 17) and by negative N-subalternation (Fact 18). □

Fact 124: The premise pairs **ae,ee-1-QX**, **ao,eo-1-QX**, **ie,oe-1-QX**, **ie,oe-2-QX**, **ae,ee-3-QX**, **ie,oe-3-QX**, and **ao,eo-3-QX** are inconcludent in the predicable semantics.

Proof. Follows from the inconcludence of the corresponding premise pairs of the form QN (Facts 110 and 123) by N-X-subordination (Fact 19). □

Fact 125: The premise pairs **ia,oa-1-QN**, **oa,ia-2-QN**, **ia,oa-1-QX**, and **oa,ia-2-QX** are inconcludent in the predicable semantics.

Proof. Follows from the inconcludence of aa-2-QN (Fact 114) by a_Q- and i_Q-conversion (Fact 17) and by N-X-subordination (Fact 19). □

Fact 126: The premise pairs **ia,ie-1-NQ** and **ie,ia-2-NQ** are inconcludent in the predicable semantics.

Proof. Follows from the inconcludence of aa-2-NQ (Fact 115) by a_N- and i_N-conversion (Fact 17). □

Fact 127: The premise pairs **ia,ie-1-XQ**, **ie,ia-2-XQ**, and **oa,oe-1-XQ** are inconcludent in the predicable semantics.

Proof. Follows from the inconcludence of the corresponding premise pairs of the form NQ (Facts 126 and 113) by N-X-subordination (Fact 19). □

Fact 128: The premise pairs **ai,ao-2-NQ**, **aa,ae-2-XQ**, and **ai,ao-2-XQ** are inconcludent in the predicable semantics.

Proof. Follows from the inconcludence of aa-2-NQ (Fact 115) by a_Q- and i_Q-conversion (Fact 17) and by N-X-subordination (Fact 19). □

Fact 129: The premise pairs **ai,ei-2-QN**, **aa,ea-2-QX**, and **ai,ei-2-QX** are inconcludent in the predicable semantics.

Proof. Follows from the inconcludence of aa-2-QN (Fact 114) by a_N- and i_N-conversion (Fact 17) and by N-X-subordination (Fact 19). □

Fact 130: The premise pair **oa,oe-2-XQ** is inconcludent in the predicable semantics.

Proof. Follows from the inconcludence of oa-2-NQ (Fact 112) by N-X-subordination (Fact 19). □

Fact 131: The premise pairs **ii,oi-1-QN**, **io,oo-1-QN**, **ii,io-1-NQ**, and **oi,oo-1-NQ** are inconcludent in the predicable semantics.

Proof. Follows from the inconcludence of ia-1-QN (Fact 125), ie-1-QN (Fact 123), ia-1-NQ (Fact 126), and oa-1-NQ (Fact 113) by subalternation (Fact 18). □

Fact 132: The premise pairs **ii,oi-1-QX**, **io,oo-1-QX**, **ii,io-1-XQ**, and **oi,oo-1-XQ** are inconcludent in the predicable semantics.

Proof. Follows from Fact 131 by N-X-subordination (Fact 19). □

Fact 133: The premise pairs **ii,oi-2-QN**, **io,oo-2-QN**, **ii,io-2-NQ**, and **oi,oo-2-NQ** are inconcludent in the predicable semantics.

Proof. Follows from the inconcludence of ia-2-QN (Fact 125), ie-2-QN (Fact 123), ia-2-NQ (Fact 126), and oa-2-NQ (Fact 112) by subalternation (Fact 18). □

Fact 134: The premise pairs **ii,oi-2-QX**, **io,oo-2-QX**, **ii,io-2-XQ**, and **oi,oo-2-XQ** are inconcludent in the predicable semantics.

Proof. Follows from Fact 133 by N-X-subordination (Fact 19). □

Fact 135: The premise pairs **ii,oi-3-QN**, **ii,io-3-NQ**, **io,oo-3-QN**, and **oi,oo-3-NQ** are inconcludent in the predicable semantics.

Proof. The inconcludence of ii-3-QN and ii-3-NQ follows from that of aa-2-QN and aa-2-NQ (Facts 114 and 115) by a_N- and a_Q-conversion (Fact 17). The inconcludence of io-3-QN follows from that of ao-3-QN (Fact 123) by subalternation (Fact 18). The inconcludence oi-3-NQ follows from that of oi-1-NQ (Fact 131) by i_Q-conversion (Fact 17). □

Fact 136: The premise pairs **ii,oi-3-QX**, **ii,io-3-XQ**, **io,oo-3-QX**, and **oi,oo-3-XQ** are inconcludent in the predicable semantics.

Proof. Follows from Fact 135 by N-X-subordination (Fact 19). □

Finally, here are some moods that are not mentioned by Aristotle and whose invalidity in the predicable semantics follows from the foregoing Facts:

Fact 137: The moods **eao,eeo-3-NQQ**, **eio,eoo-3-NQQ**, and **oao,oeo-3-NQQ** are invalid in the predicable semantics.

Proof. Follows from the inconcludence of ae-1-QN (Fact 110) by e_N-, a_Q-, and i_Q-conversion (Fact 17), and by negative N-subalternation

(Fact 18). If eao-3-NQQ were valid, the premise pair ae-1-QN would yield an o_Q-conclusion. □

Fact 138: The moods **eao,eeo-3-XQQ**, **eio,eoo-3-XQQ**, and **oao,oeo-3-XQQ** are invalid in the predicable semantics.

Proof. Follows from Fact 137 by N-X-subordination (Fact 19). □

The premise pairs listed in Facts 105–136 include all premise pairs whose inconcludence Aristotle asserts in *Prior Analytics* 1.1–22. This can be checked by means of the synopsis in Appendix A. Thus it has been shown that all premise pairs held to be inconcludent by Aristotle are inconcludent in the predicable semantics. Moreover, it has been shown in Facts 116–136 that, given the rules of conversion, subalternation, and modal subordination, all of Aristotle's claims of inconcludence follow from the primitive claims of inconcludence listed in Table 15.3 on p. 230 above (Facts 105–115).

In sum, then, it has been shown in this appendix that the predicable semantics is adequate with respect to Aristotle's modal syllogistic. In other words, the semantics has been shown to match all of Aristotle's claims of validity, invalidity, and inconcludence in *Prior Analytics* 1.1–22.

As a corollary, it has also been shown that the deductive system given in Table 15.1 on p. 225 is adequate with respect to Aristotle's modal syllogistic. For, on the one hand, all moods held to be valid by Aristotle are deducible in this deductive system. At the same time, no mood or conversion rule held to be invalid by him is deducible in it; for, as we saw, every mood and conversion rule deducible in the system is valid in the predicable semantics, and hence not held to be invalid by Aristotle.

Finally, it has been shown that, given the rules of conversion, subalternation, and modal subordination (Facts 17–21), all of Aristotle's claims of invalidity and inconcludence in *Prior Analytics* 1.1–22 follow from the primitive claims of invalidity and inconcludence given in Table 15.3 on p. 230.

Miscellany

This section contains some miscellaneous results that were used in the main text of the study but that are not needed to establish the adequacy of the predicable semantics.

Fact 139: Given a complement operator satisfying the law of obversion $\mathbf{A}ab \leftrightarrow \forall z(\mathbf{A}bz \supset \neg\mathbf{A}\bar{a}z)$, the following holds:
$\neg\mathbf{A}ab \supset \exists z(\mathbf{A}bz \wedge \forall u(\mathbf{A}zu \supset \neg\mathbf{A}au))$. (see p. 99)

Proof. Assume $\neg\mathbf{A}ab$. By obversion, this implies $\mathbf{A}bz \wedge \mathbf{A}\bar{a}z$. For *reductio,* assume $\forall z(\mathbf{A}bz \supset \exists u(\mathbf{A}zu \wedge \mathbf{A}au))$. This together with $\mathbf{A}bz$ implies $\mathbf{A}zu \wedge \mathbf{A}au$. $\mathbf{A}\bar{a}z$ and $\mathbf{A}zu$ imply $\mathbf{A}\bar{a}u$ by (ax_2). So we have $\mathbf{A}\bar{a}u \wedge \mathbf{A}au$. By obversion, this implies $\neg\mathbf{A}aa$, contradicting (ax_1). □

Fact 140: $\neg\mathbf{\Pi}ab \supset (\widehat{\mathbf{\Sigma}}a \vee \widehat{\mathbf{\Sigma}}b)$ (see p. 259)

Proof. For *reductio,* assume $\neg\widehat{\mathbf{\Sigma}}a \wedge \neg\widehat{\mathbf{\Sigma}}b$. First, this implies $\neg(\widehat{\mathbf{\Sigma}}a \wedge \widehat{\mathbf{\Sigma}}b)$. Second, $\neg\widehat{\mathbf{\Sigma}}a \wedge \neg\widehat{\mathbf{\Sigma}}b$ implies $\neg\widehat{\mathbf{N}}ab \wedge \neg\widehat{\mathbf{N}}ba$ by ($df_{\widehat{\mathbf{\Sigma}}}$). Third, $\neg\widehat{\mathbf{\Sigma}}a \wedge \neg\widehat{\mathbf{\Sigma}}b$ implies $\neg\mathbf{\Gamma}ab$ by ($df_{\mathbf{\Gamma}}$). These three consequences imply $\mathbf{\Pi}ab$ by ($df_{\mathbf{\Pi}}$). □

Fact 141: $\exists z(\mathbf{A}bz \wedge \overline{\mathbf{\Pi}}az)$ is equivalent to $\neg\mathbf{\Gamma}ab$ (see p. 262)

Proof. First, from left to right. For *reductio*, assume $\mathbf{\Gamma}ab$. This implies $\neg\exists z(\mathbf{A}bz \wedge \mathbf{A}az)$ by ($df_{\mathbf{\Gamma}}$). By ($df_{\overline{\mathbf{\Pi}}}$), $\exists z(\mathbf{A}bz \wedge \overline{\mathbf{\Pi}}az)$ and $\neg\exists z(\mathbf{A}bz \wedge \mathbf{A}az)$ imply $\mathbf{A}bz \wedge \mathbf{\Pi}az$. Now, $\mathbf{\Gamma}ab$ and $\mathbf{A}bz$ imply $\mathbf{\Gamma}az$ by Fact 15. This contradicts $\mathbf{\Pi}az$ and ($df_{\mathbf{\Pi}}$). Second, from right to left. For proof by contraposition, assume $\neg\exists z(\mathbf{A}bz \wedge \overline{\mathbf{\Pi}}az)$. This implies $\neg\exists z(\mathbf{A}bz \wedge \mathbf{A}az)$ by ($df_{\overline{\mathbf{\Pi}}}$). By ($ax_1$), $\neg\exists z(\mathbf{A}bz \wedge \overline{\mathbf{\Pi}}az)$ implies $\neg\overline{\mathbf{\Pi}}ab$, hence $\neg\mathbf{\Pi}ab$ by ($df_{\overline{\mathbf{\Pi}}}$). By Fact 140, $\neg\mathbf{\Pi}ab$ implies $\widehat{\mathbf{\Sigma}}a \vee \widehat{\mathbf{\Sigma}}b$. So we have $\widehat{\mathbf{\Sigma}}a \vee \widehat{\mathbf{\Sigma}}b$ and $\neg\exists z(\mathbf{A}bz \wedge \mathbf{A}az)$, hence $\mathbf{\Gamma}ab$ by ($df_{\mathbf{\Gamma}}$). □

Fact 142: $\mathbf{\Gamma}ab \wedge \forall z(\mathbf{A}cz \supset \mathbf{\Pi}bz) \vdash \mathbf{\Gamma}ac$ (see p. 266)

Proof. $\forall z(\mathbf{A}cz \supset \mathbf{\Pi}bz)$ implies $\neg\widehat{\mathbf{\Sigma}}b$ by Fact 14. By ($df_{\mathbf{\Gamma}}$), $\mathbf{\Gamma}ab$ and $\neg\widehat{\mathbf{\Sigma}}b$ imply $\widehat{\mathbf{\Sigma}}a \wedge \forall z(\mathbf{A}bz \supset \neg\mathbf{A}az)$. For *reductio,* assume $\neg\mathbf{\Gamma}ac$. By ($df_{\mathbf{\Gamma}}$), $\widehat{\mathbf{\Sigma}}a$ and $\neg\mathbf{\Gamma}ac$ imply $\mathbf{A}cz \wedge \mathbf{A}az$. By Fact 3, $\widehat{\mathbf{\Sigma}}a$ and $\mathbf{A}az$ imply $\widehat{\mathbf{\Sigma}}z$. $\mathbf{A}cz$ and $\forall z(\mathbf{A}cz \supset \mathbf{\Pi}bz)$ imply $\mathbf{\Pi}bz$. By Fact 13, $\mathbf{\Pi}bz$ and $\widehat{\mathbf{\Sigma}}z$ imply $\mathbf{A}zu \wedge \mathbf{A}bu$. By ($ax_2$), $\mathbf{A}az$ and $\mathbf{A}zu$ imply $\mathbf{A}au$. This contradicts $\mathbf{A}bu$ and $\forall z(\mathbf{A}bz \supset \neg\mathbf{A}az)$. □

Fact 143: $\mathbf{\Pi}ab \vee \neg\mathbf{A}ab$ is equivalent to $\neg\overline{\mathbf{N}}ab$ (see p. 263)

Proof. First, from left to right. Assume $\mathbf{\Pi}ab \vee \neg\mathbf{A}ab$. This implies $\neg\widehat{\mathbf{N}}ab$ by ($df_{\mathbf{\Pi}}$) and Fact 1. Moreover, $\mathbf{\Pi}ab \vee \neg\mathbf{A}ab$ implies $\neg(\widehat{\mathbf{\Sigma}}a \wedge \mathbf{A}ab)$.

Otherwise, we would have $\widehat{\mathbf{\Sigma}}a \wedge \widehat{\mathbf{\Sigma}}b \wedge \mathbf{A}ab$ by Fact 3, contradicting $\mathbf{\Pi}ab \vee \neg\mathbf{A}ab$ and ($\mathrm{df}_{\mathbf{\Pi}}$). So we have $\neg\widehat{\mathbf{N}}ab$ and $\neg(\widehat{\mathbf{\Sigma}}a \wedge \mathbf{A}ab)$, hence $\neg\overline{\mathbf{N}}ab$ by ($\mathrm{df}_{\overline{\mathbf{N}}}$). Second, from right to left. For proof by contraposition, assume $\neg\mathbf{\Pi}ab \wedge \mathbf{A}ab$. $\mathbf{A}ab$ implies $\neg\mathbf{\Gamma}ab$ by (ax_1) and ($\mathrm{df}_{\mathbf{\Gamma}}$). By ($\mathrm{df}_{\mathbf{\Pi}}$), $\neg\mathbf{\Gamma}ab$ and $\neg\mathbf{\Pi}ab$ imply that one of the following three cases obtains: $\widehat{\mathbf{\Sigma}}a \wedge \widehat{\mathbf{\Sigma}}b$ or $\widehat{\mathbf{N}}ab$ or $\widehat{\mathbf{N}}ba$. In the first case we have $\widehat{\mathbf{\Sigma}}a \wedge \mathbf{A}ab$, in the second case we have $\widehat{\mathbf{N}}ab$, and in the third case we have $\widehat{\mathbf{\Sigma}}a \wedge \mathbf{A}ab$ by ($\mathrm{df}_{\widehat{\mathbf{\Sigma}}}$). In each of the three cases we have $\overline{\mathbf{N}}ab$ by ($\mathrm{df}_{\overline{\mathbf{N}}}$). □

Fact 144: The mood **oao-3-MXM** is valid in the predicable semantics:
$\neg\overline{\mathbf{N}}ab \wedge \mathbf{A}cb \vdash \neg\overline{\mathbf{N}}ac$ (see pp. 208–209 and 229)

Proof. For *reductio,* assume $\overline{\mathbf{N}}ac$ and $\mathbf{A}cb$. By ($\mathrm{df}_{\overline{\mathbf{N}}}$), $\overline{\mathbf{N}}ac$ implies $\widehat{\mathbf{N}}ac$ or $\widehat{\mathbf{\Sigma}}a \wedge \mathbf{A}ac$. In the first case, $\widehat{\mathbf{N}}ac$ and $\mathbf{A}cb$ imply $\widehat{\mathbf{N}}ab$ by (ax_4), and hence $\overline{\mathbf{N}}ab$ by ($\mathrm{df}_{\overline{\mathbf{N}}}$). In the second case, $\widehat{\mathbf{\Sigma}}a \wedge \mathbf{A}ac$ and $\mathbf{A}cb$ imply $\widehat{\mathbf{\Sigma}}a \wedge \mathbf{A}ab$ by (ax_2), and hence $\overline{\mathbf{N}}ab$ by ($\mathrm{df}_{\overline{\mathbf{N}}}$). □

I should like to conclude with a comment on how the predicable semantics relates to the *Topics*' theory of predicables and categories. This theory and its connection to $\mathrm{a_X}$- and $\mathrm{a_N}$-predication were characterized by means of the statements S1–25 on pp. 116–159 above. Two of these statements figure as axioms in the predicable semantics, namely, (ax_3) and (ax_5), stating the validity of Barbara NXN and of N-X-subordination for a-propositions.[4] Most of the other statements are not included in the predicable semantics, either as axioms or as derivable theorems. Some of these statements concern relations specific to the theory of predicables, such as the relations of essential predication and of being a genus of something. Other statements concern the relations of $\mathrm{a_X}$- and $\mathrm{a_N}$-predication. Two examples of such statements that are not included in the predicable semantics are the following:[5]

$$\mathbf{\Sigma}a \supset \mathbf{N}aa$$
$$(\mathbf{A}ab \wedge \mathbf{\Sigma}a \wedge \widehat{\mathbf{\Sigma}}b) \supset \widehat{\mathbf{\Sigma}}a$$

4. These are S15 and S16 on pp. 129 and 131.
5. See S21, p. 140, and S24, p. 151.

The statements S1–25 constitute the interpretive background for the predicable semantics and should be understood to be true in the intended models of this semantics. For ease of exposition, most of these statements are not explicitly included in the predicable semantics, because they are not needed to establish the validity of the moods and conversion rules held to be valid by Aristotle. Nevertheless, S1–25 could all be added to the list of axioms of the predicable semantics; for none of them is violated in the models used above to establish the truth in the predicable semantics of Aristotle's claims of invalidity and inconcludence. In other words, S1–25 are consistent with all models used in the proofs of Facts 73–86 and Facts 105–115.[6]

Among the models used in the present appendix, there is only one that violates a statement in S1–25: the model in Fact 100 violates the second of the two statements mentioned above, according to which non-substance essence terms are not a_X-predicated of substance terms. This model is used to establish the invalidity of moods such as Barbara NQQ and Darii NQQ (see Facts 100 and 102). Aristotle does not discuss these moods, neither claiming them to be valid nor claiming them to be invalid. If S1–25 were added to the predicable semantics, these moods would be valid in the predicable semantics while everything else would remain as it is.[7] Thus, the predicable semantics accounts for all of Aristotle's claims of invalidity and inconcludence without violating S1–25.

6. The only potentially problematic statement is S8 (p. 123), according to which every subject of an essential predication has a genus. The statement implies that every subject of an a_N-predication has a genus (see S11, p. 126). In order for this to be true in the models used in Facts 73–115 (along with the other statements in S1–25), it must be assumed that in some cases something can be a genus of itself (see pp. 146–147n24). Alternatively, one could assume that every genus is distinct from that of which it is a genus and replace S8 by S8*, according to which every subject of an essential predication either has a genus or is a genus of something (p. 147n24). On this latter strategy, some of the models used above would need to be slightly modified so that the formula $\exists z(a \neq z \wedge (\mathbf{N}za \vee \mathbf{N}az))$ is true in them for every a for which $\mathbf{N}aa$ is true.

7. Facts 100 and 102 are used in Facts 101 and 103 to establish the invalidity of moods such as Barbara XQQ and Darii XQQ. Aristotle does not mention these moods either. Their invalidity in the predicable semantics could also be established by models that do not violate S1–25.

Appendix C

Aristotle's Terms

In order to establish the invalidity of moods and the inconcludence of premise pairs, Aristotle uses counterexamples consisting of concrete terms such as 'animal', 'white', and 'walking'. The present appendix gives an overview of the categorical propositions used by Aristotle in these counterexamples. The purpose of this is to show what kinds of terms he uses as subjects and predicates of true or false propositions of a given kind. For example, it will become clear that Aristotle does not employ any true a_Q-propositions whose predicate is a substance term, and that he usually does not employ any true a_X-propositions whose predicate is a substance term while the subject is a nonsubstance term.

In order to prove the inconcludence of a purely assertoric premise pair, Aristotle usually gives two counterexamples. In both of them the premise pair is true, but in one of them the major term is a_X-predicated of the minor term, and in the other the major term is e_X-predicated of the minor term (see pp. 211–212). In particular, his counterexamples in the assertoric syllogistic assume the truth of the propositions shown in Table C.1.[1] Table C.2 contains the categorical propositions that are assumed to be true as premises of counterexamples in the modal syllogistic. Table C.3 contains the categorical propositions that are assumed to be false in the modal syllogistic in order to show that a certain conclusion does not follow from a given premise pair.

In counterexamples establishing the invalidity of a mood, the conclusion of the mood is assumed to be false. For instance, in order to establish

(text continues on page 334)

1. Similar overviews for the assertoric syllogistic are given by Thom (1981: 60–2) and Boger (2004: 198–200).

	Predicate term	Subject term	Passages
ax	animal	man	26a8–9, 26b7, 27a19, 27a22, 27b38, 28a32–3, 28b37, 29a2, 29a9–10
	animal	horse	26a8–9, 26b25, 28a32, 28a34
	science	medicine	26a11–12
	condition	wisdom	26a35
	good	wisdom	26a35
	condition	ignorance	26a35–6
	white	swan	26a38, 26b8, 26b13, 27b26–7, 27b34
	white	snow	26b8, 26b13, 27b33
	animal	swan	26b9, 27b34
	inanimate	snow	26b14
	substance	animal	27a19–20
	substance	man	27a19
	substance	number	27a20
	animal	raven	27b5–6, 27b31–2
	substance	raven	27b5
	substance	unit	27b7–8
	animate	animal	28b24
	animate	man	28b24
ix	good	condition	26a35
	white	horse	26a38, 26b25
	animal	white	26b25, 29a9–10
	white	stone	26b25, 27b26–7
	animal	substance	27b7–8
	white	animal	27b33–4, 27b38–9
	white	man	27b38
	white	inanimate	27b39
	animal	wild	28b37–8
	man	white	29a10
	inanimate	white	29a10
ex	man	horse	26a8–9, 28a35
	man	stone	26a9

Table C.1. Propositions assumed to be true in the assertoric syllogistic (*APr.* 1.4–7)

	Predicate term	Subject term	Passages
ex	animal	stone	26a9, 26b25, 27a22–3
	science	line	26a11–12
	line	medicine	26a12
	line	unit	26a12
	science	unit	26a12
	good	ignorance	26a35–6
	horse	swan	26a38
	horse	raven	26a38–9
	white	raven	26a38–9, 27b5–6, 27b31–2
	animal	snow	26b9, 27b33
	inanimate	man	26b12, 28a33
	inanimate	swan	26b14
	animal	number	27a20
	line	animal	27a22–3
	line	man	27a22
	line	stone	27a22–3
	animal	unit	27b7
	animal	science	27b7–8, 28b37–8, 29a1–2
	substance	science	27b7–8
	black	snow	27b16
	snow	animal	27b16
	swan	stone	27b26–7
	stone	raven	27b32
	animal	inanimate	27b39, 28a32–3, 28a34–5, 29a10
	horse	man	28a32
	horse	inanimate	28a34–5
	man	inanimate	28a35
	man	wild	28b37, 29a2
	science	wild	28b38, 29a1–2
	raven	white	29a3
	raven	snow	29a3
ox	good	condition	26a35

Table C.1. *(continued)*

	Predicate term	Subject term	Passages
ox	white	horse	26a38, 26b25
	man	white	26b7, 26b12, 29a10
	animal	white	26b25, 27b5–6, 29a9–10
	white	stone	26b25, 27b32
	animal	substance	27b5
	black	animal	27b16
	white	animal	27b31–2, 27b38–9
	white	man	27b38
	white	inanimate	27b39
	man	animal	28b24
	animal	wild	29a1–2
	snow	white	29a3
	inanimate	white	29a10

Table C.1. *(continued)*

	Predicate term	Subject term	Passages
ax	motion	animal	30a29–30, 30b5–6
	animal	man	31a17
	moving	animal	32a5
	wakefulness	animal	31b28–9, 31b31–2
	white	man	37b22–3
	white	horse	37b22–3
	health	animal	37b37–8, 38a2
	health	man	37b37–8, 38a2
	health	horse	37b37–8, 38a2
ix	animal	white	35b18–19, 40a2–3
	white	cloak	35b19
	white	man	35b18–19
	health	animal	38a2, 38a12
	health	man	38a2, 38a12

Table C.2. Propositions assumed to be true in the modal syllogistic (*APr.* 1.8–22)

	Predicate term	Subject term	Passages
i_X	health	horse	38a2, 38a12
	man	white	40a2–3
	horse	white	40a2–3
e_X	motion	animal	30a33, 30b5–6
	animal	white	30b33–4
	good	horse	31b5
	wakefulness	white	32a1–2
	raven	reasoning	34b33
	moving	knowledge	34b38–9
	animal	snow	35a24
	animal	pitch	35a24
o_X	wakefulness	man	31b41
	animal	snow	35b10
	animal	pitch	35b10
	animal	white	31a14–15, 35b18–19, 40a2–3
	health	animal	38a12
	health	man	38a12
	health	horse	38a12
	man	white	40a2–3
	horse	white	40a2–3
	white	cloak	35b19
	white	man	35b18–19
a_N	animal	man	30a30, 30a33, 30b33–4, 31a14–15, 31b41
	animal	horse	31b5
	white	swan	36b11–12, 38a31–2, 38b4–5, 38b20, 38b28–9
	white	snow	36b11–12
	motion	awake	38a41–2, 38b4–5
i_N	animal	white	30b6, 32a1–2, 36b6–7, 36b14–15

Table C.2. *(continued)*

	Predicate term	Subject term	Passages
i_N	biped	animal	31b28–9, 31b31–2
	white	man	36b14–15
	white	inanimate	36b14–15
	white	swan	38b37
e_N	animal	snow	36a30–1
	animal	pitch	36a30–1
	white	raven	36b9–10
	white	pitch	36b9–10
	sleeping horse	man	40a37–8, 40b12
	waking horse	man	40a38, 40b12
o_N	biped	animal	32a5
	animal	white	31a17, 36b6–7, 36b14–15
	white	man	36b14–15
	white	inanimate	36b14–15
a_Q/e_Q	reasoning	man	34b33–4
	knowledge	man	34b38–9
	white	animal	35a24, 35b10, 36a30–1
	white	man	33b7, 36b6–7, 37a5–6, 37b4, 37b13, 37b16, 37b22–3, 38a31–2, 38b4–5, 38b20, 38b28–9
	white	cloak	33b7–8, 36b6–7
	white	horse	37b4, 37b13, 37b16, 37b22–3
	health	animal	37b37–8, 38a2, 38a12
	health	man	37b37–8, 38a2, 38a12
	health	horse	37b37–8, 38a2, 38a12
	motion	animal	38a41–2, 38b4–5

Table C.2. *(continued)*

	Predicate term	Subject term	Passages
a_Q/e_Q	sleep	man	40a37–8
i_Q/o_Q	animal	white	33b7, 35b18–19, 36b9–10, 36b11–12, 36b14–15, 39b5, 40a2–3
	white	cloak	35b19
	white	man	33b7, 35b18–19, 36b14–15, 37b16, 38b37
	white	horse	37b16
	white	cloak	33b7–8
	white	inanimate	36b14–15
	man	white	39b5, 40a2–3
	horse	white	39b5, 40a2–3
	health	animal	38a12
	health	man	38a12
	health	horse	38a12
	sleep	man	40b12

Table C.2. *(continued)*

	Predicate term	Subject term	Passages
not: a_N	motion	man	30a29–30
not: i_N	motion	white	30b5–6
	wakefulness	biped	31b28–9
	biped	wakefulness	31b31–2
not: e_N	motion	man	30a33
	man	white	30b33–4
	moving	man	34b38–9
not: o_N	motion	white	30b5–6
	man	white	31a14–15, 31a17
	good	animal	31b5

Table C.3. Propositions assumed to be false in the modal syllogistic (*APr.* 1.8–22)

	Predicate term	Subject term	Passages
not: o_N	wakefulness	animal	31b41, 32a1–2
	biped	moving	32a5
not: e_Q	raven	man	34b33–4
	man	white	37a7
not: i_M	animal	cloak	33b7, 35b18–19, 36b6–7
	raven	man	34b33–4
	white	pitch	35a24, 35b10, 36a30–1
	animal	pitch	36b9–10
	animal	snow	36b11–12
	animal	inanimate	36b14–15
	man	horse	37b4, 37b13, 37b16, 37b22–3
	horse	man	37b37–8, 38a2, 38a12, 39b5, 40a2–3
	man	swan	38a31–2, 38b20, 38b28–9, 38b37
	swan	man	38b4–5, 38b20, 38b28–9, 38b37
	sleep	waking horse	40a38, 40b12
not: o_M	animal	man	33b7, 35b18–19, 36b6–7, 36b14–15, 37b37–8, 38a2, 38a12, 39b5, 40a2–3
	white	snow	35a24, 35b10, 36a30–1
	animal	raven	36b9–10
	animal	swan	36b11–12
	animal	awake	38a41–2, 38b4–5
	sleep	sleeping horse	40a37–8, 40b12

Table C.3. *(continued)*

the invalidity of Barbara XNN, Aristotle gives a counterexample in which the premise pair is true while the a_N-conclusion is false. In counterexamples establishing the inconcludence of a premise pair, Aristotle can be taken to assume that certain o_M- and i_M-propositions are false (see pp. 211–214). Thus, I take him to assume that the major term is not o_M- or i_M-predicated of the minor term in these counterexamples. In the modal syllogistic, Aristotle sometimes does not spell out a given counterexample, but indicates it only by referring back to an earlier counterexample, saying that "the proof is through the same terms" or "the proof is the same." The above overview includes counterexamples indicated by such back references. It is not always entirely clear to which counterexample Aristotle refers. Tables C.1, C.2, and C.3 are based on the following interpretation of these references.

1.9 30a33: invalidity of Celarent XNN. Refers to the counterexample to Barbara XNN at 30a28–32.

1.10 31a11–15: invalidity of Baroco NXN. Refers to the counterexample to Camestres NXN at 30b33–40.

1.10 31a15–17: invalidity of Baroco XNN. Apparently refers to the counterexample to Camestres NXN at 30b33–40, but this reference is problematic (cf. p. 185). For the sake of concreteness, I have treated the passage as if it were a standard back reference indicating a counterexample with the same terms and with the same order of terms used in the counterexample to Camestres NXN (that is, with 'man' as major term and 'white' as minor term).

1.11 31b31–3: invalidity of Disamis NXN. Refers to the counterexample to Datisi XNN at 31b27–31. The major term and the minor term are interchanged when the counterexample to Datisi XNN is transformed into a counterexample to Disamis NXN; cf. Alexander *in APr.* 149.13–20 and p. 219n15 above.

1.17 37b10–13: inconcludence of ae-2-QQ, aa-2-QQ, and ee-2-QQ. Refers to the proof of the inconcludence of ea-2-QQ at 37b3–8.

1.17 37b13–16: inconcludence of ie-2-QQ, ei-2-XQ, io-2-QQ, and so on. Refers to the proof of the inconcludence of ea-2-QQ at 37b3–8.

1.18 37b19–23: inconcludence of ea-2-QX and ae-2-XQ. Refers to the proof of the inconcludence of ea-2-QQ at 37b3–8.

1.18 37b40–38a2: inconcludence of ei-2-QX and ie-2-XQ. Refers to the proof of the inconcludence of aa-2-QX and aa-2-XQ at 37b35–8.

1.18 38a8–12: inconcludence of ao-2-QX, oa-2-XQ, io-2-QX, oi-2-XQ, oi-2-QX, and io-2-XQ. Refers to the proof of the inconcludence of aa-2-XQ at 37b35–8.

1.19 38b4–5: inconcludence of ae-2-NQ. Refers to the two counterexamples at 38a28–b4, which are supposed to show the inconcludence of ea-2-QN. The major term and the minor term are interchanged when transforming the two counterexamples for ea-2-QN into counterexamples for ae-2-NQ; cf. Alexander *in APr.* 238.11–19 and p. 219n15 above.

1.19 38b27–37: inconcludence of ao-2-NQ, oa-2-QN, io-2-NQ, and oi-2-QN. I take this passage to refer to the single counterexample to aa-2-QN and aa-2-NQ at 38b13–20. Alternatively, it may also be taken to refer to the two counterexamples for ea-2-QN at 38a28–b4.

1.21 40a1–3: inconcludence of io-3-QX, io-3-XQ, oi-3-QX, and oi-3-XQ. Refers to the proof of the inconcludence of io-3-QQ at 39b2–6.

1.22 40b10–12: inconcludence of ie-3-QN. Refers to the proof of the inconcludence of ae-3-QN at 40a35–7.

Bibliography

Ackrill, J. L. (1963). *Aristotle's Categories and De Interpretatione.* Oxford: Clarendon Press.

Angelelli, I. (1979). "The Aristotelian Modal Syllogistic in Modern Modal Logic." In *Konstruktionen versus Positionen,* vol. 1: *Spezielle Wissenschaftstheorie,* edited by K. Lorenz. Berlin: de Gruyter, 176–215.

Angell, R. B. (1986). "Truth-Functional Conditionals and Modern vs. Traditional Syllogistic." *Mind* 95, 210–23.

Arpe, C. (1938). *Das τί ἦν εἶναι bei Aristoteles.* Hamburg: Friederichsen.

Bäck, A. T. (2000). *Aristotle's Theory of Predication.* Leiden: Brill.

Balme, D. (1992). *Aristotle's De Partibus Animalium I and De Generatione Animalium I.* Oxford: Clarendon Press.

Barnes, J. (1970). "Property in Aristotle's Topics." *Archiv für Geschichte der Philosophie* 52, 136–55.

—— (1984). *The Complete Works of Aristotle: The Revised Oxford Translation.* Vol. 1, edited by J. Barnes. Princeton, N. J.: Princeton University Press.

—— (1994). *Aristotle's Posterior Analytics.* 2nd ed. Oxford: Clarendon Press.

—— (2003). *Porphyry: Introduction.* Oxford: Oxford University Press.

—— (2007). *Truth, etc.: Six Lectures on Ancient Logic.* Oxford: Clarendon Press.

Becker, A. (1933). *Die Aristotelische Theorie der Möglichkeitsschlüsse.* Berlin: Junker und Dünnhaupt.

Bekker, I. (1831). *Aristoteles graece.* Vol. 1. Berlin: Reimer.

Bobzien, S. (2002). "The Development of Modus Ponens in Antiquity: From Aristotle to the 2nd Century AD." *Phronesis* 47, 359–94.

Bocheński, J. M. (1956). *Formale Logik.* Freiburg: Alber.

Boger, G. (2004). "Aristotle's Underlying Logic." In *Handbook of the History of Logic*. Vol. 1: *Greek, Indian and Arabic Logic*, edited by D. M. Gabbay and J. Woods. Amsterdam: Elsevier, 101–246.

Bostock, D. (2004). "An Aristotelian Theory of Predication?" *Oxford Studies in Ancient Philosophy* 27, 141–75.

Brennan, T. (1997). "Aristotle's Modal Syllogistic: A Discussion of R. Patterson: *Aristotle's Modal Logic*." *Oxford Studies in Ancient Philosophy* 15, 207–31.

Brenner, H. (2000). "Eine vollständige Formalisierung der aristotelischen Notwendigkeitssyllogistik." In *Beiträge zum Satz vom Widerspruch und zur Aristotelischen Prädikationstheorie,* edited by N. Öffenberger and M. Skarica. Hildesheim: Olms, 333–56. Preprint (1993) in *Arbeitsberichte des Arbeitskreises Mathematische Logik.* Bochum: Ruhr Universität.

Brogan, A. D. (1967). "Aristotle's Logic of Statements about Contingency." *Mind* 76, 49–61.

Brunschwig, J. (1967). *Aristote: Topiques.* Vol. 1. Paris: Les Belles Lettres.

Buddensiek, F. (1994). *Die Modallogik des Aristoteles in den* Analytica Priora A. Hildesheim: Olms.

Byrne, P. H. (1997). *Analysis and Science in Aristotle.* Albany: SUNY Press.

Charles, D. (2000). *Aristotle on Meaning and Essence.* Oxford: Clarendon Press.

Code, A. (1986). "Aristotle: Essence and Accident." In *Philosophical Grounds of Rationality: Intentions, Categories, Ends,* edited by R. Grandy and R. Warner. Oxford: Clarendon Press, 411–40.

—— (2010). "An Aristotelian Puzzle about Definition: *Metaphysics* Z.12." In *Being, Nature, and Life in Aristotle: Essays in Honor of Allan Gotthelf,* edited by J. G. Lennox and R. Bolton. Cambridge: Cambridge University Press, 78–96.

Colli, G. (1955). *Aristotele: Organon.* Torino: Einaudi.

Corcoran, J. (1972). "Completeness of an Ancient Logic." *Journal of Symbolic Logic* 37, 696–702.

—— (1973). "A Mathematical Model of Aristotle's Syllogistic." *Archiv für Geschichte der Philosophie* 55, 191–219.

—— (1974). "Aristotle's Natural Deduction System." In *Ancient Logic and its Modern Interpretations*, edited by J. Corcoran. Dordrecht: Reidel, 85–131.

—— (2003). "Aristotle's *Prior Analytics* and Boole's *Laws of Thought.*" *History and Philosophy of Logic* 24, 261–88.

Crivelli, P. (2004). *Aristotle on Truth.* Cambridge: Cambridge University Press.

—— (2012). "Aristotle's Logic." In *The Oxford Handbook of Aristotle,* edited by C. Shields. Oxford: Oxford University Press, 113–49.

Crivelli, P., and Charles, D. (2011). "'ΠΡΟΤΑΣΙΣ' in Aristotle's *Prior Analytics.*" *Phronesis* 56, 193–203.

de Rijk, L. M. (2002). *Aristotle: Semantics and Ontology.* Vol. 1: *General Introduction: The Works on Logic.* Leiden: Brill.

Detel, W. (1993). *Aristoteles: Analytica Posteriora, erster Halbband.* Berlin: Akademie Verlag.

Drechsler, M. (2005). *Interpretationen der Beweismethoden in der Syllogistik des Aristoteles.* Frankfurt a. M.: Lang.

Ebert, T. (1985). "Gattungen der Prädikate und Gattungen des Seienden bei Aristoteles: Zum Verhältnis von *Cat.* 4 und *Top.* I 9." *Archiv für Geschichte der Philosophie* 67, 113–38.

—— (1995). "Was ist ein vollkommener Syllogismus des Aristoteles?" *Archiv für Geschichte der Philosophie* 77, 221–47.

Ebert, T., and Nortmann, U. (2007). *Aristoteles: Analytica Priora, Buch 1.* Berlin: Akademie Verlag.

Fait, P. (1999). "Fragments of Aristotle's Modal Syllogistic in the Late Medieval Theory of Consequences: The Case of *consequentia ut nunc.*" In *Medieval Analyses in Language and Cognition,* edited by S. Ebbesen and R. L. Friedman. Copenhagen: Reitzel, 139–61.

Fine, K. (1994). "Essence and Modality." *Philosophical Perspectives* 8, 1–16.

Flannery, K. L. (1987). "A Rationale for Aristotle's Notion of Perfect Syllogisms." *Notre Dame Journal of Formal Logic* 28, 455–71.

Frede, M. (1981). "Categories in Aristotle." In *Studies in Aristotle,* edited by D. J. O'Meara. Washington, D.C.: Catholic University Press, 1–24.

—— (1987). *Essays in Ancient Philosophy.* Minneapolis: University of Minnesota Press.

Frege, G. (1879). *Begriffsschrift, eine der arithmetischen nachgebildete Formelsprache des reinen Denkens.* Halle: Nebert.

—— (2004). *Frege's Lectures on Logic: Carnap's Student Notes, 1910–1914,* translated and edited by E. H. Reck and S. Awodey. Chicago: Open Court.

Furth, M. (1988). *Substance, Form and Psyche: An Aristotelean Metaphysics.* Cambridge: Cambridge University Press.

Gohlke, P. (1936). *Die Entstehung der Aristotelischen Logik.* Berlin: Junker und Dünnhaupt.

Granger, H. (1984). "Aristotle on Genus and Differentia." *Journal of the History of Philosophy* 22, 1–23.

Henle, P. (1949). "On the Fourth Figure of the Syllogism." *Philosophy of Science* 16, 94–104.

Hintikka, J. (1973). *Time and Necessity: Studies in Aristotle's Theory of Modality.* Oxford: Clarendon Press.

—— (1977). *Aristotle on Modality and Determinism,* in collaboration with U. Remes and S. Knuuttila, Acta Philosophica Fennica 29.1. Amsterdam: North-Holland.

—— (2004). *Analyses of Aristotle: Selected Papers.* Vol. 6. Dordrecht: Kluwer.

Hodges, W. (1998). "The Laws of Distribution for Syllogisms." *Notre Dame Journal of Formal Logic* 39, 221–30.

Huby, P. M. (2002). "Did Aristotle Reply to Eudemus and Theophrastus on Some Logical Issues?" In *Eudemus of Rhodes,* Rutgers University Studies in Classical Humanities, vol. 11, edited by I. Bodnár and W. W. Fortenbaugh. New Brunswick, N. J.: Transaction Publishers, 85–106.

—— (2007). *Theophrastus of Eresus: Sources for His Life, Writings, Thought and Influence: Commentary.* Vol. 2: *Logic.* Leiden: Brill.

Irwin, T. (1980). "The Metaphysical and Psychological Basis of Aristotle's Ethics." In *Essays on Aristotle's Ethics,* edited by A. Oksenberg Rorty. Berkeley: University of California Press, 35–53.

—— (1988). *Aristotle's First Principles.* Oxford: Clarendon Press.

Jenkinson, A. J. (1928). *The Works of Aristotle.* Vol. 1, translated under the editorship of W. D. Ross (only the translation of *Prior Analytics* is by Jenkinson), Oxford: Oxford University Press.

Johnson, F. (1989). "Models for Modal Syllogisms." *Notre Dame Journal of Formal Logic* 30, 271–84.

—— (1993). "Modal Ecthesis." *History and Philosophy of Logic* 14, 171–82.

—— (1995). "Apodeictic Syllogisms: Deductions and Decision Procedures." *History and Philosophy of Logic* 16, 1–18.

—— (2004). "Aristotle's Modal Syllogisms." In *Handbook of the History of Logic,* vol. 1: *Greek, Indian and Arabic Logic,* edited by D. M. Gabbay and J. Woods. Amsterdam: Elsevier, 247–307.

Jones, R. E. (2010). "Truth and Contradiction in Aristotle's *De Interpretatione* 6–9." *Phronesis* 55, 26–67.

Kahn, C. H. (1978). "Questions and Categories: Aristotle's Doctrine of Categories in the Light of Modern Research." In *Questions,* edited by H. Hiż. Dordrecht: Reidel, 227–78.

Kapp, E. (1968). *Ausgewählte Schriften,* edited by H. and I. Diller. Berlin: de Gruyter.

Kneale, W., and Kneale, M. (1962). *The Development of Logic.* Oxford: Clarendon Press.

—— (1972). "Prosleptic Propositions and Arguments." In *Islamic Philosophy and the Classical Tradition,* edited by S. M. Stern, A. Hourani, and V. Brown. Oxford: Cassirer, 189–207.

Kosman, A. (1990). "Necessity and Explanation in Aristotle's *Analytics.*" In *Biologie, Logique et Métaphysique chez Aristote,* edited by D. Devereux and P. Pellegrin. Paris: Éditions du CNRS, 349–64.

Kung, J. (1977). "Aristotle on Essence and Explanation." *Philosophical Studies* 31, 361–83.

—— (1978). "Can Substance Be Predicated of Matter?" *Archiv für Geschichte der Philosophie* 60, 140–59.

Lagerlund, H. (2000). *Modal Syllogistics in the Middle Ages.* Leiden: Brill.

Lear, J. (1980). *Aristotle and Logical Theory.* Cambridge: Cambridge University Press.

Lejewski, C. (1976). "On Prosleptic Premisses." *Notre Dame Journal of Formal Logic* 17, 1–18.

Loux, M. J. (1991). *Primary Ousia: An Essay on Aristotle's* Metaphysics *Z and H.* Ithaca, N.Y.: Cornell University Press.

Łukasiewicz, J. (1957). *Aristotle's Syllogistic from the Standpoint of Modern Formal Logic.* 2nd ed. Oxford: Clarendon Press.

Maier, H. (1900a). *Die Syllogistik des Aristoteles II.1: Formenlehre und Technik des Syllogismus.* Tübingen: Laupp.

—— (1900b). *Die Syllogistik des Aristoteles II.2: Die Entstehung der Aristotelischen Logik.* Tübingen: Laupp.

Malcolm, J. (1981). "On the Generation and Corruption of the Categories." *Review of Metaphysics* 34, 662–81.

Malink, M. (2006). "A Reconstruction of Aristotle's Modal Syllogistic." *History and Philosophy of Logic* 27, 95–141.

—— (2007). "Categories in *Topics* I.9." *Rhizai: A Journal for Ancient Philosophy and Science* 4, 271–94.

—— (2008). "*TΩI* vs *TΩN* in *Prior Analytics* 1.1–22." *Classical Quarterly* 58, 519–36.
—— (2009a). "A Non-Extensional Notion of Conversion in the *Organon*." *Oxford Studies in Ancient Philosophy* 37, 105–41.
—— (2009b). Review of T. Ebert and U. Nortmann: *Aristoteles: Analytica Priora, Buch 1*. In *History and Philosophy of Logic* 30, 389–96.
Malink, M., and Rosen, J. (forthcoming). "Proof by Assumption of the Possible in *Prior Analytics* 1.15." *Mind.*
Mansion, S. (1968). "Notes sur la doctrine des catégories dans les *Topiques*." In *Aristotle on Dialectic: The* Topics, edited by G. E. L. Owen. Oxford: Clarendon Press, 189–201.
Martin, J. N. (1997). "Aristotle's Natural Deduction System Reconsidered." *History and Philosophy of Logic* 18, 1–15.
—— (2004). *Themes in Neoplatonic and Aristotelian Logic.* Aldershot: Ashgate.
McCall, S. (1963). *Aristotle's Modal Syllogisms.* Amsterdam: North-Holland.
—— (1966). "Connexive Implication." *Journal of Symbolic Logic* 31, 415–33.
—— (1967). "Connexive Implication and the Syllogism." *Mind* 76, 346–56.
McKirahan, R. D. (1992). *Principles and Proofs: Aristotle's Theory of Demonstrative Science.* Princeton, N. J.: Princeton University Press.
Mignucci, M. (1969). *Aristotele: Gli Analitici Primi.* Napoli: Loffredo.
—— (1991). "Expository Proofs in Aristotle." *Oxford Studies in Ancient Philosophy,* suppl. vol., 9–28.
—— (1996). "Aristotle's Theory of Predication." In *Studies in the History of Logic,* edited by I. Angelelli and M. Cerezo. Berlin: de Gruyter, 1–20.
—— (1997). "Remarks on Aristotle's Theory of Predication." In *Beiträge zur antiken Philosophie: Festschrift für Wolfgang Kullmann,* edited by H.-C. Günther and A. Rengakos. Stuttgart: Steiner, 135–51.
—— (1998). "Theophrastus' Logic." In *Theophrastus: Reappraising the Sources,* Rutgers University Studies in Classical Humanities, vol. 8, edited by J. M. van Ophuijsen and M. van Raalte. New Brunswick, N. J.: Transaction Publishers, 39–65.
—— (2000). "Parts, Quantification and Aristotelian Predication." *The Monist* 83, 3–21.

Minio-Paluello, L. (1964). "Appendix." In *Aristotelis Analytica Priora et Posteriora,* edited by W. D. Ross, Oxford Classical Texts. Oxford: Clarendon Press, 184–8.

Moody, E. A. (1953). *Truth and Consequence in Mediaeval Logic.* Amsterdam: North-Holland.

Moraux, P., and Wiesner, J. (2001). *Der Aristotelismus bei den Griechen.* Vol. 3: *Alexander von Aphrodisias,* by P. Moraux, edited by J. Wiesner. Berlin: de Gruyter.

Morison, B. (2008). "Aristotle, etc." *Phronesis* 53, 209–22.

Morrison, D. (1992). "The Taxonomical Interpretation of Aristotle's Categories." In *Essays in Ancient Greek Philosophy V: Aristotle's Ontology,* edited by A. Preus and J. P. Anton. Albany: SUNY Press, 19–46.

Mueller, I. (1999a). *Alexander of Aphrodisias: On Aristotle's* Prior Analytics 1.8–13, translated by I. Mueller with J. Gould; introduction, notes, and appendixes by I. Mueller. Ithaca, N. Y.: Cornell University Press.

—— (1999b). *Alexander of Aphrodisias: On Aristotle's* Prior Analytics 1.14–22, translated by I. Mueller with J. Gould; introduction, notes, and appendixes by I. Mueller. Ithaca, N. Y.: Cornell University Press.

Nortmann, U. (1990). "Über die Stärke der aristotelischen Modallogik." *Erkenntnis* 32, 61–82.

—— (1996). *Modale Syllogismen, mögliche Welten, Essentialismus: Eine Analyse der aristotelischen Modallogik.* Berlin: de Gruyter.

—— (2005). "Ekthesis." In *Aristoteles-Lexikon,* edited by O. Höffe. Stuttgart: Kröner, 163–5.

Oehler, K. (1986). *Aristoteles: Kategorien.* 2nd ed. Berlin: Akademie Verlag.

Owen, G. E. L. (1965). "Aristotle on the Snares of Ontology." In *New Essays on Plato and Aristotle,* edited by R. Bambrough. London: Routledge and Kegan Paul.

Owen, O. F. (1889). *The Organon, or Logical Treatises, of Aristotle.* Vol. 1. London: Bell.

Pacius, J. (1597). *In Porphyrii isagogen et Aristotelis Organum commentarius analyticus.* Frankfurt a. M.: Heredes Wecheli. Reprinted Hildesheim: Olms, 1966.

Patterson, R. (1989). "The Case of the Two Barbaras: Basic Approaches to Aristotle's Modal Logic." *Oxford Studies in Ancient Philosophy* 7, 1–40.

—— (1990). "Conversion Principles and the Basis of Aristotle's Modal Logic." *History and Philosophy of Logic* 11, 151–72.

—— (1993). "Aristotle's Perfect Syllogisms, Predication, and the *dictum de omni.*" *Synthese* 96, 359–78.

—— (1995). *Aristotle's Modal Logic: Essence and Entailment in the Organon.* Cambridge: Cambridge University Press.

Patzig, G. (1959). "Aristotle and Syllogisms from False Premisses." *Mind* 68, 186–92.

—— (1968). *Aristotle's Theory of the Syllogism,* translated by J. Barnes. Dordrecht: Reidel.

Price, R. (1969). "Ferison (eio-3) LQX and Bokardo (oao-3) LQX." *Mind* 78, 609–10.

Priest, G. (2006). *Doubt Truth to Be a Liar.* Oxford: Clarendon Press.

—— (2007). "Paraconsistency and Dialetheism." In *Handbook of the History of Logic,* vol. 8: *The Many Valued and Nonmonotonic Turn in Logic,* edited by D. M. Gabbay and J. Woods. Amsterdam: Elsevier, 129–204.

Primavesi, O. (1996). *Die aristotelische Topik.* München: Beck.

Prior, A. (1962). *Formal Logic.* 2nd ed. Oxford: Clarendon Press.

Raymond, D. (2010). "Polarity and Inseparability: The Foundation of the Apodictic Portion of Aristotle's Modal Logic." *History and Philosophy of Logic* 31, 193–218.

Rescher, N. (1964). "Aristotle's Theory of Modal Syllogisms and Its Interpretation." In *The Critical Approach to Science and Philosophy: Essays in Honor of Karl R. Popper,* edited by M. Bunge. London: The Free Press.

—— (1974). *Studies in Modality.* American Philosophical Quarterly, Monograph Series 8. Oxford: Blackwell.

Rescher, N., and Parks, Z. (1971). "A New Approach to Aristotle's Apodeictic Syllogisms." *Review of Metaphysics* 24, 678–89.

Rini, A. A. (1998). "Is There a Modal Syllogistic?" *Notre Dame Journal of Formal Logic* 39, 554–72.

—— (2011). *Aristotle's Modal Proofs:* Prior Analytics *A8–22 in Predicate Logic.* Dordrecht: Springer.

Rolfes, E. (1921). *Aristoteles: Erste Analytik.* Leipzig: Meiner.

Rosen, J., and Malink, M. (2012). "A Method of Modal Proof in Aristotle." *Oxford Studies in Ancient Philosophy* 42, 179–261.

Ross, D. W. (1936). *Aristotle's Physics: A Revised Text with Introduction and Commentary.* Oxford: Clarendon Press.

—— (1949). *Aristotle's Prior and Posterior Analytics: A Revised Text with Introduction and Commentary.* Oxford: Clarendon Press.

Sainati, V. (1988). "Zu einer neuen Lesart der modalen aristotelischen Syllogistik." In *Modallogik und Mehrwertigkeit,* edited by A. Menne and N. Öffenberger. Hildesheim: Olms, 33–51.

Schmidt, K. J. (1989). "Eine modal prädikatenlogische Interpretation der modalen Syllogistik des Aristoteles." *Phronesis* 34, 80–106.

—— (2000). *Die modale Syllogistik des Aristoteles: Eine modal-prädikatenlogische Interpretation.* Paderborn: Mentis.

Seel, G. (1982). *Die Aristotelische Modaltheorie.* Berlin: de Gruyter.

Simons, P. M. (1987). *Parts: A Study in Ontology.* Oxford: Clarendon Press.

Slomkowski, P. (1997). *Aristotle's Topics.* Leiden: Brill.

Smiley, T. J. (1962). "Syllogism and Quantification." *Journal of Symbolic Logic* 27, 58–72.

—— (1973). "What Is a Syllogism?" *Journal of Philosophical Logic* 2, 136–54.

Smith, R. (1982a). "What Is Aristotelian Ecthesis?" *History and Philosophy of Logic* 3, 113–27.

—— (1982b). "The Syllogism in *Posterior Analytics I.*" *Archiv für Geschichte der Philosophie* 64, 113–35.

—— (1983). "Completeness of an Ecthetic Syllogistic." *Notre Dame Journal of Formal Logic* 24, 224–32.

—— (1989). *Aristotle: Prior Analytics.* Indianapolis: Hackett.

—— (1991). "Predication and Deduction in Aristotle: Aspirations to Completeness." *Topoi* 10, 43–52.

—— (1995). "Logic." In *The Cambridge Companion to Aristotle,* edited by J. Barnes. Cambridge: Cambridge University Press, 27–65.

—— (1997). *Aristotle's Topics: Books 1 and 7.* Oxford: Clarendon Press.

—— (2011). Review of G. Striker: *Aristotle's* Prior Analytics: Book 1. In *Ancient Philosophy* 31, 417–24.

Sorabji, R. (1980). *Necessity, Cause and Blame: Perspectives on Aristotle's Theory.* London: Duckworth.

Stekeler-Weithofer, P. (1986). *Grundprobleme der Logik: Elemente einer Kritik der formalen Vernunft.* Berlin: de Gruyter.

Striker, G. (1994). "Modal vs. Assertoric Syllogistic." *Ancient Philosophy* 14, special issue: "Logic, Dialectic, and Science in Aristotle," 39–51.

—— (1996). "Perfection and Reduction in Aristotle's *Prior Analytics.*" In *Rationality in Greek Thought,* edited by M. Frede and G. Striker. Oxford: Clarendon Press, 203–19.

—— (2009). *Aristotle's Prior Analytics: Book 1.* Oxford: Clarendon Press.

Thom, P. (1981). *The Syllogism.* München: Philosophia Verlag.

—— (1991). "The Two Barbaras." *History and Philosophy of Logic* 12, 135–49.

—— (1993). "Apodeictic Ecthesis." *Notre Dame Journal of Formal Logic* 34, 193–208.

—— (1994). "Interpreting Aristotle's Contingency-Syllogistic." *Oxford Studies in Ancient Philosophy* 12, 91–109.

—— (1996). *The Logic of Essentialism: An Interpretation of Aristotle's Modal Syllogistic.* Dordrecht: Kluwer.

Thomason, S. K. (1993). "Semantic Analysis of the Modal Syllogistic." *Journal of Philosophical Logic* 22, 111–28.

—— (1997). "Relational Models for the Modal Syllogistic." *Journal of Philosophical Logic* 26, 129–41.

Thompson, M. (1953). "On Aristotle's Square of Opposition." *Philosophical Review* 62, 251–65.

Tredennick, H. (1938). *Aristotle: The Organon I: The Categories, On Interpretation, Prior Analytics* (only the translation of *Prior Analytics* is by Tredennick), Loeb Classical Library. Cambridge, Mass: Harvard University Press.

van Benthem, J. (1986). *Essays in Logical Semantics.* Dordrecht: Reidel.

van Eijck, J. (1985). "Generalized Quantifiers and Traditional Logic." In *Generalized Quantifiers in Natural Language,* edited by J. van Benthem and A. ter Meulen. Dordrecht: Foris, 1–19.

van Rijen, J. (1989). *Aspects of Aristotle's Logic of Modalities.* Dordrecht: Kluwer.

Varzi, A. C. (1996). "Parts, Wholes, and Part-Whole Relations: The Prospects of Mereotopology." *Data and Knowledge Engineering* 20, 259–86.

von Kirchmann, J. H. (1877). *Erläuterungen zu den ersten Analytiken des Aristoteles.* Leipzig: Koschny.

Wagner, T., and Rapp, C. (2004). *Aristoteles: Topik.* Stuttgart: Reclam.

Waitz, T. (1844). *Aristotelis organon graece I.* Leipzig: Hahn.

—— (1846). *Aristotelis organon graece II.* Leipzig: Hahn.

Wallies, M. (1917/18). "Zur Textgeschichte der Ersten Analytik." *Rheinisches Museum für klassische Philologie* 72, 626–32.

Waterlow, S. (1982). *Passage and Possibility: A Study of Aristotle's Modal Concepts.* Oxford: Clarendon Press.

Wedin, M. V. (1990). "Negation and Quantification in Aristotle." *History and Philosophy of Logic* 11, 131–50.

Weidemann, H. (2002). *Aristoteles: Peri Hermeneias.* 2nd ed. Berlin: Akademie Verlag.

Whitaker, C. W. A. (1996). *Aristotle's* De Interpretatione: *Contradiction and Dialectic.* Oxford: Clarendon Press.

Wieland, W. (1966a). "Die Aristotelische Theorie der Notwendigkeitsschlüsse." *Phronesis* 11, 35–60.

—— (1966b). Review of G. Patzig: *Die Aristotelische Syllogistik.* In *Philosophische Rundschau* 14, 1–27.

—— (1972). "Die aristotelische Theorie der Möglichkeitsschlüsse." *Phronesis* 17, 124–52.

—— (1975). "Die aristotelische Theorie der Syllogismen mit modal gemischten Prämissen." *Phronesis* 20, 77–92.

—— (1980). "Die aristotelische Theorie der Konversion von Modalaussagen." *Phronesis* 25, 109–16.

Williams, M. (1984). *Studies in the Manuscript Tradition of Aristotle's* Analytica. Königstein: Hain.

Wolff, M. (1998). "Prinzipien und expositorische Beweise in Aristoteles' Syllogistik." *Philosophiegeschichte und logische Analyse* 1, 131–69.

Zalta, E. N. (2006). "Essence and Modality." *Mind* 115, 659–93.

Index of Names

Index of Passages

Index of Subjects

www.ingramcontent.com/pod-product-compliance
Ingram Content Group UK Ltd.
Pitfield, Milton Keynes, MK11 3LW, UK
UKHW040706260926
475585UK00006B/28/J

9 780674 724549